THE ENCYCLOPEDIA OF

Mass Spectrometry

THE ENCYCLOPEDIA OF
Mass Spectrometry

Volume 9
Historical Perspectives
Part B: Notable People in Mass Spectrometry

Edited by

MICHAEL L. GROSS
Washington University, St. Louis, USA

RICHARD M. CAPRIOLI
Vanderbilt University School of Medicine, Nashville, USA

Volume Editors

KEITH A. NIER
Independent Historian, Madison, NJ, USA

ALFRED L. YERGEY
National Institute of Child Health and Human Development, National Institutes of Health, Bethesda, MD, USA

P. JANE GALE
Waters Corp., Milford, MA, USA

ELSEVIER

AMSTERDAM – BOSTON – HEIDELBERG – LONDON – NEW YORK – OXFORD
PARIS – SAN DIEGO – SAN FRANCISCO – SINGAPORE – SYDNEY – TOKYO

Elsevier
Radarweg 29, PO Box 211, 1000 AE Amsterdam, Netherlands
The Boulevard, Langford Lane, Kidlington, Oxford OX5 1GB, UK
225 Wyman Street, Waltham, MA 02451, USA

British Library Cataloguing in Publication Data
A catalogue record for this book is available from the British Library

Library of Congress Cataloging-in-Publication Data
A catalog record for this book is available from the Library of Congress

ISBN: 978-0-08-100379-4

Publisher: John Fedor
Acquisitions Editor: Katey Bircher
Editorial Project Manager: Jill Cetel
Project Manager: Paul Prasad Chandramohan
Cover Designer: Alan Studholme

Contents

Foreword

Truly phenomenal changes have taken place in the field of mass spectrometry over the past several decades. By 1990, mass spectrometry had evolved as a discipline and an important tool for solving problems in organic and inorganic chemistry. It stood alongside nuclear magnetic resonance and optical spectroscopy as a means for the identification and structure proof of organic and organometallic compounds. Indeed, many journals required a mass spectrum or a measurement of an accurate mass before a description of a new compound could be published. Gas chromatography/mass spectrometry had become the premier analytical method for complex mixture analysis such as in environmental chemistry, flavors, aromas, petroleum and other energy materials, and in small molecule metabolism, as in drug research. Despite these advances, mass spectrometry did not yet play a major role in polar and large-molecule analysis. Nevertheless, the seeds for a revolution had been sown ten years or more earlier. Field desorption (FD), fast atom bombardment (FAB), and Cf-252 plasma desorption (PDMS) began to push mass spectrometry into applications such as peptide sequencing and molecular weight determination of large biological and synthetic polymers. Before long, FAB became routine and was an integral part of nearly all mass spectrometry laboratories, and Cf-252 PDMS and FDMS were utilized in more specialized applications. Tandem mass spectrometry was established as a vital tool for complex mixture analysis. The subject of mass spectrometry, both as a discipline and as a measurement method, began to flourish thereafter. The development of new instrumentation and methods positioned mass spectrometry as an invaluable tool in bioanalytical and biophysical chemistry that would extend its reach into the research laboratories of a vast array of disciplines.

The invention of electrospray ionization (ESI) and matrix-assisted laser desorption ionization (MALDI) in the late 1980s opened the door for mass spectrometry to assume a primary role in virtually all of the chemical sciences, including biological, medical, organic, organometallic, and elemental analysis. The discipline of mass spectrometry experienced a leap in activity with the chemical–physical studies of proteins, oligodeoxynucleotides, and other biological compounds in the gas phase.

The broad applicability of mass spectrometry to a multitude of chemical, physical, and biological problems has created the need for an encyclopedia on the subject. To permit a full and fruitful expansion in other disciplines, this encyclopedia is intended as a learning tool to the many newcomers who do not have the theoretical and practical background needed to take full advantage of the technology. Further, the field is now so broad that even the specialists are in need of a resource to allow them to explore the vast reaches of mass spectrometry and help them teach this discipline to scientists and students new to the field.

The Editors began discussion with staff at Elsevier aimed at establishing a comprehensive work in the field of mass spectrometry, a work that covers the theory of gas-phase ion chemistry, principles and designs of ionization and mass analysis, instrument developments and techniques, and a wide variety of applications. The Encyclopedia consists of nine volumes. To permit searching this literature, an index is provided in Volume 10. We estimate that the work will consist of c. 6500 pages in 600 articles, by 1000 authors. Included will be over 15,000 figures and 20,000 references. There is extensive cross-referencing and a subject index in every volume. The articles are intended to cover a specific subject field in three areas; (i) original or pioneering work in the area, (ii) seminal contributions to the area, including appropriate examples illustrating a given application, and (iii) citations to useful review articles and additional reading. Articles are intended to be tutorial in nature, citing both advantages and disadvantages of a technique or method. Where possible, articles are divided into primary (basic) considerations, and advanced topics, the first aimed at the novice and the second to the experienced practitioner who is not necessarily an expert in that specific subject.

Volume 1 is devoted to fundamentals and to chemical physics. Work in these areas is aimed at understanding ionization phenomena and the properties of gas-phase ions. The research interacts productively with theory because often the first outcome of theoretical inquiry is the intrinsic property of a substance; that is, the property in the absence of solvent. Insight into the properties of gas-phase ions, the central subject of mass spectrometry, is illuminated by collision theory, molecular orbital theory, potential energy surfaces, ion spectroscopy, thermochemistry, equilibrium constants, and rate constants. Other ideas, some born in the past such as ion mobility, have been renewed because they can be applied to biomolecules in the gas phase.

The discipline of mass spectrometry has opened the door to new substances, some of which were unknown or only subjects of speculation in condensed-phase chemistry. The early research to produce CH_5^+, a species that violates the rule that carbon binds four times, is but one example. Distonic ions, ion-neutral complexes, clusters, isolated free radicals, and even highly reactive neutral species have become the subjects of mass spectrometry. The discovery of a new allotrope of carbon, the fullerenes, was made possible by workers in the discipline of mass spectrometry who were interested in the intrinsic properties of gas-phase clusters.

Perhaps the most exciting opportunity for mass spectrometry is in the area of biological, biochemical, and biomedical applications (**Volumes 2 and 3**). The end of the 20th century saw the technology of DNA sequencing expand and be successfully applied to a variety of genomes including human. The genome is, however, relatively static, and it is clear that the next big challenge is to understand the proteome. Here, mass spectrometry has assumed the premier role for the identification of proteins, post-translational modifications, protein–protein and other molecular interactions in solution and ultimately in the cell. Ongoing research into the immune system (e.g., antigenic peptides), brain chemistry, and imaging of tissues and cells are part of exciting ongoing work. These opportunities will challenge mass spectrometrists to build new instruments, to develop new and powerful methods, and to understand the underlying ion chemistry.

Other areas of application of mass spectrometry to biology and biochemistry are also blossoming. Indeed mass spectrometry plays a role in DNA and oligonucleotide sequencing, in understanding DNA damage and the implications in cancer and other diseases. Structure elucidation of carbohydrates and glycoproteins, lipids, drugs, and drug metabolism is also greatly facilitated by mass spectrometry.

The applications of mass spectrometry to understand the properties of organic substances constitute subjects in **Volume 4**. Molecular mass spectrometry had its beginnings in the 1940s with applications in petroleum chemistry. These applications were quickly followed by efforts to understand fragmentation of organic molecules in the gas phase. Pioneering figures such as Fred McLafferty, John Beynon, Klaus Biemann, and Carl Djerassi saw the opportunity to use mass spectrometry to determine the structure of organic compounds. This capability helped develop areas such as organic and organometallic synthesis and natural product identification. Building on the early discoveries, mass spectrometry spread out to applied areas such as drug chemistry, environmental chemistry, food, flavor, and aroma chemistry, combinatorial chemistry, and geochemistry. Advances in macromolecular determination quickly impacted the area of synthetic polymers. These applications remain active today in the chemical and related industries, government laboratories, and academe.

Elemental analysis, one of the earliest areas touched by mass spectrometry, is the subject of **Volume 5**. The discovery of stable isotopes alone is one of the fundamental accomplishments of mass spectrometry. This important discovery by J.J. Thomson approximately 100 years ago set the stage for instrument development and measurement of the isotopic composition of the elements. The area of elemental analysis remains important today. Inductively coupled plasma, when interfaced to mass spectrometry, gives science the opportunity to measure nearly all of the elements of the periodic table. It is compatible with a number of ionization methods such as spark source, thermal ionization, electrospray, laser desorption, and secondary ionization mass spectrometry. Today, isotope ratio mass spectrometry is important in biomedicine, drug disposition, archeology, forensics, and materials dating.

Key to every area and application of mass spectrometry are developments in molecular ionization, mass analysis, and detection. **Volume 6** begins with ionization methods and reviews not only those that are used today but also those that were important in the past and that served as a foundation for new methods. Subjects include the venerable electron ionization, field ionization, and chemical ionization and the associated subjects of ion thermochemistry (e.g., proton affinities, electron affinities, gas-phase acidities, etc.). The development of these subjects has impacted not only chemical analysis but also our understanding of physics and physical organic chemistry. Developments of the desorption methods (field, laser, Cf-252 plasma, and FAB) allowed mass spectrometrists to make in-depth studies in biochemistry, biophysics, and biomedicine. Further, their need helped to drive new technologies and helped the development of ESI and MALDI.

Mass analysis and detection are the subjects of **Volume 7**. These fields continue to occupy the attention of mass spectrometrists in all areas involving mass spectrometry. Although the early instruments were based mainly on magnetic sectors and double focusing spectrometers, the renaissance of the time-of-flight mass analyzer, made possible by modern electronics and computers, has significantly changed the face of mass spectrometry. The advent of the quadrupole, the quadrupole ion trap, and the Penning trap have had a major impact on contemporary mass spectrometry. The latter instrument is the basis for Fourier transform mass spectrometry (ion cyclotron resonance mass spectrometry). Many of today's inquiries into the basic properties of gas phase ions and applications in biochemistry and biomedicine also rely on these spectrometers, again demonstrating the close connection between the discipline and the technique.

The advent of soft (desorption) ionization was the stimulus for the development of the field of tandem mass spectrometry. Soft ionization often gives little or no structural information, and mass spectrometrists, starting in the 1980s, explored many combinations of analyzers to produce instrumental arrangements that are appropriate for structural chemistry and biochemistry. These instruments grew from earlier spectrometers in the 1970s that were used for ion chemistry and ion physics. A description of these tandem spectrometers appears in Volume 7 and also in Volume 1.

Combining mass spectrometry with separations methods or other types of fractionation techniques has provided major advances in performance with respect to targeted analyses and analysis of complex mixtures. This combination of techniques, such as liquid chromatography (LC-MS), capillary electrophoresis (CE-MS), supercritical fluid (SCF-MS), as well as others, is sometimes referred to as "hyphenated" techniques. **Volume 8** is devoted to the instrumental setup, operational parameters, and applications of such combined techniques.

As a part of this technology, the volume covers important aspects of the separation science as well in order to fully inform the reader of pertinent aspects involving its combination with mass spectrometry. Perhaps no other class of instruments has made a greater impact on critical issues of the past few decades than this instrumentation, particularly in the field of environmental and biological sciences. It is because of its extraordinary value that it is appropriate to dedicate a volume to this topic.

The final technical volume (**Volume 9**) is historical in nature, and is divided into Part A, *The Development of Mass Spectrometry* and Part B, *Notable People in Mass Spectrometry*. The history of mass spectrometry is rich, encompassing the early days where physics played an important role, through its important progression through many areas of chemistry and biochemistry, and today through nearly every area of science where molecular measurements are required. Indeed, it is still growing and is seeing new applications in areas of clinical significance.

In conclusion, this encyclopedia represents a grand tutorial of the subject of mass spectrometry. It will be a strong beginning for those who are new to the field and wish a centralized work and additional reading in order to learn the various subjects in more detail. By the same token, it is also intended to refresh and bring up to date the existing practitioner in areas apart or adjacent to his immediate area of interest. Finally, through some of the articles that detail today's cutting edge advances in mass spectrometry, the encyclopedia will present a glimpse of tomorrow's technology and applications. We wish readers good fortune in their travels through this work and the hope that they experience some of the inspiration and fascination that mass spectrometry has brought to the world of science and industry.

Richard Caprioli and Michael Gross (Series Editors)

Preface to Volume 9, Part B

From the beginning, this ninth volume of *The Encyclopedia of Mass Spectrometry* covering *Historical Perspectives* was designed and developed as an integrated whole, principally composed of two mutually supporting types of material grouped in two Parts. Accounts of crucial aspects of mass spectrometry are in the first one, *Part A*: *The Development of Mass Spectrometry*. Brief biographical sketches relating the lives and mass spectrometry-related work of many of the influential people involved fill this second one, *Part B*: *Notable People in Mass Spectrometry*.

Interconnections among all these articles are manifold, yet obviously each Part has a coherence of its own. Thus it has been a straight-forward matter to accommodate production convenience by issuing the two Parts under separate covers. Each is readily understood and enlightening on its own while together they provide the more comprehensive exploration of the subject that has always been the goal. We are confident that interested readers will find that this biographical collection can be not only informative in its parts, but also, taken together, can be enlightening about the range and character of the world of mass spectrometry as a whole.

Please note that in some instances autobiographical material was put into paragraph form by the editors from information contributed in other formats.

While publication of the Parts separately has allowed this biographical collection to appear more quickly than otherwise would be the case, it also has resulted in it appearing without an index. Full indexing will be provided at the end of Part A (forthcoming), online, and in the comprehensive index for the whole *Encyclopedia*.

We need not discuss here the general background of the *Historical Perspectives* project, but we must emphasize one fundamental point. We are acutely aware of how much we have depended upon the sixty-plus scientists and historians who have provided one or more of the biographical sketches in this work. Even the shortest autobiographical article takes thought, effort, care, and time to prepare, and these authors have generously provided these essentials. We thank all of them, along with all the personnel at Elsevier, for their help in producing this book.

Keith A. Nier
Alfred L. Yergey
P. Jane Gale

Introduction to Part B: Criteria for Inclusion in these Biographies of Notable People in Mass Spectrometry

People advance mass spectrometry, and while they almost always do it through working with still more people, specific individuals stand out as crucial in all manner of ways to the development of the world of mass spectrometry. Any history of mass spectrometry must devote substantial attention to the people responsible for the devices, techniques, and applications that moved the field forward. The following biographical articles highlight some of the key individuals who have made this area of science and technology grow the way it has.

The volume editors and the general editors of the whole encyclopedia agreed that the selection of individuals for biographical entries should reflect as much as possible the judgment of the scientific community rather than the personal views of the editors. In selecting individuals for inclusion, our goal has been to identify people who have earned major scientific honors for work specifically on aspects of mass spectrometry or who have advanced applications that centrally rely upon mass spectrometry. Efforts were made to obtain an autobiographical or biographical article for all those people whose advances of the techniques, devices, or applications of mass spectrometers brought them such recognition.

The scientific honors in question are the International Mass Spectrometry Conference's Thomson Medal, the ASMS's Distinguished Contribution to Mass Spectrometry award, the American Chemical Society's Field and Franklin award, and the following honors – the Nobel Prize, Crafoord Prize, membership in the National Academy of Sciences, fellowship in the Royal Society of London, the ACS Division of Analytical Chemistry Award in Chemical Instrumentation, and the Pittsburgh Spectroscopy Award – in those cases when mass spectrometric work was central to the selection. Beyond the cases in which the scientific community has declared who was notable through award recognition, the editors added after careful consultation some major pioneering figures in the history of mass spectrometry, most of whom retired or died before many of these awards were well established.

Some other honors were considered but it turned out they rarely added any names to the composite list. Awards specifically limited to younger researchers were not included. The editors note that various people involved with mass spectrometry may have achieved high honors for work that did not involve mass spectrometry and thus would be included here only if their contributions to mass spectrometry were so important that they must be regarded as major pioneers. Obvious instances include Wilhelm Wien, J. J. Thomson, and Harold Urey, each of whom won a Nobel Prize, but are included here not because of those honors but rather for their crucial work in the creation and development of mass spectrometry.

Many more scientists have achieved major honors for work in which mass spectrometry was important, though not really central. Two cases illustrate this kind of relationship. The work of Roger Guillemin and colleagues and of Andrew V. Schally and colleagues that in 1969 finally settled the nature of thyrotropin releasing factor, and which led to their sharing the Nobel Prize in medicine and physiology in 1977, did depend crucially upon mass spectrometry. However, that was only in the last stage of years of research, for most of which mass spectrometry was anything but central (*1*,*2*). The 1985 experiments that led to the 1996 Nobel Prize in chemistry going to Robert F. Curl, Jr., Harold Kroto, and Richard E. Smalley for discovering fullerenes did depend on mass spectrometry to identify what was produced, but it was essentially just a portion of the experimental arrangement in which the focus was on other matters (*3*). Thus these scientists and others in similar cases have not been included here despite making use of mass spectrometry in important ways and being highly honored by their peers.

It is important to stress that there are many more people than we have been able to cover in this volume whose work with mass spectrometry was of considerable importance. In some cases it has proven impossible to find an author to prepare the biographical entry; where possible, we have supplied what information we could to make up in part for the gap. For some people the editors do not know enough to be sure that the person should qualify. It is likely that more complete knowledge would have led to inclusion of additional people from outside the US. Finally, our decision to minimize our subjective views, as well as considerations of space, led us to leave out many people who have advanced the world of mass spectrometry significantly but in ways that are not usually rewarded with major prizes, such as providing education, running businesses, organizing conferences, running associations, editing journals, spreading information, and arranging funding.

Despite these and other limitations, it is clear that the 122 good and great scientists who are included here deserve recognition for their important contributions to the improvement and spread of mass spectrometry.

References

(*1*) Guillemin, R. Peptides in the Brian: The New Endocrinology of the Neuron. Nobel Lecture, 8 December 1977. In: *Nobel Lectures, Physiology or*

Medicine 1971–1980; Lindsten, J., Ed.; World Scientific Publishing Co.: Singapore, 1992; pp 364–397.
(*2*) Schally, A. V. Aspects of Hypothalamic Regulation of the Pituitary Gland with Major Emphasis on Its Implications for the Control of Reproductive Processes. Nobel Lecture, 8 December 1977. In: *Nobel Lectures, Physiology or Medicine 1971–1980*; Lindsten, J., Ed.; World Scientific Publishing Co.: Singapore, 1992; pp 405–438.
(*3*) Smalley, R. E. Discovering the Fullerenes. Nobel Lecture, December 7, 1996. In: *Nobel Lectures, Chemistry 1996–2000*; Ingmar Grenthe, I., Ed.; World Scientific Publishing Co.: Singapore, 2003; pp 89–103.

Keith A. Nier
Independent Historian, Madison, NJ, USA

Alfred L. Yergey
National Institute of Child Health and Human Development, NIH, Bethesda, MD, USA

P. Jane Gale
Waters Corp., Milford, MA, USA

Arthur John Ahearn

b. 20 June, 1902
d. 12 June, 1990

Arthur Ahearn was an American physicist who greatly improved the capabilities of spark source mass spectrometry and significantly advanced the techniques of trace analysis.

Ahearn was born in New London, WI, US, in 1902. He graduated from Ripon College in Wisconsin in 1923, taught high school for a year, and began graduate studies in physics at the University of Minnesota in 1924. He began work for AT&T's Bell Labs in New Jersey in 1929 and completed his Minnesota Ph.D. while employed, finishing in 1931. He worked as a research physicist at Bell Labs until retirement in 1966. He then took a position as research physicist in the Spectrochemical Analysis Section of the National Bureau of Standards in Washington, DC, until 1970.

At Bell Labs, Ahearn's research involved many areas including electron emission, electron optics and microscopy, thermionics, and mass spectrometry. His most dramatic results came in work with N. B. Hannay, developing the first modern spark source mass spectrometer based on Mattauch-Herzog geometry. They showed that spark source mass spectrometry could measure semiconductor dopants at levels below the detection limit of emission spectroscopy and generally could detect sub-ppm impurity levels directly in a solid. The subject became of ever-greater importance as the role of semiconductor electronics increased in modern society.

Ahearn's work was recognized with the Spectroscopy award at Pittcon. Arthur and Ella Ahearn had two children. When his health deteriorated seriously in 1987 they moved to Georgetown, SC, US. He died in 1990. Examples of his work are listed here (*1–5*).

References

(*1*) Hannay, N. B.; Ahearn, A. J. Mass Spectrometric Analysis of Solids. *Anal. Chem.* **1954**, *26*, 1056–1058.

(*2*) Ahearn, A. J. Mass Spectrographic Analysis of Insulators Using a Vacuum Spark Positive Ion Source. *J. Appl. Phys.* **1961**, *32*, 1195–1197.

(*3*) Ahearn, A. J., Ed. *Mass Spectrometric Analysis of Solids*; Elsevier: Amsterdam, London, New York, 1966.

(*4*) Ahearn, A. J.; Trumbore, F. A.; Frosch, C. J.; Luke, C. L.; Malm, D. L. Spark Source Mass Spectrometric Measurements of Dopants of Known Concentrations in Gallium Phosphide. *Anal. Chem.* **1967**, *39*, 350–352.

(*5*) Ahearn, A. J., Ed. *Trace Analysis by Mass Spectrometry*; Academic Press: New York, 1972.

Keith A. Nier
Independent Historian, Madison, NJ, USA

Claude Jean Allègre

b. 1937

Claude Allègre is a highly honored French geochemist/geophysicist/geologist who has also become a political figure (*1*). He was a cowinner of the Crafoord Prize (1986) and has received most of the other major awards in the earth and planetary sciences. He was the French Minister of Education, Research, and Technology in 1997–2000. He has been prominent in several public controversies. In addition to his research and his active role in the scientific community, he has written well-received books for nonspecialists that present a view of the development of this wide area of science and that put considerable emphasis on instrumentation, particularly mass spectrometry (*2*,*3*).

Claude Allègre was born in Paris in 1937 into an academic family; his mother was a school principal, his father a professor of natural science. He began his graduate scientific work in Paris under Louis Barrabé, professor of structural and applied geology in the Paris Faculty of Sciences, shortly before Barrabé died. After some years as an assistant in chemistry in the Faculties of Sciences of Reims and then Paris, in 1967 he founded the Laboratory of Geochemistry and Cosmochemistry at IPGP, the Institut de Physique du Globe de Paris. He became professor at the University of Paris VII in 1970 and director of the IPGP in 1976. He also served as head of the Bureau of Geological and Mining Research and in various other governmental and political positions. He became professor emeritus in 2007.

Allègre's scientific career has centered on the use of mass spectrometry of greater and greater sensitivity and precision in expanding the scope and power of isotope geology to questions throughout the earth and planetary sciences. He and his students and colleagues have introduced or developed a significant number of techniques and approaches including radiogenic isotopic and trace element and stable isotopic studies of Pb, Xe, Ar, He, Os, Th, and particularly the revolutionary Sm-Nd (Samarium and Neodymium) method of dating and the study of Nd isotope systematics. He has maintained a fruitful competitive and cooperative interaction with the other main groups in these fields and promoted the spread of these methods to new centers. These techniques provide effective methods for investigating many aspects of our planet's interior and for revealing the effects it has on processes near, on, and above the Earth's surface (*4*–*7*).

Allègre's work has produced a better understanding of the natures of and complex interactions between the crust of the Earth and the different layers of its mantle. This has illuminated the changing rate and sources of continental formation over the Earth's history. His studies have produced evidence of very early degassing of part of the mantle and thus helped clarify the development of the atmosphere and ocean. He has generated sophisticated models of many geological processes and helped produce an overall geochemical perspective on geodynamics and on the comparative and interconnected histories of the various components of our solar system. Some notable publications are listed below (*8*–*14*).

In addition to the Crafoord Prize, Allègre's honors include membership in the Académie des Sciences, foreign membership in the National Academy of Sciences and the Royal Society of London, and receipt of the Goldschmidt Medal (Geochemical Society of America), Wollaston Medal (Geological Society of London), Day Medal (Geological Society of America), Holmes Medal (European Union of Geosciences), and Bowie Medal (American Geophysical Union).

References

(*1*) Allègre, C. *Wikipédia*. fr.wikipedia.org/wiki/Claude_All%C3%A8gre.

(*2*) Allègre, C. *The Behavior of the Earth: Continental and Seafloor Mobility*. Harvard University Press: Cambridge, MA, and London, 1988. Van Dam, D. K., Trans., French orig. 1983.

(*3*) Allègre, C. *From Stone to Star: A View of Modern Geology*. Harvard University Press: Cambridge, MA, and London, 1992. Van Dam, D. K., Trans., French orig. 1985.

(*4*) National Academy of Sciences, Member Directory, s.v. Allegre, Claude J, nas.nasonline.org/site/Dir/2071041173?pg=vprof&mbr=1001486&returl=http%3A%2F%2Fnas.nasonline.org%2Fsite%2FDir%2F2071041173%3Fpg%3Dsrch%26view%3Dbasic&retmk=search_again_link.

(*5*) Wasserburg, G. J. Isotopic Adventures-Geological, Planetological, and Cosmic. *Annu. Rev. Earth Planet. Sci.* **2003**, *31*, 1–74.

(*6*) The Swedish Royal Academy of Sciences, *The Crafoord Prize*, Crafoord Prize Press, 1986. www.crafoordprize.se/press/arkivpressreleases/thecrafoordprize1986.5.32d4db7210df50fec2d800016948.html
(*7*) Allègre, C. J. *Isotope Geology*. Cambridge University Press: Cambridge, etc., 2008; French orig. 2005.
(*8*) Allègre, C. J. Méthode de discussion géochronologique concordia généralisée. *Earth Planet. Sci. Lett.* **1967**, *2*, 57–66.
(*9*) Allègre, C. J.; Ben Othman, D.; Polvé, M.; Richard, P. The Nd–Sr Isotopic Correlation in Mantle Materials and Geodynamic Consequences. *Phys. Earth Planet. Int.* **1979**, *19*, 293–306.
(*10*) Allègre, C. J.; Brévard, O.; Dupré, B.; Minster, F. F. Isotopic and Chemical Effects Produced in a Continuously Differentiating Convecting Earth Mantle. *Phil. Trans. Roy. Soc. Lond.* **1980**, *297*, 447–477.
(*11*) Allègre, C. J. Chemical Geodynamics. *Tectonophysics* **1982**, *81*, 109–132.
(*12*) Allègre, C. J.; Turcotte, D. Implications of a Two-Component Marble-Cake Mantle. *Nature* **1985**, *323*, 123–127.
(*13*) Allègre, C. J.; Hoffmann, A.; O'Nions, R. K. The Argon Constraints on Mantle Structure. *Geophys. Res. Lett.* **1996**, *23*, 3555–3557.
(*14*) Allègre, C. J.; Moreira, M. Rare Gas Systematics and the Origin of Oceanic Islands: The Key Role of Entrainment at the 670 km Boundary Layer. *Earth Planet. Sci. Lett.* **2004**, *228*, 85–92.

Keith A. Nier
Independent Historian, Madison, NJ, USA

Francis William Aston

b. September 1, 1877
d. November 20, 1945

Francis William Aston was an accomplished physicist who built the first mass spectrographs in England the early 1900s (*1–3*). He received the Nobel Prize in Chemistry in 1922 for his discovery, by means of his mass spectrograph, of isotopes in a large number of non-radioactive elements, and for his enunciation of the whole number rule (Fig. 1).

Aston was born in Harborne near Birmingham in 1877. Both his father William and mother Fanny Charlotte (nee Hollis) had connections with the metal industry; his father worked as a metal broker. Largely as a result of his mother's family wealth, Francis received a private school education. He attended the Harborne Vicarage School from 1889 and was enrolled at Malvern College in Worcestershire in 1891 where he boarded for two years. Here Aston demonstrated his abilities in mathematics and science. With his father's encouragement, Aston honed his experimental and fabrication skills in a home workshop in a stable, and later a disused loft, at the family property.

Figure 1
Francis William Aston *ca.* 1930s.

In 1894, Aston attended the Sir Josiah Mason Science College where he studied chemistry and physics under professors William Augustus Tilden, Percy Faraday Frankland, and John Henry Poynting. His training with Poynting was particularly influential and would ultimately lead to his appointment at the Cavendish Laboratory in Cambridge as Joseph John Thomson's research assistant (*1–7*). Aston was awarded a Bachelor of Science degree in 1898 and received a Forster scholarship to conduct research with Frankland on the optical rotation of a tartartic acid derivative that resulted in his first publication in 1901 (*8*). He was encouraged to study fermentation chemistry at a new School of Brewing at Birmingham that led to his employment years as a brewer's chemist at the W. Butler & Co. Springfield brewery in Wolverhampton for three years.

With Poynting's support, he returned to the then University of Birmingham in 1903 as an Associate. His research on electrical discharges in vacuum tubes enabled Aston to accurately measure the length of the chamber volume in what was described as cathode dark space and now bears his name (*9*). After extended world travel in 1908 with monies left to him upon the death of his father, Aston returned to Birmingham for a term before being recommended by Poynting to Sir J. J. Thomson for a research assistant position at the Cavendish Laboratory, Cambridge.

Thomson had been studying positive rays originating in evacuated discharge tubes from residual gases present (see this volume, *Joseph John Thomson*). The rays followed parabolic paths when deflected with crossed electric and magnetic fields as evident by their image on a fluorescent screen. His results with neon gas, gave rise to two parabolas that suggested either the existence of two forms of the element, one dubbed meta-neon, or that the second parabola resulted from neon hydride. Aston was charged with separating and examining the properties of the two forms of neon. He did so by measuring their density using a quartz microbalance specifically designed by him for that purpose. He showed that although the differences in their density were small, they were too large to be attributed to experimental error. Aston made a failed attempt to separate the constituents through their fractional distillation in charcoal cooled with liquid air and had limited success separating them by diffusion through pipeclay after repeated cycles. His greatest success in this endeavor, however, came after World War I during which he was stationed at the Royal Aircraft Establishment in Farnborough. His period at the RAE enabled Aston to consolidate his thoughts and he benefitted from his interactions with other noted scientists, among them Frederick Lindemann.

Aston realized that the approach with the best chance of success involved constructing an instrument using a sequence of electric and magnetic fields in which the constituents were separated by their mass. In 1919, on his return to Cambridge, Lindemann and Aston published on the possibility of separating isotopes of the elements by a range of approaches. Of them they wrote that "*the positive ray method of electrical separation is the only one offering hopes of a pure product and that only in microscopic quantity*" (*10*). Aston built his first mass spectrograph at the Cavendish laboratory, then directed by Ernest Rutherford, and successfully separated the forms of neon by mass with sufficient accuracy to prove that atmospheric neon is a mixture of two isotopes of atomic weight 20.00 and 22.00 correct to about 0.1% (*11–13*). The instrument quickly established the existence of isotopes of other elements including the alkali metals, nickel, chlorine and the other noble gases. His work resulted in the enunciation of the whole number rule in which all atoms have masses that are integer multiples of the mass of a hydrogen atom. His achievements, and their importance to our understanding of atoms and molecules, were recognized in 1922 with his award of the Nobel Prize in Chemistry (see this volume, *Magnetic and Electrostatic Analyzers before 1960, Mass Spectrometry and the Separation and Enrichment of Isotopes*, and *A History of Mass Spectrometry in Nuclear Science*).

After studying more than 50 elements with his first instrument, the flight tube of which now resides at the Science Museum in London, Aston began to construct his second mass spectrograph in order to deal with the greater number of isotopes evident for a number of metals and the minute variations in their masses (*14,15*). This instrument, reported on fully in 1927, could achieve a mass accuracy of 1 in 10,000 (*16*). The ^{16}O atom was chosen as the standard reference mass and the percentage divergence, expressed in parts per 10,000, was reported as a 'packing fraction' for many other elements (*17*).

Aston's third mass spectrograph (Fig. 2) continued to refine the method and introduced finer collimating slits to achieve mass accuracies of 1 in 100,000 and a mass resolution of 1:2000, values that are still very acceptable to this day (*18*). Using this mass spectrograph, Aston reported a mass for ^{12}C of 12.00355 ± 0.00015 that was to eventually become the

Figure 2
Francis Aston working at his third mass spectrograph, Cavendish laboratory, *ca.* mid 1930s.

preferred reference mass for all other isotopes. By the mid 1930s, Aston had almost single-handedly recorded the stable isotopes of most of the chemical elements, not withstanding the contributions of Arthur Dempster in Chicago (*19*, and see this volume, *Arthur Jeffrey Dempster*). He was made Chairman of the International Union of Chemistry and President of the Cambridge Philosophical Society in 1935 (*20*). He ceased his research around the outbreak of World War II.

Apart from receiving the Nobel Prize, Aston received many accolades and awards throughout his career. Among them, he received a D.Sc. from the universities of Birmingham and Cambridge in 1914 and 1922 respectively, was made a Fellow of Trinity college in 1920, received both the Hughes medal of the Royal Society of London and the John Scott medal from the Franklin Institute in 1922, the Duddell medal from the Institute of Physics in 1944, honorary degrees from Freiburg and Calcutta universities, and gave the Bakerian lecture to the Royal Society of London in 1927 and the Liversidge Memorial lecture to the Chemical Society of London in 1932. His two monographs on *Isotopes*, and *Mass Spectra and Isotopes*, were first published in 1922 (*14*) and 1933 (*17*).

Away from the laboratory, Aston occupied his time by playing sport (cycling, swimming, tennis, rock climbing and golf) throughout most of his life. He also enjoyed music, played the cello, and traveled extensively combining his scientific activities with his interests in astronomy and photography. He resided for most of his adult life at Trinity College largely in rooms in Nevile's Court (*7*). Aston never married and was survived upon his death in Cambridge in November of 1945 by his sisters.

His outstanding research studying isotopes and positive rays using several mass spectrographs, combined with the work of Canadian Arthur Dempster, is responsible for the essential role that mass spectrometry now plays in chemical and biochemical analysis.

References

(*1*) Hevesy, G. Francis William Aston 1877–1945. *J. Chem. Soc.* **1948**, 1468–1475.

(*2*) Thomson, G. P. Dr Francis William Aston FRS. *Nature* **1946**, *157*, 290–292.

(*3*) Squires, G. Francis Aston and the Mass Spectrograph. *J. Chem. Soc. Dalton Trans.* **1998**, *23*, 3893–3900.

(*4*) Brock, W. H. Aston, Francis William. In: *Dictionary of Scientific Biography*; Gillispie, C. C., Ed.; Charles Scribner's Sons: New York, 1971; vol. 1, pp 320–322.

(*5*) Feather, N., Osborne, P. Aston, Francis William (1877–1945). In: *Dictionary of National Biography*; Oxford University Press, 2004.

(*6*) Downard, K. M. Cavendish's Crocodile and Dark Horse: The Lives of Rutherford and Aston in Parallel. *Mass Spectrom. Rev.* **2007**, *26*, 713–723.

(*7*) Downard, K. M. Francis William Aston: The Man Behind the Mass Spectrograph. *Eur. J. Mass Spectrom.* **2007**, *13*, 177–197.

(*8*) Frankland, P. F.; Aston, F. W. Influence of a Heterocyclic Group on Rotatory Power; the Ethyl and Methyl Esters of Dipyromucyltartaric Acid. *J. Chem. Soc. Trans.* **1901**, *79*, 511–520.

(*9*) Aston, F. W. Experiments on the Length of the Cathode Dark Space with Varying Current Densities and Pressures in Different Gases. *Proc. R. Soc. A* **1907**, *79*, 80–95.

(*10*) Lindemann, F. A.; Aston, F. W. The Possibility of Separating Isotopes. *Phil. Mag.* **1919**, *37*, 523–534.

(*11*) Aston, F. W. A Positive-Ray Spectrograph. *Phil. Mag.* **1919**, *38*, 707–715.

(*12*) Aston, F. W. Neon. *Nature* **1919**, *104*, 334–335.

(*13*) Aston, F. W. Constitution of Atmospheric Neon. *Phil. Mag.* **1920**, *39*, 449–455.

(*14*) Aston, F. W. *Isotopes*, Edward Arnold & Co.: London, 1922.

(*15*) Aston, F. W. Atomic Weights and Isotopes. *J. Franklin Inst.* **1922**, *193*, 581–608.

(*16*) Aston, F. W. A New Mass Spectrograph and the Whole-Number Rule. *Proc. R. Soc. A* **1927**, *115*, 487–514.

(*17*) Aston, F. W. *Mass Spectra and Isotopes*; Longmans, Green & Co.: New York, 1933.

(*18*) Aston, F. W. A Second-Order Focusing Mass Spectrograph and Isotopic Weights by the Doublet Method. *Proc. R. Soc. A* **1937**, *163*, 391–404.

(*19*) Aston, F. W. The Story of Isotopes. *Science* **1935**, *82*, 235–240.

(*20*) Aston, F. W., *et al.* International Table of Stable Isotopes, Sixth Report of the Committee on Atoms of the International Union of Chemistry. *Nature* **1942**, *150*, 515.

Kevin M. Downard
University of Sydney, Australia

Kenneth Tompkins Bainbridge

b. July 27, 1904
d. July 14, 1996

Harvard University Archives, call # HUP Bainbridge, K. (1)

Kenneth Tompkins Bainbridge was born on July 27, 1904, in Cooperstown, New York and grew up in New York City where he attended the Horace Mann High School. Like so many of his contemporaries, Bainbridge became interested in radio while still in high school. The proximity of his home to the naval shipyard provided an opportunity to try his hand at communicating with ship's operators returning from Europe following World War I. He was one of the earliest 'ham' radio operators, with the call letters 2WN.

Bainbridge entered the Massachusetts Institute of Technology in 1921 as a 'co-op' student with General Electric. This program required Bainbridge to work alternate semesters, thereby providing funds for his education. This extended his undergraduate program by one year, but he earned both S. B. (*Scientiæ Baccalaureus*) and S. M. (*Scientiæ Magister*) degrees. Despite having filed patents (*1,2*) on photoelectric tubes while working at General Electric during this period, he opted to pursue a graduate degree in physics at Princeton, beginning in 1926.

At Princeton, his search for the then undetected element with atomic weight 87 in the periodic table, *eka*-cesium (*3*), now known as francium, led him to the use of mass spectrometry (*3*). He built a 180° sector mass spectrograph after the design of Dempster (*4*). Although unsuccessful in his effort to find evidence of *eka*-cesium, his interest in mass spectrometry grew from those early Princeton experiments. After graduating from Princeton in 1929, Bainbridge performed research at the Bartol Research Foundation of The Franklin Institute in Philadelphia for the next four years. There he used mass spectrometry to determine the exact masses of the lithium isotopes. Combining his work with data from lithium disintegration by proton bombardment, he was able to confirm the mass–energy equivalence proposed by Einstein, $E = Mc^2$ (*5*). Upon departure from the Bartol Research Foundation, he published a review article on the state of 'The Masses of Atoms and the Structure of Atomic Nuclei' (*6*).

The following year Bainbridge performed research at the Cambridge Cavendish Laboratory as a Guggenheim fellow. During this period, he was able to interact with the world's leading scientists in the development and use of mass spectrometry as a tool for the study of the elements and their isotopes. While at the Cavendish, he designed a mass spectrograph (*7*) which he later constructed after joining the Physics faculty at Harvard in 1934 (see this volume, *Magnetic and Electrostatic Analyzers before 1960*). This instrument had a resolving power of 10,000 and the interesting property that the mass scale (*8*) was linear over the mass range of the photoplate to one part in 7000! In addition to the mass spectrograph, he led the development of the Harvard Cyclotron, using both instruments in his research into the physics of the nucleus. With these instruments, he made a number of fundamental measurements of the light elements through argon. It was during this period that a young scientist by the name of Alfred Nier, with a recent physics doctorate from the University of Minnesota, came to Harvard under a National Research Council Fellowship to conduct postdoctoral research in Bainbridge's lab (see this volume, *Alfred Otto Carl Nier*).

On leave of absence from Harvard in 1940, Bainbridge was the first physicist to work at the Massachusetts Institute of Technology Radiation Laboratory in Cambridge, Massachusetts, where he contributed to early research on radar in the United States. In this capacity, he visited England to coordinate American efforts with those in Great Britain. While making significant contributions to the radar effort, he was drafted into the Manhattan project in May of 1943, on the recommendation of E. O. Lawrence. He joined the research group at Los Alamos, and in 1945 he was made Director of the Trinity Project, the first test of the atomic bomb. In addition to that responsibility, he was intimately involved in the design of the explosive assemblies to trigger the A-bomb. Thus Bainbridge had a hand in two of the most prominent wartime projects that affected the Allied War Effort, radar and the atomic bomb. He was one of only five Manhattan Project scientists who had the authority to stop the first test of the A-bomb at his own discretion.

Returning to Harvard in the fall of 1945 Bainbridge undertook the design of a large mass spectrograph with even higher resolving power. With this instrument, he continued his investigations into the exact masses of the elements and their isotopes, as well as their modification when undergoing high energy bombardment in a new cyclotron of his own design, the synchrocyclotron. From this research he was able to establish the existence of the neutrino. Bainbridge continued his mass spectrometric studies of the elements and nuclear physics until his retirement from Harvard in 1975. In addition to his research, he developed new courses for the physics curriculum, taught both undergraduate and graduate courses, and served as Chairman of the Physics Department. During his tenure, he facilitated the establishment of the Morris Loeb Lectureship. He was awarded the George Vasmer Leverett Chair in Physics (*9*,*10*).

References

(*1*) Bainbridge, K. T. *Photoelectric Tubes*. Patent # 364006, (The British Thomson-Houston Co., Ltd.). Great Britain, 1929.

(*2*) Bainbridge, K. T. *Photoelectric Tube*. Patent # 1901577, (General Electric Co.). United States, 1933.

(*3*) Bainbridge, K. T. A Search for Element 87 by Analysis of Positive Rays. *Phys. Rev.* **1929**, *34*, 752–762.

(*4*) Dempster, A. J. A New Method of Positive Ray Analysis. *Phys. Rev.* **1918**, *11*, 316–324.

(*5*) Bainbridge, K. T. Equivalence of Mass and Energy. *Phys. Rev.* **1933**, *44*, 123.

(*6*) Bainbridge, K. T. Masses of Atoms and the Structure of Atomic Nuclei. *J. Franklin Inst.* **1933**, *215*, 509–534.

(*7*) Bainbridge, K. T.; Jordan, E. B. Mass-Spectrum Analysis. 1. The Mass Spectrograph. 2. The Existence of Isobars of Adjacent Elements. *Phys. Rev.* **1936**, *50*, 282–296.

(*8*) Bainbridge, K. T.; Jordan, E. B. A High-Dispersion, Linear-Scale Mass Spectrograph with Velocity and Direction Focusing. *Phys. Rev.* **1936**, *49*, 421.

(*9*) Grandinetti, D. P., *Bainbridge, Kenneth T.* (*Kenneth Tompkins*) *1904–1996. Papers of Kenneth Tompkins Bainbridge: An Inventory*. Harvard University Archives, Call #: HUGFP152 http://oasis.lib.harvard.edu/oasis/deliver/~hua23003 **2004** (accessed August 24, 2014).

(*10*) Pound, R. V.; Ramsey, N. F. Kenneth Tompkins Bainbridge: July 27, 1904–July 14, 1996; *Biographical Memoirs*, vol. 76. National Academy Press: Washington, DC, 1999. pp. 18–34.

Michael A. Grayson
Washington University in St. Louis, St. Louis, MO, USA (ret.)

Michael Barber

b. November 3, 1934
d. May 8, 1991

Michael (Mickey) Barber was an English chemist and mass spectrometrist whose work in the exploration of molecular-level surface phenomena with photoelectron spectroscopy and secondary ion mass spectrometry led to the invention of Fast Atom Bombardment (FAB) in 1980 (*1*; and see this volume, *Static and Continuous-Flow Fast Atom Bombardment [FAB]*). Predating the invention of electrospray and MALDI by several years, FAB made possible the use of ion bombardment to desorb intact thermally labile and polar compounds dissolved in a liquid matrix. The sudden availability of molecular weight and structural information for species that either from chemical structure or molecular weight were previously outside the domain of mass spectrometry opened the door to what would over the next 10 years become the rich and varied world of biological mass spectrometry (*2,3*).

Mickey was born, raised, and spent all his adult life in and around the city of Manchester, England. After completing his early education at Manchester Grammar School, he attended Queen's College Oxford. There, under the direction of Jack Linnett, FRS, he designed and built a mass spectrometer for the analysis of flames, work for which he received a D.Phil. He joined Associated Electrical Industries (AEI) in 1961, working in both surface science and mass spectrometry. He became a lecturer at the University of Manchester Institute of Science and Technology (UMIST) in 1973 and was promoted to professor in 1985, the same year he was elected a Fellow of the Royal Society.

In 1964, while still at AEI, Barber and Martin Elliott developed a method (later known as the Barber–Elliott method) for studying the fragmentation of ions within the analyzer of a mass spectrometer. Returning to his first love, surface science, in the late 1960s and early 1970s, Mickey designed an X-ray photoelectron spectrometer (PES) with which he and collaborators John Thomas and Eurwyn Evans (among others) were able to identify the state of bonding of heteroatoms at the surfaces of carbon fibers and other solids – proof that PES could determine submonolayer amounts of oxygen and other light elements adsorbed on a solid surface. He continued this interest in PES at UMIST and also began working with John Vickerman's group on secondary ion mass spectrometry (SIMS).

As Brian Green of Micromass (later Waters) tells the FAB story (*3*), at UMIST Mickey began to explore the effect of a beam of primary ions or neutrals with energies of a few keV bombarding the surface of a sample stage coated with solid organic compounds. Unfortunately, he found that the molecules of interest so rapidly decomposed under this energetic bombardment "that any secondary ions emitted from the intact molecule were transitory and hence of little practical use. Mickey's key inventive step in 1980 was to dissolve the organic compound in a relatively involatile liquid substrate such as glycerol, so that the surface was continuously replenished with fresh sample. In this way, substantial [secondary] ion beams of the intact molecule were produced over several minutes. This simple yet so elegant amalgamation of ideas revolutionized the analysis of labile polar molecules by mass spectrometry and increased the molecular mass of accessible compounds to over 10,000 a.m.u."

In 1990, Mickey founded the Centre for Mass Spectrometry at UMIST, a joint venture among UMIST, other universities, and industry in the Manchester area. He was to have become its first director, had not his untimely death intervened. The Centre was renamed the Michael Barber Centre for Mass Spectrometry in 1994 in recognition of Mickey's many contributions to the development of the science of mass spectrometry.

Mickey's work was recognized with his receipt of the Royal Society of Chemistry Award for Analytical Chemistry 1979 and The Strock Medal and Award of the American Society for Applied Spectroscopy in 1983. In May, 1991 at the annual conference of the American Society for Mass Spectrometry (ASMS), the award for Distinguished Contributions to Mass Spectrometry was given to Mickey posthumously. Made aware of his receipt of the award in April, Mickey was at work on his acceptance speech when he died in early May of that year.

Mickey's interests outside chemistry were famously broad and wide. He was a sportsman of some repute; in addition to being a crack shot, he gained a half-blue for lacrosse while at Oxford. He was fluent in Russian, having spent his 2 years of National Service as a Russian translator. His duties included serving as the official translator for the visit of the General Secretary of the Russian Communist Party to Oxford

University. He is also said to have read 'War and Peace' cover-to-cover in Tolstoy's original Russian language. Mickey was an accomplished musician, singing in his youth and playing the pipe organ for his local village parish as an adult. So strong were his views on the subject of pipe organs that when time came for a new organ, he oversaw specification and construction of the instrument. Mike, as she preferred him to be called, and his wife Joan raised two daughters.

References

(*1*) Barber, M.; Bordoli, R. S.; Sedgwick, R. D.; Tyler, A. N. Fast Atom Bombardment of Solids (F.A.B.): A New Ion Source for Mass Spectrometry. *J. Chem. Soc. Chem. Commun.* **1981**, 325–327.

(*2*) Brenton, A. G.; Monaghan, J. J.; Munn, R. W.; Thomas, J. M. Professor Michael Barber FRS. *Rapid Commun. Mass Spectrom.* **1991**, *5*, 340–342.

(*3*) Green, N. In Memory: Professor Michael Barber FRS. *Org. Mass Spectrom.* **1992**, *27*, 67.

Simon J. Gaskell
Queen Mary's College, London, UK

P. Jane Gale
Waters Corp., Milford, MA, USA

Hans-Dieter Beckey

b. June 8, 1921
d. June 18, 1992

Hans-Dieter Beckey was a German scientist widely noted for his work in combining field ionization with mass spectrometry and for a seminal achievement of introducing field desorption into mass spectrometry (see this volume, *A Historical Perspective on Field Ionization (FI) and Field Desorption (FD) Mass Spectrometry*). The field ionization method represented the first 'soft' ionization method developed for volatile organic compounds, while the later-invented field desorption method was the first 'soft' ionization method for nonvolatile organic compounds.

H. D. Beckey was born in Hamburg, Germany, in 1921 and he attended the Wilhelm high school there. As a result of World War II and a severe injury he suffered during its last days, he didn't start his university studies until 1947. He graduated in physics from the University of Hamburg (advisor P. Hartek) and moved to the University of Bonn in 1950 as assistant to Wilhelm Groth, where he received his PhD in physical chemistry in 1952 with Groth as his advisor. His habilitation was in March 1959. He became full professor of physical chemistry at the University of Bonn in 1966. He served as chairman of the 4th International Mass Spectrometry Conference in Berlin in 1976.

H. D. Beckey married Gudrun Kischke in 1951 and the couple had three children. In 1969, Beckey became seriously ill but was able to continue his scientific work and academic duties until worsening illness forced his resignation in 1979. H. D. Beckey was a respected academic teacher. He enjoyed playing the violin and was a member of a string quartet for almost three decades. He died in Bonn in 1992.

H. D. Beckey started his scientific career with investigations on isotope separation in the transition region between Knudsen and Poiseuille flow. For several years he worked on gas diffusion, isotope separation, and gas phase radical reactions.

In 1958 he began using field ionization (FI) with mass spectrometry (MS), an analytical method that transfers more or less volatile substances into the gas phase prior to ionization (*1*). Based on earlier work by Inghram and Gomer, he used high electric fields of 10^7–10^8 V/cm to ionize organic molecules for subsequent mass spectrometric analysis (*2,3*, see this volume, *Mark Gordon Inghram II*). These high electric fields were originally generated using fine metal tips, later thin wires, to which a potential of up to 15 kV with respect to a counterelectrode was applied (*1,4*). To enhance the field strength and to achieve good ion currents, microneedles were grown onto these wire emitters by a field polymerization process in the presence of suitable organic gases (originally acetone, later benzonitrile or related compounds) (*5,6*). FI provides mass spectra dominated by molecular ions (or protonated molecules) with little fragmentation, even in cases where under the classical electron impact (EI) ionization little if any molecular ion is produced, as with many labile organic molecules. In this manner Beckey demonstrated that EI and FI provide complementary information useful for structure elucidation of organic compounds.

A further very important discovery by Beckey was that with an FI source the fast unimolecular decomposition of organic ions in the gas phase could be studied in timeframes ranging from pico- to microseconds (*7,8*). This early work represents the basis f the method later termed 'field ionization kinetics' used by several investigators to study the mechanisms of the unimolecular decay of organic ions in the gas phase, for example, the study of competing rearrangements and direct dissociations as well as isomerization reactions or to elucidate reaction mechanisms (*9*).

Beckey also used FI-MS successfully to detect thermally induced free radicals (*10*). Direct mixture analysis of petroleum or higher boiling oil fractions, where quantitative mixture analysis is also possible, is another important area of analytical application of the FI-MS method that continues to be of importance in the petroleum industry (*11*). The applications of field ionization in structure determination and quantitative analysis were reviewed by Beckey in 1969 (*12*).

Beckey's most important contribution to mass spectrometry of organic compounds, however, was the invention of field desorption, 'soft' ionization for nonvolatile and very polar compounds including larger biopolymers (*13*). His success with field ion emitters covered with long microneedles that provide a large adsorption area led Beckey to dip these 'activated emitters' directly into an aqueous solution of D-glucose. He then introduced the emitters into the FI source and (after evaporation of the aqueous solvent and applying a high electric field), he was able to desorb the intact protonated glucose molecule for subsequent mass analysis. It is remarkable and also typical of Beckey that he carried out this crucial experiment himself. In the following years this new, soft ionization method was applied to the mass spectrometric analysis of a large variety of biologically important compounds such as peptides, pyrolysis products of bacteria and DNA (*14,15*), and nucleosides and nucleotides (*16*). When Beckey became seriously ill, his coworker Hans-Rolf Schulten continued the development of field desorption mass spectrometry (*17*).

Beckey published two monographs on field ionization and field desorption (*18,19*). He received the Nernst Prize (now the Nernst–Haber–Bodenstein Prize) from the German Bunsen Society for Physical Chemistry in 1964.

References

(*1*) Beckey, H. D. Massenspektrometrische Untersuchung con Chemisorptions- und Assoziationsreaktionen mit Hilfe einer neuen Feldemissionen Quelle. *Naturwissenschaften* **1958**, *45*, 259.

(*2*) Inghram, M. G.; Gomer, R. Mass Spectrometric Analysis of Ions from the Field Microscope. *J. Chem. Phys.* **1954**, *22*, 1279–1280.

(*3*) Inghram, M. G.; Gomer, R. Mass Spectrometric Investigation of the Field Emission of Positive Ions. *Z. Naturforsch.* **1955**, *10a*, 863–872.

(*4*) Beckey, H. D. Massenspektrometrische Untersuchungen mit Hilfe einer Feldemissions-Ionenquelle. *Z. Naturforsch.* **1959**, *14a*, 712–721.

(*5*) Beckey, H. D.; Hilt, E.; Maas, A.; Migahed, M. D.; Ochterbeck, E. A Method for Strong Activation of Field Ion Emitters. *Int. J. Mass Spectrom. Ion Phys.* **1969**, *3*, 161–165.

(*6*) Beckey, H. D.; Heising, H.; Hey, H.; Metzinger, H. G. Use of Thin Wires in Field Ionisation Mass Spectrometry. In: *Advances in Mass Spectrometry, 4*; Kendrick, E., Ed.; Institute of Petroleum: London, 1968; pp. 817–830.

(*7*) Beckey, H. D.; Knöppel, H. Zur Kinetik des Monomolekularen Zerfalls organischer Ionen in hohen elektrischen Feldern. *Z. Naturforsch.* **1966**, *21a*, 1920–1930.

(*8*) Beckey, H. D.; Hey, H.; Levsen, K.; Tenschert, G. Study of the Kinetics of Fast Unimolecular Decomposition Processes and of Organic Rearranged Reactions by Field Ionisation Mass Spectrometry. *Int. J. Mass Spectrom. Ion Phys.* **1969**, *2*, 101–123.

(*9*) Schwarz, H.; Levsen, K. Feldionisationskinetik zur Aufklärung von Reaktionsmechanismen. *Nachr. Chem. Tech. Lab.* **1978**, *26*, 136–141.

(*10*) Butzert, H.; Beckey, H. D. Untersuchung von freien Radikalen aus heterogenen Zersetzungreaktionen mit einem Feldionen-Massenspektrometer. *Z. Phys. Chem., NF* **1968**, *62*, 83–102.

(*11*) Beckey, H. D. Mass Spectrometric Analysis with a New Source for the Production of Field Ions at Thin Wires or Metal Edges. *Fresenius Z. Anal. Chem.* **1963**, *197*, 80–90.

(*12*) Beckey, H. D. Determination of the Structures of Organic Molecules and Quantitative Analyses with the Field Ionization Mass Spectrometer. *Angew. Chemie. Int. Ed.* **1969**, *8*, 623–639.

(*13*) Beckey, H. D. Field Desorption Mass Spectrometry: A Technique for the Study of Thermally Unstable Substances of Low Volatility. *Int. J. Mass Spectrom. Ion Phys.* **1969**, *2*, 500–503.

(*14*) Beckey, H. D.; Schulten, H. R. Felddesorptions-Massenspektrometrie biologisch wichtiger Substanzen. *Z. Anal. Chem.* **1975**, *273*, 345–358.

(*15*) Schulten, H. R.; Beckey, H. D.; Meuzelaar, H. L. C.; Boerboom, A. J. H. High Resolution Field Ionization Mass Spectrometry of Bacterial Pyrolysis Products. *Anal. Chem.* **1973**, *45*, 191–195.

(*16*) Schulten, H. R.; Beckey, H. D. High Resolution Field Desorption Mass Spectrometry I: Nucleosides and Nucleotides. *Org. Mass Spectrom.* **1973**, *7*, 861–867.

(*17*) Schulten, H. R. Biochemical, Medical, and Environmental Applications of Field-Ionization and Field-Desorption Mass Spectrometry. *Int. J. Mass Spectrom. Ion Phys.* **1979**, *32*, 97–283.

(*18*) Beckey, H. D. *Field Ionization Mass Spectrometry*; Pergamon: Oxford, 1971.

(*19*) Beckey, H. D. *Principles of Field Ionization and Field Desorption Mass Spectrometry*; Pergamon: Oxford, 1977.

Karsten Levsen

Fraunhofer Institute for Toxicology and Experimental Medicine (ITEM), Hannover, Germany

Alfred Benninghoven

b. February 8, 1932.

Photo taken from University of Munster website, reference (1)

German physicist Alfred Benninghoven is best known for his pioneering work on static secondary ion mass spectrometry (SIMS) and more particularly for its extension to the characterization of monolayers of organic molecules adsorbed on solid surfaces (see this volume, *The Development of SIMS for Imaging* and *The Development of Imaging Mass Spectrometry*)

Alfred Benninghoven was born in Frankfurt, Germany, in February, 1932. After studying physics at the Universities of Paris and Cologne, he was graduated in 1961 as Dr. rer. nat. He completed his habilitation at the University of Cologne in surface physics two years later. In 1972 he was appointed to a chair at Münster University in Experimental Physics, where he spent the next 25 years as the Director of the Physics Institute of the University of Münster. Becoming emeritus in 1997, he has continued to work there on secondary ion mass spectrometry (SIMS). On a number of occasions between 1978 and 1990 Prof Benninghoven was the Dean or Vice-Dean of the Department of Physics (*1*).

Benninghoven's approach to surface science has been to probe the different ways that ions and molecules interact with solid surfaces. Through the many inter-disciplinary and international collaborations in which he has been involved, he turned SIMS into a recognized standard method of surface analysis, applicable in areas as diverse as microelectronics, biosciences, and medicine. With his group at Münster, he not only pursued applications of static SIMS but also the development of instrumentation with which to further their efforts. In 1989, he and colleagues from the Physics Institute founded the company IonTOF to produce the specialized tools needed for organic surface analysis. IonTOF quickly became a leader in commercially-produced TOF-SIMS instrumentation.

Benninghoven's work has been described in more than 300 scientific papers. He has also written and/or edited a dozen or so books on the subject. A sampling of these publications, many of which have become reference works on SIMS, is given below (*2–9*). Benninghoven has been awarded a number of national and international prizes for his work in SIMS, among them the Technology Transfer prize awarded by the German Ministry of Education and Research and the 1984 Gaede-Langmuir Prize awarded by the American Vacuum Society "for the development of concepts and instrumentation in static secondary ion mass spectrometry and the demonstration of its usefulness in manifold applications" (*10*).

Ever the ambassador for SIMS, Benninghoven organized the first international SIMS conference in Münster in 1987, drawing practitioners of SIMS from around the world. So great was the enthusiasm of participants of this first conference that similar gatherings have been held in venues across Europe, the US, and Asia biennially (*11*). Benninghoven has continued to be a member of the Organizing Committee and is a coeditor of the proceedings from these meetings, published under the title, *Secondary Ion Mass Spectrometry SIMS N*, where *N* indicates the number of this conference in the sequences of such conferences (*8*). In 1981, he initiated a second biennial conference with a focus (as its name suggests) on organic surfaces: Ion Formation from Organic Solids or IFOS. This conference is held in Münster, and Benninghoven edits the published proceedings (*9*). From 1977 to 1983, Benninghoven was president of the German Vacuum Society, a technical arm of the German Physical Society.

References

(*1*) Anon. Press Release, University of Münster, summer 2002. http://www.unimuenster.de/Rektorat/exec/upm.php?rubrik = Alle&neu = 0&monat = 200202&nummer = 03199

(*2*) Benninghoven, A.; Kirchner, F. The Energy Distribution of Atomized Neutral and Charged Particles. *Z. Naturforsch* **1963**, *18A*, 1008–1010.

(*3*) Benninghoven, A. Analysis of Submonolayers on Silver by Negative Secondary Ion Emission. *Phys. Stat. Sol.* **1969**, *34*, K169–K171.

(*4*) Benninghoven, A.; Jaspers, D.; Sichtermann, W. Secondary Ion Emission of Amino Acids. *J. Appl. Phys* **1976**, *11*, 35.

(*5*) Benninghoven, A.; Sichtermann, W. Secondary Ion Mass Spectrometry: A New Analytical Technique for Biologically Important Compounds. *Org. Mass Spectrom* **1977**, *12*, 595–597.

(*6*) Benninghoven, A. Some Aspects of Secondary Ion Mass Spectrometry of Organic Compounds. *Int J. Mass Spectrom. Ion Phys.* **1983**, *53*, 85–89.
(*7*) Benninghoven, A.; Rudenauer, F. G.; Werner, H. W. *Secondary Ion Mass Spectrometry: Basic Concepts, Instrumental Aspects, Applications and Trends*; John Wiley and Sons, 1987.
(*8*) See the series published, for example, as Benninghoven, A.; Colton, R. J.; Simons, D. S.; Werner, H. W. *Secondary Ion Mass Spectrometry SIMS V*; Springer-Verlag: Berlin and Heidelberg, 1986, 592 pp.
(*9*) See the series published, for example, as Benninghoven, A. *Ion Formation from Organic Solids (IFOS III): Mass Spectrometry of Involatile Material: Proceedings of the Third International Conference,* Münster, Fed. Rep. of Germany, September 16–18, 1985. Springer-Verlag Berlin and Heidelberg, 1986, 219 pp.
(*10*) http://www2.avs.org/historybook/links/glwincit.htm (accessed August 25, 2014).
(*11*) International SIMS Conference website: http://www.simssociety.org/ (accessed August 25, 2014).

P. Jane Gale
Waters Corp., Milford, MA, USA

John Herbert Beynon

b. December 29, 1923

Photo from Boyd RCM ref. 22

John Beynon is a British physicist whose pioneering work in organic mass spectrometry in the 1950s and '60s and his now classic book of 1960 *Mass Spectrometry and its Applications to Organic Chemistry* set the stage for many of the subsequent developments in the field. Beynon's instrumentation designs were incorporated into very successful commercial high resolution mass spectrometers. He transformed the study of metastable ions, contributing both to the experimentation and theory related to their production and energetics. As a founding member of several mass spectrometry societies, he encouraged the formal and informal exchange of information among colleagues.

John Beynon was born in Ystalyfera, a coal mining town in Wales, on December 29, 1923, the older of two sons of parents who did not finish secondary school. He was educated at Ystalyfera Grammar School, finishing his A Levels when he was only 16. He attended the University of Wales at Swansea on a scholarship and received a B.S. in Physics in 1943. With Britain in the throes of World War II when he graduated, Beynon rejected graduate school and instead joined the Fighting Vehicles Research Establishment, where he was engaged in tank armament research from 1943–1947. In 1947, free at last to leave tank warfare, he married Yvonne Fryer, an English girl of Salisbury, and joined the Dyestuffs Division of Imperial Chemical Industries (ICI), Ltd. in Manchester, UK. He was ICI's Manager of Physics and Physical, Polymer and Analytical Chemistry from 1947–1969. He was made a Senior Research Associate at ICI in 1964, a post created to give selected men the opportunity to carry out personal research. Availing himself of opportunities to explore academia, he was a Boomer Memorial Fellow at the University of Minnesota in 1965. Finding that the grass was indeed greener, he accepted the positions of Professor of Chemistry and Director of the Mass Spectrometry Center at Purdue University in 1969, posts he held until 1975. Already an honorary Fellow of Swansea University (1964–1969), in 1974 he became Royal Society Research Professor and Director of the Mass Spectrometry Research Unit at Swansea, a post he held until 1986. He has been Research Professor of Physics and Chemistry there since 1976. Between 1972–1975 and 1982–1985, he was a visiting professor at Essex University, UK. In addition, he has been a Research Associate at the Institut Jozef Stefan in Ljubljana, Slovenia, since 1976.

As the now famous story goes, Beynon's introduction to mass spectrometry began with his arrival at ICI, when his supervisor, A.J. Hailwood, gave him the initial task of building a mass spectrometer, a device about which, by his own admission, Beynon knew virtually nothing. Learning the physics and creating the technology, however, proved to be pivotal in his career: Beynon and his work became central to the development of organic mass spectrometry as a field of study and as a tool of chemical analysis and knowledge (*1*). The proposed mass spectrometer was intended for studying isotopically-labeled metabolites of pharmaceuticals being produced by the small pharmaceutical unit of the research department at ICI. Understanding that resolution would be important to this task, Beynon based his design on one recently published by Thode and coworkers that included a 6"-90° sector (*2*, and see this volume, *Henry George Thode*). He and A. E. (Bert) Williams, who would become his long-term collaborator, literally built the apparatus from the ground up with a few things that were bought and many things that were scavenged, including equipment abandoned by the US military in its post-WWII departure from Britain. Two years in construction, the instrument produced its first spectrum from a sample of coal gas in 1949. Beynon's identification of benzene in that sample so impressed ICI management that the focus of the instrument thereafter was on identifying impurities in the many products of the Dyestuffs Division, not following metabolites of pharmaceuticals (*3*).

Beynon's instrument was the first mass spectrometer designed specifically to study organic compounds unrelated to petroleum. The accurate mass measurements (50 ppm) he was able to make with it allowed him to determine the atomic composition of ions he observed in the spectra of the profusion of ICI samples that soon came his way. His 1954 paper in *Nature* reporting this work, entitled, 'Qualitative Analysis of Organic Compounds by Mass Spectrometry' (*4*) has been cited as 'the birth of organic mass spectrometry' (*5*). Building on the successes of the first instrument, Beynon was then able to collaborate with J.D. Waldron at Metropolitan Vickers (later AEI and later still Kratos) on the design of a double-focusing instrument for studying organic compounds at high resolution. The instrument, at Beynon's insistence, incorporated Nier-Johnson geometry with a radius of 6" in the magnetic field, giving it a resolving power (10% valley definition) of 13,500. Produced specifically for Beynon by Metropolitan Vickers, the MS8 was so successful in doing what Beynon had envisioned that it soon became the prototype for the MS9 (see below), an instrument famous for its later ubiquity in organic mass spectrometry labs in the UK and US (see this volume, *Mass Spectrometry Manufacturers and Markets from the Beginning to 1960, The Beginnings of Organic Mass Spectrometry*, and *Alfred Otto Carl Nier*).

As Bert Williams later observed, "John was a person of vision surrounded by an aura of urgency; the vision was, of course, unique to John but the sense of urgency was communicated to those who worked with him" (*6*). Using Beynon's research to fuel commercial development of high resolution magnetic sector instruments became a pattern, allowing the two major British instrument manufacturers of the day, Metropolitan Vickers and VG Organic, Ltd to create organic mass spectrometers that became the workhorses of their eras. The MS8 that worked so well for Beynon at ICI was redeveloped into the MS9, the first commercial high resolution mass spectrometer, of which more than 200 were eventually sold. Some years later, based on work he had started at Purdue, Beynon collaborated with VG Organic Ltd. to design and build the ZAB (Zero alpha and beta coefficients). Beynon took delivery of the first of these instruments at Purdue and his group went on over the next six years to produce more than 100 papers and the monograph *Metastable Ions* from work done with it (*7*).

The time at Purdue was productive, launching Beynon in the research in which he would be involved for many years to come in various academic venues. Metastable ions – ions formed by electron ionization that have internal energy in excess of that required to break chemical bonds but are sufficiently long-lived to fragment only after leaving the source – became the focus of much of his thinking, instrument design, and experimental work. In 1969, during his sabbatical from ICI at Purdue, he developed ideas for studying metastables by what would become known as ion kinetic energy spectroscopy (IKES) from his work on the prototype Hitachi RMH2 there. Returning to Purdue two years later, he built a reverse geometry mass spectrometer (i.e. inverting the Nier–Johnson order of the electrostatic and magnetic analyzers) mass spectrometer that allowed him to probe metastables in a complementary way that he termed mass-analyzed ion kinetic energy spectroscopy (MIKES) (*7,8*, and see this volume, *Ion Kinetic Energy Spectra [IKES] and Mass-Analyzed Ion Kinetic Energy Spectra [MIKES]*).

Over the course of his long and productive career, Beynon published more than 350 papers and wrote a number of books, many of them related to the topic of obtaining tandem mass spectrometry (MS/MS) data with various kinds of double-focusing instruments. As late as 1985, he was still interested in instrument design: a translational energy spectrometer created at Swansea was described in the *International Journal of Mass Spectrometry and Ion Processes* (*9*). A sampling of his publications is given below (*10–18*); a complete listing can be found in his curriculum vitae (*19*).

While a member-at-large of the American Society for Testing and Materials (ASTM) E-14 Committee on mass spectrometry (the precursor organization to the American Society for Mass Spectrometry, ASMS), Beynon participated in discussions and then produced recommendations on symbolism and nomenclature for mass spectrometry that were adopted by the IUPAC (*20*). He was an energetic advocate for mass spectrometry as an analytical discipline in which free and open exchanges of ideas would foster new developments. In 1960 he founded the British Mass Spectrometry Society and became its first chairman. He was a founding member of ASMS when it separated from ASTM, the founding president of the European Mass Spectrometry Society (1993), and an honorary member of the Japanese and Yugoslavian mass spectrometry societies. In 1987, perceiving the need for a vehicle for publication of fast-breaking developments in what he saw as a phase of explosive growth about to begin for mass spectrometry, Beynon secured the support of publisher John Wiley and Sons for a new journal called *Rapid Communications in Mass Spectrometry*. It was to be comprised of short articles on 'hot' topics published on a short timeline. Starting in 1987 as a very thin production (7 issues, 135 pages), by the turn of the century each volume contained some 2000 pages. In recognition of Beynon's leadership role in promoting publication of mass spectrometry research work, in 2003 Wiley funded an annual award named the RCM Beynon Award (*21*).

Beynon became a Fellow of the Royal Society of London in 1978. He has been the recipient of many medals and prizes, some of which include the Jozef Stefan Medal in 1980 and the Medal of the Serbian Chemical Society in 1981; the Jan Marcus Marci Medal of the Czechoslovak Spectroscopic Society in 1984, the Thomson Medal of the International Mass Spectrometry Conference in 1985, the Frank H. Field and Joe L. Franklin Award for Outstanding Work in Mass Spectrometry of the American Chemical Society in 1987, the Gold Medal of the British Mass Spectrometry Society in 1988, and the Gold Medal, Italian Mass Spectrometry Society in 1990.

Outside mass spectrometry, Beynon has had a lifelong interest in the photography of insects and flowers, and, in later years, in the creation of a wholly different kind of art through computer-generated composite images of his photographs (*22*). He was a field hockey player in his youth, and he occasionally figured as "Rhythm Man" in a jazz band. While John Beynon the scientist will certainly be remembered for his enormous contributions to mass spectrometry, John Beynon the man will be remembered for being the consummate raconteur, whose telling of stories and jokes was a legendarily large part of his personality (*22–24*).

References

(*1*) Beynon, J. H. Interview by Michael A Grayson at Swansea. Wales, United Kingdom, 22 April 2008 (Philadelphia: Chemical Heritage Foundation, Oral History Transcript # 0420). http://www.chemheritage.org/Oral-Histories/Documents/Beynon (accessed January 11, 2013).

(*2*) Graham, R. L.; Harkness, A. L.; Thode, H. G. An a.c. Operated Mass Spectrometer for Isotope Abundance Measurements. *J. Sci. Instrum.* **1947**, *24*, 119–128.

(*3*) Beynon, J. H. Letters to the Editor. *Mass Spectrom. Rev.* **1994**, *13*, 183–185.

(*4*) Beynon, J. H. Qualitative Analysis of Organic Compounds by Mass Spectrometry. *Nature* **1954**, *174*, 735–737.

(*5*) Maccoll, A. Biographical Note on J. H. Beynon. *Org. Mass Spectrom.* **1983**, *18*, 503–505.

(*6*) Williams, A. E. My Years with John (1947–73). *Org. Mass Spectrom.* **1983**, *18*, 506–508.

(*7*) Beynon, J. H.; Cooks, R. G.; Caprioli, R. M.; Williams, A. E. *Metastable Ions*; Elsevier Publishing Company: Amsterdam, 1973; pp 298.

(*8*) Jennings, K. R.; Mason, R. S. Tandem Mass Spectrometry Utilizing Linked Scanning of Double Focusing Instruments. *Org. Mass Spectrom.* **1977**, *12*, 163.

(*9*) Guilhaus, M.; Boyd, R. K.; Brenton, A. G.; Beynon, J. H. Advantages of a Second Electric Sector on a Double-Focusing Mass Spectrometer of Reversed Configuration. *Int. J. Mass Spectrom. Ion Proc.* **1985**, *67*, 209–225.

(*10*) Beynon, J. H. *Mass Spectrometry and its Applications to Organic Chemistry*; Elsevier Publishing Company: Amsterdam, 1960; pp 640.

(*11*) Beynon, J. H.; Williams, A. E. *Mass and Abundance Tables for Use in Mass Spectrometry*; Elsevier Publishing Company: Amsterdam, 1963; pp 570.

(*12*) Beynon, J. H. *Table on Meta-Stable Transitions for Use in Mass Spectrometry*; Elsevier Publishing Company: Amsterdam, 1965; pp 783.

(*13*) Beynon, J. H.; Saunders, R. A.; Williams, A. E. *Mass Spectrometry of Organic Molecules*; Elsevier Publishing Company: Amsterdam, 1968.

(*14*) Beynon, J. H., Ed. *Table of Ion Energies for Metastable Transitions in Mass Spectrometry*; Elsevier Publishing Company: Amsterdam, 1970; pp 509.

(*15*) Beynon, J. H. *Recommendations for Symbolism and Nomenclature for Mass Spectroscopy*; Elsevier Publishing Company: Amsterdam, 1978; pp 10.

(*16*) Beynon, J. H.; Brenton, A. G. *An Introduction to Mass Spectrometry*; University of Wales Press: Swansea, 1982; pp 57.

(*17*) Beynon, J. H.; Gilbert, J. R. *Applications of Transition State Theory to Unimolecular Reactions: An Introduction*; John Wiley & Sons: London, 1984; pp 94.

(*18*) Waddell, D. S.; Boyd, R. K.; Brenton, A. G.; Beynon, J. H. Experimental and Theoretical Investigations of Angular Distributions of Fragments from Dissociations of Polyatomic Ions. *Int. J. Mass Spectrom. Ion Proc.* **1986**, *68*, 71–90.

(*19*) Beynon, J. H. John Herbert Beynon: Curriculum Vitae. *Int. J. Mass Spectrom.* **2003**, *230*, xxix–xxx.

(*20*) International Union of Pure and Applied Chemistry, Analytical Chemistry Division, Commission on Analytical Nomenclature. Recommendations for Nomenclature of Mass Spectrometry. *Pure Appl. Chem.* **1974**, *37*, 469–480.

(*21*) Boyd, R. K. The RCM Beynon Prize. *Rapid Commun. in Mass Spectrom.* **2003**, *18*, iv.

(*22*) Boyd, R. K. John H. Beynon. *Rapid Commun. in Mass Spectrom* **2004**, *18*, 1–6.

(*23*) Boyd, R. K. John H. Beynon. *Int. J. Mass Spectrom. Ion Proc.* **1988**, *86*, ix–xii.

(*24*) Bowers, M. T.; McLafferty, F. W.; Cooks, R. G.; Ast, T.; Brenton, G. Special Issue in Honor of John H. Beynon. *Int. J. Mass Spectrom. Ion Proc.* **2003**, *230*, vii–xxvii.

P. Jane Gale
Waters Corp., Milford, MA, USA

Klaus Biemann

b. November 2, 1926

Klaus Biemann was born in Innsbruck, Austria, and received his primary and high school education in Klosterneuburg, a suburb of Vienna. His father was a pharmacist and it was more or less understood that Biemann would follow his father into that profession. His family owned the pharmacy and lived above it, so Biemann was immersed in the profession from an early age. To help his sister after the death of their father, he began an apprenticeship in pharmacy. It was during high school that he became interested in chemistry, receiving the highest grades in the subject. After World War II, his family moved back to Innsbruck where Biemann continued on an academic course to a master's degree in Pharmacy from the University of Innsbruck in1948. By that time, however, the family's pharmacy business had been lost, leaving Biemann free to continue his studies in chemistry.

He received his doctorate in organic chemistry from the University of Innsbruck under the tutelage of Professor Hermann Bretschneider in 1951 and was immediately appointed as an instructor. He embarked on the traditional Austrian path into academia by starting his *habilitation* but was intrigued by an announcement of a summer program for foreign academics at the Massachusetts Institute of Technology (MIT). He applied for and received the award in 1954. After arriving at MIT, he worked with Professor George Buchi on a variety of structural problems in organic chemistry. In Buchi's laboratory, Biemann had his first exposure to analytical instrumentation based on physical methods, such as ultraviolet and infra-red spectrophotometry. After returning to Innsbruck late in 1954, Biemann decided that he would rather pursue a career in the United States and returned in the fall of 1955 as a postdoctoral appointee under Professor Buchi. After a two year stint in that position, Professor Arthur C. Cope, then Chair of the Chemistry Department, offered Biemann a position on the faculty. With his background in organic chemistry and with the analytical training he received while working towards the pharmacy degree, Biemann landed a position in the Analytical Chemistry division of the department.

In this role, Biemann had to create a research program in analytical chemistry, despite his strong background in organic synthesis. He chose to develop a method for amino acid sequencing of small peptides commencing from the C-terminus, which he thought promised to complement the method previously designed by Sanger working from the N-terminus. While still in Innsbruck, Biemann had developed a synthesis reaction, which he now planned to use for this work on a microchemical level. He was able to obtain a grant from the National Institutes of Health for his peptide research; however, before he started on the project, a serendipitous event changed the direction of his research forever: at a conference in Chicago on flavors and fragrances, Biemann learned of mass spectrometry from a paper presented by William H. Stahl of the U.S. Army Quartermaster Corps.

After researching the literature on this relatively new analytical technique, Biemann saw how, coupled with his synthetic chemistry skills, it could be used as a powerful tool in his proposed study of peptides. With support from Cope he purchased a Consolidated Electrodynamics Corporation 21–103C mass spectrometer, at that time a workhorse in the petrochemical industry (see this volume, *Mass Spectrometry Manufacturers and Markets from the Beginning to 1960*). Since the only viable ionization technique was electron ionization, it was mandatory that samples be in the vapor phase for introduction to the ion source. Peptides, even small ones, are not volatile, however, and here Biemann's training in organic synthesis came to the fore: he decided to convert the peptides to the corresponding polyamino alcohols. These more volatile derivatives of the peptides could be introduced into the ion source and were found to fragment under electron ionization in ways that would reveal their amino acid sequence (*1,2*). In addition, he turned his newly acquired mass spectrometer to the study of the structure of natural products, particularly alkaloids (*3,4*). Based on the early results of his research and the need in 1962 for such a text, Biemann wrote *Mass Spectrometry: Organic Chemical Applications* which quickly became a standard in the field (*5* and see this volume, *The Beginnings of Organic Mass Spectrometry* and *A Perspective on Books in Mass Spectrometry in Chemistry*).

This was a period of rapid development in all forms of analytical instrumentation, mass spectrometry not excepted. Biemann's laboratory became a cutting edge center for mass spectrometric instrumentation and the ancillary analytical and computational tools necessary to attack the most complicated problems. Gas chromatography (GC) was becoming a common instrument for the separation of complex mixtures in laboratories throughout the world, and Biemann's lab was no different (see this volume, *A History of Gas Chromatography Mass Spectrometry (GS/MS)*). Besides using GC to separate the increasingly complex mixtures of compounds his group encountered in their research, they rapidly adopted the combination of the two instruments with the molecular effusion separator developed by Biemann and then graduate student J. T. Watson (*6*). With the addition of a high resolving power mass spectrometer in 1962, his group demonstrated the value of accurate mass determination in the analysis and structure determination of organic compounds and devised new ways of presenting complex data sets for ease of interpretation with the 'element map' (*7,8*). Seeing the copious amount of mass spectral data that could be generated in the most modest combined GC/MS analysis, his group pioneered the application of digital computers to mass spectral data analysis for both nominal mass and high resolving power mass spectral data (*9*, see this volume, *Reflections on the History of Computers in Mass Spectrometry*).

When the Apollo program brought back soil and rocks from the Moon in 1969, Biemann's lab was one of the first to receive such lunar material for analysis. Later, when Project Viking, a trip to the surface of Mars, was proposed, Biemann was chosen to head up a team to design an experiment for the detection and analysis of organic compounds in the Martian soil. While analyzing material from the Moon posed problems, here on Earth in terms of avoiding sample contamination and working in isolation labs to guard against possible exposure to unknown lunar microbes, the Viking program required the analysis to be performed remotely on the surface of Mars, a far more difficult and challenging problem (see this volume, *The Development of Mass Spectrometry in the Earth and Planetary Sciences*). The experiment involved heating the soil sample to volatilize, and finally pyrolize, any organic compounds present into a miniaturized GC/MS instrument. Although the instrument worked perfectly, the data indicated the absence of organic material at both landing sites (*10*). These results disappointed many of those who had hoped for any signs indicating the possibility of life on the Red Planet.

When Biemann began his career at MIT, there were no academic graduate programs for training analytical chemists in the art and science of mass spectrometry. His lab became the first. While other chemistry departments did establish programs in the years that followed, Biemann's was always considered the premier lab, both in the United States and abroad. During his illustrious career, Professor Biemann nurtured a total of 140 graduate research students and postdoctoral fellows and in the process published with them over 350 papers in the scientific literature (*11*, and see, e.g., this volume, *John Michael Hayes, David Michael Hercules, Donald Frederick Hunt*, and *James A. McCloskey*).

Biemann has been honored with numerous awards, among them are a Fullbright Fellowship in 1954, Fellowship in the American Academy of Arts and Sciences in 1962, the Tricentennial Medal of the University of Innsbruck in 1970, the Exceptional Scientific Achievement Medal from the National Aeronautics and Space Administration in 1977, Honorary Member of the Japanese Society for Medical Mass Spectrometry in 1981, a Guggenheim Fellowship in 1983–84, Frank H. Field and Joe L. Franklin Award for Outstanding Achievement in Mass Spectrometry of the American Chemical Society, membership in the National Academy of Sciences in 1993, the Award in Analytical Chemistry from the American Chemical Society in 2001, and the Benjamin Franklin Medal in Chemistry in 2007.

Despite all the acclaim, however, from Biemann's point of view the most notable was not an award he received but an award that his students created in his name: the 'Biemann-Medal' bestowed annually by the American Society for Mass Spectrometry on a young investigator for exceptional achievements early in his or her research career.

Biemann's interview for Chemical Heritage Foundation Oral History Project is available online at http://www.chemheritage.org/discover/collections/oral-histories/details/biemann-klaus.aspx (accessed November 16, 2014).

References

(*1*) Biemann, K.; Gapp, F.; Seibl, J. Application of Mass Spectrometry to Structure Problems. I. Amino Acid Sequence in Peptides. *J. Am. Chem. Soc.* **1959**, *81*, 2274–2275.

(*2*) Biemann, K.; Seibl, J.; Gapp, F. Mass Spectrometric Identification of Amino Acids. *Biochem. Biophys. Res. Commun.* **1959**, *1*, 307–311.

(*3*) Biemann, K.; Seibl, J. Application of Mass Spectrometry to Structure Problems. II. Stereochemistry of Epimeric, Cyclic Alcohols. *J. Am. Chem. Soc.* **1959**, *81*, 3149–3150.

(*4*) Biemann, K. The Determination of Carbon Skeleton of Sarpagine by Mass Spectrometry. *Tetrahedron Letters* **1960**, *1*(36), 9–14.

(*5*) Biemann, K. *Mass Spectrometry: Organic Chemical Applications*; McGraw Hill Book Co., Inc.: New York, 1962.

(*6*) Watson, J. T.; Biemann, K. Direct Recording of High Resolution Mass Spectra of Gas Chromatographic Effluents. *Anal. Chem.* **1965**, *37*, 844–851.

(*7*) Watson, J. T.; Biemann, K. High-Resolution Mass Spectra of Compounds Emerging from a Gas Chromatograph. *Anal. Chem.* **1964**, *36*, 1136–1137.

(*8*) Biemann, K.; Bommer, P.; Desiderio, D. M. Application of Mass Spectrometry to Structure Problems. XXI. Element Mapping, a New Approach to Interpretation of High-Resolution Mass Spectra. *Tetrahedron Lett.* **1964**, 1725–1731.
(*9*) Hites, R. A.; Biemann, K. Computer Recording and Processing of Low-Resolution Mass Spectra. *Adv. Mass Spectrom.* **1968**, *4*, 37–54.
(*10*) Biemann, K.; Oro, J.; Toulmin, III, P., *et al.* The Search for Organic Substances and Inorganic Volatile Compounds in the Surface of Mars. *J. Geophys. Res.* **1977**, *82*, 4641–4658.
(*11*) Biemann, K. The Massachusetts Institute of Technology Mass Spectrometry School. *J. Am. Soc. Mass Spectrom.* **1994**, *5*, 332–338.

Michael A. Grayson
Washington University in St. Louis, St. Louis, MO, USA (ret.)

Walker Bleakney

b. February 8, 1901
d. January 15, 1992

Source: *National Academy of Sciences Biographical Memoirs, vol. 73.*

Walker Bleakney was an American physicist who spent almost his entire career in the Princeton University physics department. He was elected to the National Academy of Sciences in 1959. He made significant contributions to the development of mass spectrometry, most notably a much-improved design for an electron ionization source.

Walker Bleakney was born on a farm near Elderton, Pennsylvania in 1901; the family moved to Oregon in his childhood, first to near the town Milton, then near Echo. After working as a field hand to save money, he entered Whitman College in Walla Walla, Washington, in 1920. He clearly got a good education. Whitman's physics professor was the award-winning Benjamin H. Brown and of the four physics majors who graduated in 1924, one became a Princeton professor – Bleakney, one a college president – E. J. Workman, one the author of one of the earliest and most successful quantum mechanics texts – V. Rojansky, and one a Nobel laureate and co-inventor of the transistor – Walter Brattain (*1–3*).

After a year on scholarship studying electrical engineering at Harvard, Bleakney began graduate study in physics at the University of Minnesota with John Tate as his advisor. As part of a general research program investigating ionization phenomena, Bleakney worked on multiply-charged mercury ions created by electron ionization (*4,5*). In doing so he designed a significantly-improved electron ionization (EI) source and a new form of mass analyzer. The analyzer design was not taken up by others but the source became the basis for further improvements by Tate's group, including those incorporated in the first commercial mass spectrometers and the well-known Nier source (see this volume, *Magnetic and Electrostatic Analyzers before 1960* and *The Development of Electron Ionization*).

Bleakney received his Ph.D. in 1930 and was awarded a National Research Fellowship with which he went to Princeton. Two years later, he joined the faculty and remained for thirty-seven more years. During the 1930s, Bleakney continued research on ionization, investigated isotopic abundances, and made further innovations in mass spectrometers (*6–9*). Among these was the creation with John Hipple of a mass spectrometer with 'perfect' focus in which ions travel trochoidal paths (*10,11*, and see this volume, *John Alfred Hipple, Jr.*). Perhaps his most influential contribution to mass spectrometry however was facilitating the entries of Harold Urey and David Rittenberg into working with these instruments (*12,13*, and see this volume, *David Rittenberg* and *Harold Clayton Urey*).

With the start of World War II, Bleakney turned to the study of effectiveness of explosives and the detailed study of shock waves (*14*). He left Princeton for Santa Barbara upon retirement in 1969. Survived by his wife "Tommie," nee Dorothy Clyde Thomas, Walker Bleakney died on January 15, 1992.

References

(*1*) Bleakney, W. Reminiscences of My Youth in Oregon. *Am. J. Phys* **1972**, *40*, 953–959.

(*2*) Reynolds, G. T.; Sherr, R.; Emrich, R. J.; Mariner, T.; Panofsky, W. K. H. Walker Bleakney. *Physics Today* **1993**, *46*, 80.

(*3*) Reynolds, G. T. Walker Bleakney: 1901–1992. In: *Biographical Memoirs*; National Academies Press: Washington, DC, 1998, vol. 73, pp 3–15.

(*4*) Bleakney, W. A New Method of Positive Ray Analysis and Its Application to the Measurement of the Probability and Critical Potentials for the Formation of Multiply Charged Ions in Hg Vapor by Electron Impact. *Phys. Rev.* **1929**, *34*, 157–160.

(*5*) Bleakney, W. Probability and Critical Potentials for the Formation of Multiply Charged Ions in Hg Vapor by Electron Impact. *Phys. Rev.* **1930**, *35*, 139–148.

(*6*) Bleakney, W. Ionization Potentials and Probabilities for the Formation of Multiply Charged Ions in Helium, Neon, and Argon. *Phys. Rev.* **1930**, *36*, 1303–1308.

(*7*) Bleakney, W. A Search for Isotopes of Hydrogen and Helium. *Phys. Rev.* **1932**, *41*, 32–38.

(*8*) Sampson, M. B.; Bleakney, W. A Mass Spectrograph Study of Ba, Sr, In, Ga, Li, and Na. *Phys. Rev.* **1936**, *50*, 456–460.

(*9*) Bleakney, W. The Relative Abundance of Isotopes. *Proc. Am. Philos. Soc.* **1937**, *77*, 395–409.

(*10*) Bleakney, W.; Hipple, J. A. Jr. A New Mass Spectrometer with Improved Focusing Properties. *Phys. Rev.* **1938**, *53*, 521–529.
(*11*) Mariner, T.; Bleakney, W. A Large Mass Spectrometer Employing Crossed Electric and Magnetic Fields. *Rev. Sci. Instrum.* **1949**, *20*, 297–303.
(*12*) Rittenberg, D.; Bleakney, W.; Urey, H. C. The Equilibrium Between the Three Hydrogens. *J. Chem. Phys.* **1934**, *2*, 48–49.
(*13*) Manian, S. H.; Bleakney, W.; Urey, H. C. An Investigation of the Relative Abundance of the Oxygen Isotopes O16:O18 in Stone Meteorites. *J. Am. Chem. Soc* **1934**, *56*, 2601–2609.
(*14*) Emrich, R. J. Walker Bleakney and the Development of the Shock Tube at Princeton. *Shock Waves* **1996**, *5*, 327–339.

Keith A. Nier
Independent Historian, Madison, NJ, USA

John Hamilton Bowie

b. July 16, 1938

John Hamilton Bowie is an Australian chemist who pioneered in investigating the chemistry of gas phase negative ions and whose work has extended from the biochemistry of Australian frogs to molecules in interstellar regions.

Bowie was born on July 16, 1938, in Melbourne, Australia. He is the only child of William Hamilton Bowie and Alice Maud Bowie (nee Boyce). He was educated at Scotch College in suburban Hawthorn, where he studied science with a view to becoming a high school science teacher (*1*). Despite receiving a fairly uninspired education in chemistry, John continued his studies in this subject (*2*). He particularly enjoyed the practical aspects of chemistry that distinguished it from his other mostly rote-learnt studies.

Bowie enrolled at the University of Melbourne to study science in 1956 and was awarded a Bachelor of Science degree with honors in 1958. He was the recipient of the Cuming Major and Dixon prizes that year and opted to continue his studies in chemistry by enrolling in a newly offered master's degree rather than obtaining teaching qualifications. The master's degree involved both coursework and research and included a project that enabled Bowie to initiate his research career in natural product chemistry under the direction of Dr. Ray Cooke. The project involved characterizing the 'coloring matters' of Australian plants by infrared absorption spectroscopy and chemical analysis (*3*).

A lecture at Melbourne in 1960 by visitor Professor Alan Johnson FRS, then Head of the chemistry department at Nottingham University, prompted Bowie to ask if he could join Johnson's laboratory. Encouraged by Johnson, Bowie applied for and received a Commonwealth United Kingdom Postgraduate Award from the Australian government. He left for Nottingham in 1961 to conduct doctoral research on the structures of various antibiotics including streptolidine, viomycin, and rhodomycinone (from an unknown *Streptomyces* strain) (*4*). These studies primarily employed chemical analysis, UV-VIS and NMR spectroscopy but also exposed Bowie to mass spectrometry for the first time. His research in mass spectrometry began in earnest in Cambridge when he was awarded an Imperial Chemical Industries Fellowship to study the chemistry of aphids with Nobel Prize winner Lord (Alexander Robertus) Todd, FRS, and Dr. D. W. Cameron at Cambridge (*5*).

It was during the next 2 years of postdoctoral studies that Bowie began to collaborate with Dudley Williams, who had just been appointed a senior assistant in research and a fellow and lecturer at Churchill College, Cambridge (see this volume, *Dudley Howard Williams*). Bowie worked after hours with Williams, together with R. Graham Cooks and others, to acquire mass spectra of a host of organic compounds using an AEI MS9 double-focusing mass spectrometer (*6*, and see this volume, *R. Graham Cooks* and *Magnetic and Electrostatic Analyzers before 1960*). Impressed by the power and speed of the mass spectrometer for chemical analysis and for following chemical transformations, Bowie made mass spectrometry the centerpiece of his research when he was appointed to the academic staff at the University of Adelaide as a lecturer in mid-1966 (*7*).

The former head of the organic chemistry department at Adelaide, Athel Beckwith, had acquired a Hitachi RMU 7D mass spectrometer a few years earlier. Bowie made good use of it, at first continuing research he had began in the UK. The instrument offered the capability to study negative ions, so recognizing the dearth of negative ion mass spectra of organic compounds and the importance of negative ions in organic reaction pathways, Bowie increasingly focused on negatively charged ions from the late 1960s, a focus that would continue for the next four decades (*8–11*).

The receipt of a Dynaspec ion cyclotron resonance (ICR) 9 spectrometer in the early 1970s facilitated the study of the reactions of negative ions in the gas phase (see this volume, *A History of Ion Cyclotron Resonance (ICR) and Fourier Transform (FTICR) Mass Spectrometry*). Bowie spent a sabbatical year during this period, returning to Cambridge, visiting Keith Jennings at Sheffield University, and stopping with Graham Cooks at Purdue in 1972 (see this volume, *Keith R. Jennings*). In 1979, he visited John Beynon's lab in Swansea for an introduction to the newly developed reverse geometry instrument (see this volume, *John Herbert Beynon*). He later visited and collaborated with a number of notable mass spectrometrists and ion chemists including Nico Nibbering, Helmut Schwarz, Charles DePuy, and John Holmes (see this volume, *Nicholaas Martinus Maria Nibbering*, *Helmut Schwarz*, and *John Leonard Holmes*). A long-standing collaboration with John Sheldon at Adelaide facilitated companion theoretical studies of organic ions and reaction intermediates.

The acquisition of a VG ZAB 2HF reverse geometry instrument in the mid 1980s promoted the implementation of MIKES experiments in conjunction with the charge-reversal of gas phase anions (*12*, and see this volume, *Ion Kinetic Energy Spectra (IKES) and Mass-Analyzed Ion Kinetic Energy Spectra (MIKES)*). This instrument was used for extensive studies of negative-ion rearrangement reactions and in a modified form was among the first used to investigate interstellar molecules through neutralization–reionization approaches (*13,14*). This instrument was also the first used to initiate studies of the structures of host-defense peptides excreted from the glands of Australian frogs in collaboration with Mike Tyler (*15*). This research was significantly advanced following the purchase of a Finnigan LCQ and Micromass QTOF2 instrument (*16*). Close to 200 bioactive peptides have been characterized in these studies and their mode of action investigated with the aid of NMR spectroscopy in collaboration with John Carver and Frances Separovic.

Bowie's highly productive and successful research career promoted his rise through the academic ranks at Adelaide where he has remained since his initial return to Australia. He was appointed a professor in 1983, served as head of department in 1993/1994, Dean of Science in 1981/1982 and held the position of Pro-Vice-Chancellor and Acting Vice Chancellor (when required) from 1989 to 1992. He has received many awards and accolades, among them the Rennie (1967), H. G. Smith (1974) and A. J. Birch (2001) Medals from the Royal Australian Chemical Institute, the Sir Joseph Verco medal of the Royal Society of South Australia (2003), and a DSc from the University of Adelaide in 1969. In 2006, he received the Thomson Medal from the International Mass Spectrometry Foundation.

References

(*1*) Anonymous. Biographical Note. *Org. Mass Spectrom.* **1975**, *10*, v–vi.

(*2*) O'Hair, R. A. J. John Hamilton Bowie – An Appreciation. *Aust. J. Chem.* **2003**, *56*, 343–348.

(*3*) Bowie, J. H.; Cooke, R. G. Colouring Matters of Australian Plants. IX. Anthraquinones from Morinda Species. *Aust. J. Chem.* **1962**, *15*, 332–335.

(*4*) Bowie, J. H.; Bullock, E.; Johnson, A. W. The Structure of Streptolidine. *J. Chem. Soc.* **1963**, 4260.

(*5*) Bowie, J. H.; Cameron, D. W.; Findlay, J. A.; Quartey, J. A. K. Hemolymph Pigments of Aphids. *Nature* **1966**, *210*, 395–397.

(*6*) Bowie, J. H.; Williams, D. H.; Lawesson, S. O.; Schroll, G. Studies in Mass Spectroscopy. VI. Mass Spectra of Substituted Diethyl Malonates. *J. Org. Chem.* **1966**, *31*, 1792–1797.

(*7*) Bowie, J. H.; Duus, F.; Lawesson, S. O.; Larsson, F. C. V.; Madsen, J. O. Electron Impact Studies. XL. Negative-Ion Mass Spectra of Organousulfur Compounds: Thioglycollic Acids and Arylsulphinylamines. *Aust. J. Chem.* **1969**, *22*, 153–159.

(*8*) Bowie, J. H. Negative-ion Mass Spectrometry: A Viable Alternative? H. G. Smith Memorial Lecture. *Proc. Roy. Aust. Chem. Inst.* **1975**, *42*, 175–180.

(*9*) Bowie, J. H.; Trenerry, V. C.; Klass, G. Reactions of Negative Ions in the Gas Phase. *Mass Spectrom. Rev.* **1981**, *6*, 233–240.

(*10*) Bowie, J. H. The Formation and Fragmentation of Negative Ions Derived from Organic Molecules. *Mass Spectrom. Rev.* **1984**, *3*, 161–207.

(*11*) Bowie, J. H. Twenty Five Years of Negative-Ion Studies at Adelaide. *Org. Mass Spectrom.* **1993**, *28*, 1407–1413.

(*12*) Bowie, J. H.; Blumenthal, T. +E collision-induced mass spectra from negative ions. *J. Am. Chem. Soc.* **1975**, *28*, 559–562.

(*13*) Eichinger, P. C. H.; Bowie, J. H. The Gas-Phase Smiles Rearrangement. The Effect of Ring Substitution. An Oxygen-18 Labeling Study. *Org. Mass Spectrom.* **1992**, *27*, 995–999.

(*14*) Blanksby, S. J.; Schroeder, D.; Dua, S.; Bowie, J. H.; Schwarz, H. Conversion of Linear to Rhombic C_4 in the Gas Phase: A Joint Experimental and Theoretical Study. *J. Am. Chem. Soc.* **2000**, *122*, 7105–7113.

(*15*) Stone, D. J. M.; Bowie, J. H.; Tyler, M. J.; Wallace, J. C. The Structure Of Caerin 1.1, A Novel Antibiotic Peptide From Australian Tree Frogs. *J. Chem. Soc. Chem. Commun.* **1992**, *17*, 1224–1225.

(*16*) Wabnitz, P. A.; Bowie, J. H.; Tyler, M. J.; Wallace, J. C.; Smith, B. P. Aquatic Sex Pheromone From A Male Tree Frog. *Nature* **1999**, *401*, 444–445.

Kevin M. Downard
University of Sydney, Australia

Harrison Scott Brown

b. September 26, 1917
d. December 8, 1986

Harrison Brown was an American nuclear chemist who became a pioneer in applying the insights and instruments of nuclear chemistry and physics in astronomical and geochemical research. He guided the work that determined the general age of the earth, meteorites, and the solar system. He led the creation of a world-class program in geochemistry at the California Institute of Technology (Cal Tech) and went on to play in major role in public discussions of several controversial global issues (*1–3*).

Brown was born in 1917 in Sheridan, Wyoming, USA. After his father died when he was 10 years old, he and his mother moved to San Francisco. He earned his bachelor's degree in chemistry in 1938 at the University of California at Berkeley, where he was exposed to mass spectrometry and nuclear chemistry. He worked on mass spectrometry studies of cobalt and earned his PhD in 1941 at The Johns Hopkins University in Baltimore, MD, USA, under the direction of Robert Fowler (*4,5*).

In 1942, Brown was recruited by Glen Seaborg to the Metallurgical Laboratories, a branch of the Manhattan Project at the University of Chicago. His task was the separation of the plutonium formed in the fuel of an atomic reactor from the uranium fuel. With Orville Hill he worked out a method and, after experiments at Chicago, moved to the Oak Ridge branch of the Manhattan Project to help put it into practice.

In 1946, Brown returned to Chicago as an assistant professor of chemistry and member of the Institute for Nuclear Studies. He turned his attention to three linked topics or problem areas, through which he saw that nuclear chemistry would guide our understanding the formation of the earth and solar system. One topic area was determining the abundances of elements and isotopes in the solar system generally and meteorites in particular. A second area was determining the ages of many terrestrial rocks as clues to the course of geological developments. And the third area was determining the age of meteorites as an indicator of the age of the earth and solar system. With the improved mass spectrometer designs available in years immediately after World War II, he could make progress in the first area. For the other two it was necessary to develop methods and machines with even greater sensitivity and freedom from contamination. Brown set the general course and got graduate students (George Tilton and Clair Patterson) and colleagues (especially Mark Inghram) to work together for years to achieve results in the other two areas (see this volume, *Mark Gordon Inghram II, Clair Cameron Patterson*, and *George Robert Tilton*). Through these overlapping studies he was able to develop strong evidence that the solar system condensed from a comparatively cool cloud of gases and dust, that the process took place farther back in time than had previously been estimated, and that the earth's atmosphere was not formed during the initial consolidation of the planet but through a secondary process (*6–9*).

These projects continued through the disruptions of moves to new institutions. Brown was hired in 1951 to start a new program in geochemistry at Cal Tech with the clear expectation that he would build a world-class group in short order; he persuaded Patterson to make the same move, while Tilton went to the Carnegie Institution in Washington, DC.

At the same time that Brown was helping reshape geochemistry in particular and the earth and planetary sciences generally, he also became much involved in public affairs. He grew concerned with the issue of the control of nuclear weapons and with the more general issues of international relations and the relation of science and public affairs. He was involved in the formative years of *The Bulletin of the Atomic Scientists*, and he wrote and published *Must Destruction Be Our Destiny*? in 1946, following that with many public lectures. This began a pattern of bringing scientific results, methods, and questions to wider public audiences that he continued for the rest of his life, involving authorship of many books, some of which are listed below (*10–14*).

Brown was very successful in developing a fine research tradition in geochemistry at Cal Tech and in fostering interdisciplinary work as well. He added a professorship of science and government to his professorship of geochemistry in 1967. He became very much involved in promoting international scientific contacts as foreign secretary for the National Academy of Sciences, to which he had been elected in 1955.

His attention turned more and more to issues of population growth, economic development, technology, energy, and environment, all the time directly and sometimes controversially tackling the political and social questions as well as addressing scientific topics. He left Cal Tech in 1977 to become the founding director of the Resource Systems Institute at the East-West Center in Honolulu, Hawaii. Health problems led to his retirement in 1983, though he then found the energy to serve as editor-in-chief of *The Bulletin of the Atomic Scientists*. Harrison Brown was married three times and divorced twice. He died in Albuquerque, NM, USA, in 1986.

References

(*1*) Revelle, R. Harrison Brown: 1917–1986; *Biographical Memoirs*, vol. 65, National Academies Press: Washington, DC, 1998. pp. 40–55.

(*2*) Holdron, J. Harrison Brown: 1917–1986. *Bull. At. Sci.* **1987**, *43*, 3–8.

(*3*) Smith, K. R., Fesharaki, F., Holdren, J. P., Eds. *Earth and the Human Future: Essays in Honor of Harrison Brown*; Westview Press: Boulder, CO, 1986.

(*4*) Brown, H.; Mitchell, J. J.; Fowler, R. D. The Construction of A Mass Spectrometer for Isotope Analysis. *Rev. Sci. In strum.* **1941**, *12*, 435–442.

(*5*) Mitchell, J. J.; Brown, H.; Fowler, R. D. On The Isotopic Constitution of Cobalt. *Phys. Rev.* **1941**, *60*, 359.

(*6*) Brown, H. An Experimental Method for the Estimation of the Age Of the Elements. *Phys. Rev.* **1947**, *72*, 348.

(*7*) Inghram, M. G.; Hess, D. C.; Brown, H. S.; Goldberg, E. On the Isotopic Composition of Meteoritic and Terrestrial Gallium. *Phys. Rev.* **1948**, *74*, 343–344.

(*8*) Brown, H. Table of Relative Abundances of Nuclear Species. *Rev. Mod. Phys.* **1949**, *21*, 625–634.

(*9*) Tilton, G. R.; Patterson, C.; Brown, H., *et al.* Isotopic Composition and Distribution of Lead, Uranium, and Thorium in a Precambrian Granite. *Geol. Soc. Am. Bull.* **1955**, *66*, 1121–1148.

(*10*) Brown, H. *Must Destruction Be Our Destiny?* Simon & Schuster: New York, 1946.

(*11*) Brown, H.; *The Challenge of Man's Future*; Viking Press: New York, 1954.

(*12*) Brown, H. The Age of the Solar System. *Sci. Am.* **1957**, *196*, 80–95.

(*13*) Brown, H.; Bonner, J.; Weir, J. *The Next Hundred Years: Man's Natural and Technological Resources; A Discussion Prepared for Leaders of American Industry*; Viking Press: New York, 1957.

(*14*) Brown, H. *The Human Future Revisited: The World Predicament and Possible Solutions*; W.W. Norton: New York, 1978.

Keith A. Nier
Independent Historian, Madison, NJ, USA

Curt Brunnée

b. May 24, 1928

Curt Brunnée is a German physicist, whose name, through his distinguished career over 35 years, became synonymous with the mass spectrometry division, Mess- und Analysen Technik (MAT), of the ATLAS Werke AG in Bremen, Germany. Through Brunnée's creativity and steadfastness during many changes in the company, MAT-ATLAS and later Finnigan-MAT became the cradle for the worldwide success of mass spectrometers such as the CH4, MAT 8200, and the MAT 90 series.

Curt Brunnée grew up in Rostock, Germany (former East Germany) and discovered his passion for magic tricks and his natural curiosity in 'how things work' early in his life. He and his family narrowly escaped several life-threatening situations during the World War II years. He entered the Universität Rostock in 1946 as a physics and philosophy major. In post-war Germany, political events made university studies less and less attractive, and Brunnée headed west under life-threatening circumstances in the winter of 1947–48. He entered the Universität Marburg in 1948 and obtained the degree of 'Diplom Physiker' in 1952 and the Dr. *rer. nat.* degree in 1955 with Dr. Wilhelm Walcher, a pioneer of isotope separation beginning in 1937 and professor for experimental physics at Marburg (1947–1978).

By the time Brunnée entered the field of mass spectrometry, Germany's technology lagged behind the world. Pioneering work in mass spectrometry had been done in England (Thomson, Aston, and others) and the United States (Dempster, Nier and others), and commercial instruments of high quality were available from Metropolitan Vickers Electrical Company Ltd., England and Consolidated Engineering Corporation, USA (see this volume, *Magnetic and Electrostatic Analyzers before 1960, Mass Spectrometry Manufacturers and Markets from the Beginning to 1960, Joseph John Thomson, Francis William Aston, Arthur Jeffrey Dempster, Kenneth Tompkins Bainbridge*, and *Alfred Otto Carl Nier*). Not only was Germany not at the developmental forefront in mass spectrometry; the end of WWII had brought total destruction of Germany's economy, and resources for instrument development such as laboratories, funding, and personnel were scarce. Only very few academic institutes had home-built mass spectrometers, one of which, as it happened, was located in the physics department of the Universität Marburg. In that laboratory, under the supervision of Wilhelm Walcher, Brunnée studied ion-surface phenomena (*1*). In 1956, Walcher brought Brunnée in contact with his friend Dr. Ludolf Jenckel (1912–1990), who had founded the Mess- und Analysen Technik (MAT) as an internal group of the ATLAS Werke AG in 1947.

The ATLAS Werke AG (today ATLAS Elektronik), founded in 1902 by the northern German Lloyd as a shipyard, was and still is primarily known for construction of ships, propellers, and naval equipment (Fig. 1). After WWII, Jenckel had successfully convinced the company's leadership to invest in the development of mass spectrometry, which was considered a rather unattractive and exotic technique at the time. After the first difficult years under the direction of Jenckel, MAT was down to only two research and development scientists in 1950 but survived. ATLAS-MAT was relocated in 1956 to a different building in Bremen, filed its first German mass spectrometry patent for the analysis of gas mixtures in positive and negative mode in 1958 (Deutsches Patentamt, Auslegeschrift 1018645, Dr. Willard H. Bennett), and was made a separate company, ATLAS-MAT GmbH in 1962. In 35 years of affiliation with MAT, Brunnée subsequently witnessed the acquisition of MAT by Krupp in 1964, the takeover of the Krupp-MAT division by Varian in 1967, Finnigan in 1981, and Thermo-Electron (Thermo-Fisher since 2006).

Brunnée's contributions to the growth of MAT are countless. He was attracted to the company and initially misjudged ATLAS-MAT as a prospective and financially sound company. Indeed, before the landmark CH4 mass spectrometer in 1958, the company had barely avoided bankruptcy due to weak sales more than once. One of Brunnée's first assignments was the installation of a CH3 mass spectrometer, the successor of Jenckel's first 60° sector prototype, at the University of Osaka, Japan. Despite frequently lacking even the most basic supplies, Brunnée finally succeeded. The CH3 spectrometer was inferior to comparable commercial instruments of the time, and Brunnée was regularly involved with machining instrument parts by himself and improving them. His seminal contribution to the next generation of mass spectrometers, the CH4 (Fig. 2), was a motor-driven vacuum inlet system (Fig. 3) that

Figure 1
ATLAS Werke before 1939 (with permission from Stadtteilarchiv Kulturhaus Pusdorf, Germany).

Figure 2
CH4 Mass spectrometer, connected to a capillary gas chromatograph (1962), (John Wiley & Sons, Ltd., reproduced with permission from Brunnée, C., *50 Years of MAT in Bremen*, *Rapid Comm. Mass Spectrom.*, *11*, 694–707 (1997)).

made the analysis of 'non-volatile' and thermally labile compounds possible without chemical derivatization – a true advantage. Instrument development became Brunnée's primary focus throughout his career (*2–6*).

After the acquisition of ATLAS-MAT by Krupp in 1964 and Krupp-MAT by Varian in 1967, he became head of the 280-employee manufacturing facility in Bremen in 1970 and implemented new computerized

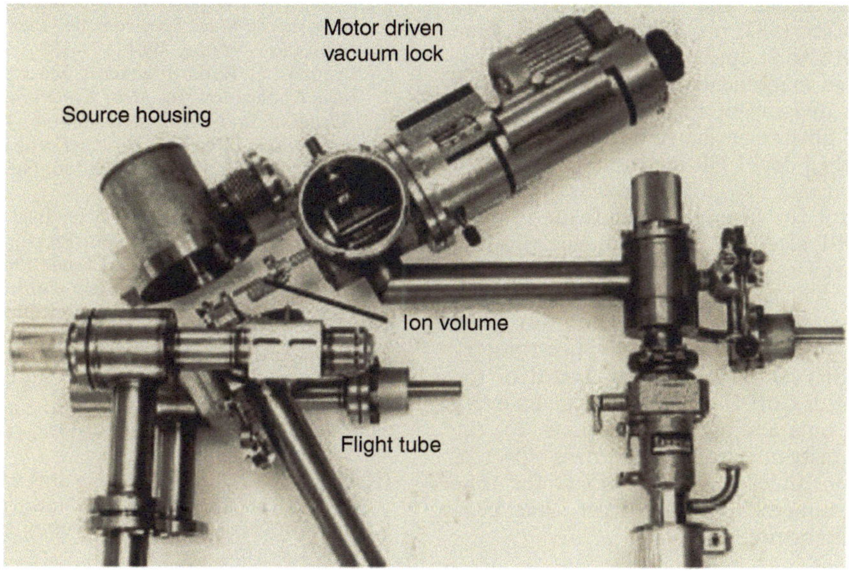

Figure 3
Vacuum lock for the direct introduction of solid samples (1960) (John Wiley & Sons, Ltd., reproduced with permission from Brunnée, C., *50 Years of MAT in Bremen, Rapid Comm. Mass Spectrom.*, *11*, 694–707 (1997).

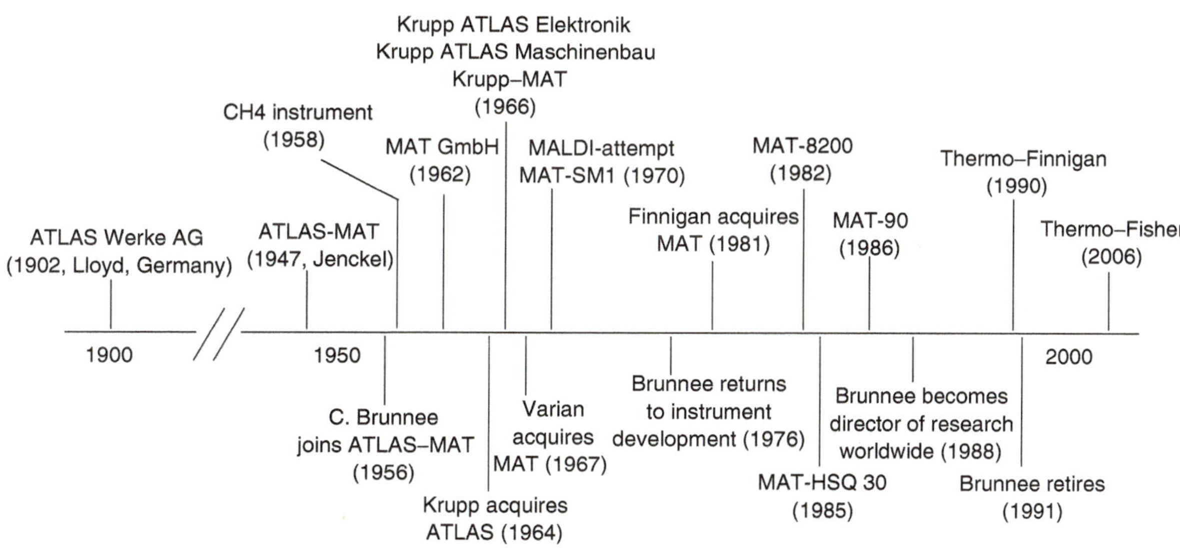

Figure 4
Significant events in the career of C. Brunnée related to Krupp-MAT (later Thermo-Fisher).

manufacturing procedures improving the quality of quadrupole rods significantly. Ahead of his time and technology in 1970, Brunnee attempted to use the Matrix-Assisted Laser Desorption/Ionization (MALDI) technique in collaboration with the Dutch company CIVO TNO in the ATLAS MAT SM1 instrument, however without significant commercial success. After six years, he returned to the development of instrumentation as director of the construction department and experimental machine shop. Over the next ten years, partly or fully-automated GC–MS instruments (*7,8*) such as the MAT 112 S, MAT 311/312, and MAT 8200, and MAT HSQ 30 were developed. Particularly the two latter instruments were extremely versatile, and the MAT HSQ 30 (Finnigan-MAT since 1981) provided for the first time high-resolution MS/MS capability in a MAT instrument. Brunnée was made head of the entire research and development division of Finnigan-MAT. In the last ten years of his career, he developed the completely computerized MAT 90 instrument (*9*) which became popular across the globe. He was promoted to research director for Finnigan worldwide in 1988 and retired in 1991 after an extraordinarily productive career in search of the ideal mass analyzer (*10*) (Fig. 4).

Brunnée was awarded one of the Thomson Gold Medals by the International Mass Spectrometry Foundation (IMSF) in 1994. Another award of the IMSF is named after Curt Brunnée and is given every three years for outstanding contributions to the development of instrumentation for mass spectrometry by a person under the age of 45 at the time of the award. Brunnée's life and career have been described in the literature (*11,12*).

References

(*1*) Brunnee, C. Ion Reflection and Secondary Electron Emission by Impact of Alkali Ions on Pure Molybdenum Surfaces. *Zeitschrift für Physik* **1957**, *147*, 161–183.

(*2*) Brunnee, C. Modifications of a Mass Spectrometer for the Analysis Of Higher Hydrocarbons. *Fresenius' Zeitschrift für Analytische Chemie* **1959**, *170*, 371–380.

(*3*) Brunnee, C.; Delgmann, L. A Modified Leak-Type Mass Spectrometer for Breathable Air Analysis and General Process Control Problems. *Fresenius' Zeitschrift für Analytische Chemie* **1963**, *197*, 51–57.

(*4*) Brunnee, C.; Delgmann, L.; Kronenberger, K. The Atlas Quadrupole Mass Spectrometer. *Publ. Group. Avan. Methodes Spectrog.* **1964**, 177–188.

(*5*) Brunnee, C. Radio-Frequency Mass Spectrometers and their Application. In: *Mass Spectrometry, a NATO Advanced Study Institute on Theory, Design and Applications, Glasgow, Scotland, August 1964*; Reed, R. I., Ed.; Academic Press: London, New York, 1965; pp 37–59.

(*6*) Brunnee, C. Mass Spectrometer Inlet System for Organic Compounds with Low Vapor Pressure. *Fresenius' Zeitschrift für Analytische Chemie* **1966**, *217*, 333–339.

(*7*) Brunnee, C. A New Computer-Controlled Mass Spectrometer for Exact Isotope Ratio Measurements on Gases. *Chemie Technik (Heidelberg)* **1978**, *7*, 111–115.

(*8*) Brunnee, C.; Kappus, G.; Rache, H.; Seiler, E. U.; Windel, B. A Computerized Mass Spectrometer for Isotopic Analyses of Solids. *American Laboratory (Shelton, CT, United States)* **1978**, *10*, 141–142, 144, 146–147, 148–149.

(*9*) Brunnee, C.; Jung, G.; Markwardt, U.; Pesch, R.; Wollnik, H. An Automated Magnetic Sector Mass Spectrometer. *American Laboratory* **1986**, *18*, 102, 104, 106, 108–110.

(*10*) Brunnee, C. The Ideal Mass Analyzer: Fact or Fiction? *Int. J. Mass Spectrom. Ion. Proc.* **1987**, *76*, 125–237.

(*11*) Brunnee, C. 50 Years of MAT in Bremen. *Rapid Commun. Mass Spectrom.* **1997**, *11*, 694–707.

(*12*) ten Noever de Brauw, M. C. A Short Story about the Life of Curt Brunnee. *Rapid Commun. Mass Spectrom.* **1997**, *11*, 708–713, reprinted in Jennings, K. R., Ed. *A History of European Mass Spectrometry*, IM Publications. Chichester, England. 2012, pp. 164–174.

Andreas H. Franz

Department of Chemistry, University of the Pacific, Stockton, CA, USA

Angus Ewan Cameron

b. October 14, 1906.
d. September 27, 1981.

Gus Cameron, as he was known, was an American chemist who worked extensively with mass spectrometers for most of his career in science, primarily with the US Atomic Energy Commission (AEC) at Oak Ridge, Tennessee, USA (*1,2*). He was also a long-serving member of the International Commission on Atomic Weights within the International Union of Pure and Applied Chemistry (IUPAC). In 1960/61, when IUPAC and the International Union of Pure and Applied Physics voted to change the atomic mass scale, he converted all mass spectrometric measurements of the atomic weight values derived from the isotopic composition of all of the chemical elements from the older scale to the new standard in which the mass of carbon-12 (^{12}C) is 12 exactly.

Angus Ewan (Gus) Cameron was born on October 14, 1906 in Sylvania, Pennsylvania. He died on September 27, 1981 in Oak Ridge, Tennessee. Cameron was married to Gray Williams and had three sons, Allan W. Cameron, Douglas Cameron, and Alexander Cameron. He was an ardent sportsman, especially enjoying the game of squash; he liked to teach young people how to play the game.

Cameron received his undergraduate degree from Oberlin College in Ohio, where he graduated magna cum laude in 1928. He then went to the University of Minnesota and received his PhD in Physical Chemistry in 1932. Cameron held a National Research Council Fellowship in Chemistry for 1932–1933 at the University of Rochester in Rochester, New York at their Institute of Applied Optics. He then joined the Eastman Kodak Company in Rochester, where he worked at the Kodak Research Laboratories.

Except for a brief period of time in 1947–1948 at the Eastman Company's Distillation Products Inc. in Rochester, NY, Cameron spent the bulk of his scientific career, beginning in 1943, at US AEC (now the US Department of Energy) sites at Oak Ridge, Tennessee later renamed the Oak Ridge National Laboratory (ORNL). These were the Y-12 (electromagnetic separation) plant, the K-25 (gaseous diffusion) plant, and the X-10 (graphite reactor) laboratory. Mass spectrometry played an important role in these installations (*3*). Due to the classified nature of the work during the War, much of Cameron's work and accomplishments were never published.

The Oak Ridge installations were part of the World-War II Manhattan Project to develop an atomic bomb. Y-12, known also as the Clinton Engineering Works, used Calutrons to enrich natural uranium in the isotope uranium-235 (^{235}U) by electromagnetic isotope separation. A Calutron was like a magnetic deflection mass spectrometer but on a very large scale employing beam currents on the order of 20–100 milli-amperes. Multiple Calutrons were arranged around magnets in massive ovals, which resembled and were called race tracks that were122 feet long, 77 feet wide, and 15 feet high. The Eastman Company transferred Cameron to Y-12, which it operated, where he was in charge of many projects, the most important of which being all of the mass spectrometers. From 1943 to 1947, he served as the Superintendent of the Production Assay Department with over 300 employees. His mass spectrometer laboratory provided measurements of the isotopes of various uranium materials employed in the weapons and related research programs (see this volume, *Mass Spectrometry and the Separation and Enrichment of Isotopes* and *A History of Mass Spectrometry in Nuclear Science*).

The K-25 plant used gaseous diffusion, a second fissionable isotope enrichment method, to separate the isotope ^{235}U from ^{238}U by differential passage through porous barriers. The K-25 building when built was the biggest building in the world. Hundreds of mass spectrometers, designed by Al Nier, were used to monitor the diffusion cascade (see this volume, *Alfred Otto Carl Nier*). Cameron worked as a physicist at K-25 from 1948 to 1954, when he became a Fulbright Fellow for a year at the Max Planck Institute in Mainz, Germany.

Following the war, the Y-12 facility was converted to the separation of stable isotopes and Cameron was the Department Head of the Stable Isotopes Division from 1955 to 1957. He then became Section Chief for mass spectrometry in the Analytical Chemistry Division at ORNL from 1957 to 1960 and Assistant Director of the Division from 1960 to his retirement in 1971. He then served as a consultant to that group until his death.

During his time at Y-12 and K-25, Cameron built a time of flight mass spectrometer with David Eggers and produced a TOF spectrum of the mercury isotopes in 1944 (see this volume, *The Development of Time-of-Flight (TOF) Mass Spectrometry*). In 1945, he and Jerry Spears built a spark source mass spectrometer using the geometry of Arthur Dempster and analyzed the ten isotopes of tin (see this volume, *Arthur Jeffrey Dempster*). Cameron developed graded glass-to-copper seals that were most useful in the vacuums in mass spectrometers. He perfected a technique for making uranium hexafluoride for gas mass spectrometry by reacting cobalt tri-fluoride with oxides of uranium. To minimize the exposure to the toxic and radiation effects of plutonium, he devised a method of confining a mass spectrometer in a glove box so that an isotopic analysis could be performed in a safe manner. This early effort was later used to establish the trans-plutonium mass spectrometer laboratory, in which ORNL built a double focusing instrument with pulse counting detection that was used for many years. Cameron also served as a consultant to Litton Systems on the development of the gas chromatograph mass spectrometer (GCMS) for the Viking mission to Mars (see this volume, *The Development of Mass Spectrometry in the Earth and Planetary Sciences*). The GCMS instrumentation has been on display at the Smithsonian Institution in Washington, DC (*4–10*).

Cameron was elected to the International Commission on Atomic Weights in 1959 and served until his death in 1981. Cameron also served on the Commission's IUPAC Mass Spectrometric Evaluation Group and Sub-committee on Assessment of Isotopic Composition. When the decades-long problem of conflicting atomic mass scales, namely the chemical one setting natural oxygen = 16 and the physical one with $^{16}O = 16$, was finally resolved during 1960–1961, he was the co-author of the Cameron-Wichers report that converted all atomic weight measurements for all chemical elements to the new carbon-12 = 12 scale (*11,12*).

References

(*1*) Jaques Cattell Press, Eds., American Men and Women of Science, Physical and Biological Science, Vol. II, s.v. "Cameron, Angus Ewan." R.R. Bowker Co., New York, London, 14th edition, 1979, p. 695.

(*2*) Walker, R. L.; Carter, J. A.; Christie, W. H. In Memoriam of A, E. Cameron. *Int. J. Mass Spectros. Ion Phys.* **1982**, *42*, 1–2.

(*3*) McLuckey, S. A.; Smith, D. H.; Compton, R. N. Mass Spectrometry R & D at Oak Ridge National Laboratory. *Rapid Commun. Mass Spectrom.* **1996**, *10*, 253–255.

(*4*) White, J. R.; Cameron, A. E. The Natural Abundance of Isotopes of Stable Elements. *Phys. Rev.* **1948**, *74*, 991–1000.

(*5*) Cameron, A. E.; Eggers, D. F. Jr. An Ion Velocitron. *Rev. Scientific Instr.* **1948**, *19*, 605–607.

(*6*) Cameron, A. E. A. Compact High-Vacuum Valve. *Rev. Scientific Instr.* **1954**, *25*, 1027–1028.

(*7*) Cameron, A. E.; Lippert, E. L. Jr. Isotopic Composition of Bromine in Nature. *Science* **1955**, *121*, 136–137.

(*8*) Wampler, J. M.; Smith, D. H.; Cameron, A. E. Isotopic Comparison of Lead in Tektites with Lead in Earth Materials. *Geochim. Cosmochim. Acta* **1969**, *33*, 1045–1055.

(*9*) Smith, D. H.; Cameron, A. E.; Dean, J. A. Mass Spectrometric Investigation of Surface Ionization. VI. Desorption of Th^+, Np^+, and Pu^+ from Tungsten Surfaces. *J. Chem. Phys.* **1971**, *54*, 170–173.

(*10*) Walton, J. R.; Cameron, A. E.; Walker, R. L.; Hebble, T. L. Determination of the Abundance of Krypton in the Earth's Atmosphere by Isotope Dilution Mass Spectrometry. *Int. J. Mass Spectros. Ion Phys.* **1973**, *12*, 439–453.

(*11*) Cameron, A. E.; Wichers, E. Report of the International Commission on Atomic Weights 1961. *J. Am. Chem. Soc.* **1962**, *84*, 4175–4197.

(*12*) Cameron, A. E. The Determination of Atomic Weights by Mass Spectrometry. *Anal. Chem.* **1963**, *35*, 23A–35A.

Norman E. Holden
National Nuclear Data Center, Brookhaven National Laboratory, Upton, NY, USA

Richard M. Caprioli

b. April 12, 1943

Dr. Richard Caprioli is an American chemist and biochemist whose general research interests lie in using mass spectrometry to discover temporal and spatial processes in biological systems. This work has included technology developments in the areas of electrospray and laser desorption ionization mass spectrometry and their applications to the study of intact tissues and other biological samples, work that has focused on achieving ultra-high sensitivity detection of endogenous compounds (e.g., neuropeptides) in intact tissues and in live animal systems.

Caprioli was born in New York City on April 12, 1943. He studied at Columbia University, receiving a B. Sc. in Pharmcy from there in 1965 and a Ph.D. in Biochemistry in 1969 under Dr. David Rittenberg. Leaving New York to take up a postdoctoral appointment in Chemistry with Professor John Beynon at Purdue University, West Lafayette, IN, he was appointed as Assistant Professor of Biochemistry at Purdue in 1970. He moved to the University of Texas Medical School in Houston in 1975, where he became Professor of Biochemistry and Molecular Biology and Director of the Analytical Chemistry Center and held a joint appointment at the M.D. Anderson Cancer Center. He moved to Vanderbilt University in 1998, where he currently holds the positions of Stanford Moore Professor of Biochemistry, Professor of Chemistry, Pharmacology and Medicine, and Director, Mass Spectrometry Research Center, Vanderbilt University, Nashville, TN, USA.

Caprioli's experience in mass spectrometry began in graduate school, where he learned procedures and instrumentation techniques related to stable isotope ratio mass spectrometry. Always with an interest in the applications for new technology, in 1986 Caprioli pioneered the development of continuous-flow fast atom bombardment (cf-FAB) mass spectrometry, a variation on the Fast Atom Bombardment (FAB) technique introduced by Michael Barber in 1980 (see this volume, *Static and Continuous – Flow Fast Atom Bombardment*) that enabled the coupling of chromatography and flow injection analysis with a desorption ionization technique for the first time. In 1994, he designed and developed a low flow rate electrospray (ESI) ionization source for LC/MS/MS that coupled small bore separations techniques such as capillary chromatography and capillary electrophoresis to a mass spectrometer.

In 1997, Caprioli published the first of a series of papers that describe the development of matrix-assisted laser desorption (MALDI) mass spectrometry for imaging of biological materials, especially for tissue sections, both for basic biological research and for clinical specimens. This technology enables imaging of peptides, proteins, drugs, and other compounds with molecular weight specificity localized at their sites of concentration in tissue sections. Investigations in specific research areas involve elucidation of spatial distributions of molecules in both health and disease. Of special interest is the molecular distribution, spatial rearrangement, and alterations of expression levels of such molecules in cancer tissues including human glioblastomas, breast, colorectal, prostate, and lung cancer.

Caprioli has been a member of the American Society for Mass Spectrometry since 1975 and served as its President from 2000–2001. He is also a member of the American Society for Biochemistry and Molecular Biology, the American Association for Cancer Research, and the American Chemical Society. He has been the Editor-in-Chief of the *Journal of Mass Spectrometry* since 1990. He is currently Series Co-Editor of *The Encyclopedia of Mass Spectrometry* and additionally has edited several volumes in this work. He has been a member of the Board of Directors of the US HUPO since its inception.

Caprioli's work is well documented in the publication of over 300 scientific papers and three books. He also holds 12 issued US patents and 7 patents pending involving mass spectrometry technologies. For these efforts, he has received numerous awards, including the Thomson Medal Award from the International Mass Spectrometry Society for outstanding achievements in mass spectrometry and for distinguished service to international mass spectrometry. He was named one of the 'Pioneers in Proteomics' by the National Institutes of Health and received the Donald H. Coffey Award from the Society for Basic Urologic Research in 2005. Also in that year, he received the honorary degree Laurea Honoris Causa, from the University of Naples in Biological Sciences. In 2006, he received the Field and Franklin Award from the American Chemical Society for Outstanding Achievement in Mass Spectrometry, the Eastern Analytical Society 2010 Award for Achievements in Mass Spectrometry and the HUPO (Human Proteome Organization) Distinguished Achievement Award in Proteomic Sciences for 2010.

The publications cited below are representative of the author's most significant contributions to the field (*1–6*).

References

(*1*) Caprioli, R. M.; Fan, T.; Cottrell, J. Continuous-Flow Sample Probe for Fast Atom Bombardment Mass Spectrometry. *Anal Chem.* **1986**, *58*(14), 2949–2954.
(*2*) Emmett, M. R.; Caprioli, R. M. Microelectrospray MS: Ultra-High-Sensitivity Analysis of Peptides and Proteins. *J. Am. Soc. Mass Spectrom.* **1994**, *5*, 605–613.
(*3*) Caprioli, R. M.; Farmer, T. B.; Gile, J. Molecular Imaging of Biological Samples: Localization of Peptides and Proteins Using MALDI-TOF MS. *Anal Chem.* **1997**, *69*, 4751–4760.
(*4*) Stoeckli, M.; Chaurand, P.; Hallahan, D. E.; Caprioli, R. M. Imaging Mass Spectrometry: A New Technique Applied to Mammalian Brain Tumor Protein Expression. *Nat. Med.* **2001**, *7*, 493–496.
(*5*) Chaurand, P.; Sanders, M. E.; Jensen, R. A.; Caprioli, R. M. Proteomics in Diagnostic Pathology: Profiling and Imaging Proteins Directly in Tissue Sections. *Am. J. Pathol.* **2004**, *165*(4), 1057–1068.
(*6*) Andersson, M.; Groseclose, M. R.; Deutch, A. Y.; Caprioli, R. M. Imaging Mass Spectrometry of Proteins and Peptides: 3D Volume Reconstruction. *Nat. Methods* **2008**, *5*(1), 101–108.

Richard M. Caprioli
Departments of Biochemistry, Chemistry, Pharmacology and Medicine and the Mass Spectrometry Research Center, Vanderbilt University School of Medicine, Vanderbilt University, Nashville, TN, USA

Brian T. Chait

b. July 6, 1949

Brian Chait was born and raised in Cape Town, South Africa. He attended the University of Cape Town, from which he received a B.Sc. in Natural Sciences in 1969 and a B.Sc. Hons. in Physics in 1970. He went on to study nuclear physics at Oxford, where he received his DPhil in 1976.

Moving to Canada to take up a post-doctoral appointment in Ken Standing's laboratory at the University of Manitoba in 1977, Chait began the work on which he would focus for much of his career: creating instrumentation with which to study biological systems, in particular proteins (see this volume, *Kenneth Graham Standing*). Chait first built a SIMS instrument in Manitoba in the Standing lab and then built another at the Rockefeller University during an appointment as a research associate in Frank Field's laboratory there (see this volume, *The Development of Secondary Ion Mass Spectrometry (SIMS) for Imaging, A history of Imaging Mass Spectrometry*, and *Frank Henry Field*). He has remained at Rockefeller throughout his career, becoming an assistant professor in 1981, Associate Professor in 1985, Professor in 1991, and the Camille and Henry Dreyfus Professor in 1995. He became Associate Director of the Mass Spectrometric Biotechnology Research Resource in 1983 and its Director in 1988.

Throughout his career at Rockefeller, Chait has used mass spectrometry as a tool for investigating a variety of biological and biochemical phenomena. His focus has been on characterizing proteins and their structure as related to a cell's activities. Along the way, however, he has also developed tools to aid in the process, devices that fill a need so simply that they are often quickly adopted in other laboratories across the world.

Chait's awards include the 2007 HUPO Discovery Award in Proteomics Sciences, the 2002 Frank H. Field and Joe L. Franklin Award for Outstanding Achievement in Mass Spectrometry from the American Chemical Society, the 2000 Bijvoet Medal from Utrecht University and the 1998 American Association for the Advancement of Science Newcombe-Cleveland Prize. He has been awarded 22 United States patents for his inventions.

The publications cited below are representative of the author's most significant contributions to the field (*1-9*).

References

(*1*) Chait, B. T.; Agosta, W. C.; Field, F. H. Fission Fragment Ionization (252Cf) Mass Spectrometry. Positive and Negative Spectra and Decomposition Mechanisms for Seven Compounds. *Int. J. Mass Spectrom. Ion Phys.* **1981**, *39*, 339–366.

(*2*) Gale, P. J.; Bentz, B. L.; Chait, B. T.; Field, F. H.; Cotter, R. J. Reduction in Liquid Secondary Ion Mass Spectrometry. Comparison of the Fission Fragment and Liquid Secondary Ion Mass Spectra of Organic Dyestuffs. *Anal. Chem.* **1986**, *58*, 1070–1076.

(*3*) Merrifield, R. B.; Singer, J.; Chait, B. T. Mass Spectrometric Evaluation of Synthetic Peptides for Deletions and Insertions. *Anal. Chem.* **1988**, *174*, 399–414.

(*4*) Beavis, R. C.; Chait, B. T. Cinnamic Acid Derivatives as Matrices for Ultraviolet Laser Desorption Mass Spectrometry of Proteins. *Rapid Commun. Mass Spectrom.* **1989**, *3*, 432–435.

(*5*) Katta, V.; Chait, B. T. Hydrogen/Deuterium Exchange Electrospray Ionization Mass Spectrometry: A Method for Probing Protein Conformational Changes in Solution. *J. Am. Chem. Soc.* **1993**, *115*, 6317–6321.

(*6*) Rout, M. P.; Aitchison, J. D.; Hjerdaas, K.; Suprapto, A.; Zhao, Y.; Chait, B. A Compositional Survey of the Nuclear Pore Complex. *Mol. Biol. Cell* **1998**, *9*, 132A.

(*7*) Zhang, W. Z.; Krutchinsky, A. N.; Chait, B. T. '*De novo*' Peptide Sequencing by MALDI-Quadrupole-Ion Trap Mass Spectrometry: A Preliminary Study. *J. Am. Soc. Mass Spectrom.* **2003**, *14*(9), 1012–1021.

(*8*) Tackett, A. J.; Dilworth, D. J.; Davey, M. J., *et al.* Proteomic and Genomic Characterization of Chromatin Complexes at a Boundary. PMCID 2171912.*J. Cell Biol.* **2005**, *169*(1), 35–47.

(*9*) Ueberheide, B. M.; Fenyo, D.; Alewood, P. F.; Chait, B. T. Rapid Sensitive Analysis of Cysteine Rich Peptide Venom Components. PMID 19380747 PMCID 2678425. *Proc. Natl. Acad. Sci. USA* **2009**, *106*(17), 6910–6915.

P. Jane Gale
Waters Corp., Milford, MA, USA

Keith A. Nier
Independent Historian, Madison, NJ, USA

Alfred L. Yergey
National Institute of Child Health and Human Development, NIH, Bethesda, MD, USA

Robert N. Clayton

b. March 20, 1930

Robert N. Clayton is a Canadian-American isotope geochemist who has used stable isotope geochemical methods to discover important aspects of the nature and development of the earth, moon, and meteorites. His research opened the field of isotope anomalies in meteorites and has led to significant insights into the nuclear and chemical history of the solar system.

Robert Clayton was born in Hamilton, Ontario, Canada, in 1930. He did his undergraduate studies at Queens University, in Kingston, Ontario, where his chemistry major included courses in geology and mineralogy. He graduated in 1951 and added an MSc in 1952. He earned his PhD at the California Institute of Technology, with Samuel Epstein as his advisor, in 1955 and spent an additional year there as a research fellow (see this volume, *Samuel Epstein*). He was an assistant professor of geochemistry for two years at Penn State University before moving in 1958 to the University of Chicago, where he has remained. At Chicago he 'inherited' Harold Urey's lab and equipment. He became full professor in 1966 and the Enrico Fermi Distinguished Service Professor in 1980. He formally retired in 2001, but adding 'Emeritus' to his title did not stop his research (*1–4*).

Clayton's research career divides into two periods, namely that with an almost entirely terrestrial focus and a second with more emphasis on an extraterrestrial orientation, divided by the preparation for and working with samples brought back from the first Apollo lunar landing. He and his students investigated the isotopic characteristics of rocks and waters in various environments to trace the movement and interactions of the water and cycles of elements in the earth's crust and mantle. Because of careful preparation for work on lunar samples, Clayton was among those ready to respond to the opportunity presented by the Allende meteorite fall in 1969. The discovery of major anomalous ratios among the three oxygen isotopes in meteorite inclusions changed the field of cosmochemistry fundamentally. From that point on, Clayton's research has broadened to include the characteristics of the solar nebula and the formation of meteorites and the terrestrial planets along with his earlier interests in various terrestrial processes such as the development of earth's atmosphere. It is worth noting that through all this, Clayton has demonstrated the continuing capabilities of what some might think of as outmoded instrumentation, for he has continued to rely upon the McKinney-Nier machine at Chicago even while employing newer techniques and devices from ion probes to resonance ionization mass spectrometry. A small sample of his research publications are listed below.

Clayton is married. He retained his Canadian identity though his whole professional life has been in the US. He is a member of the National Academy of Sciences and the Royal Society of London. He has received many other honors for his work, including the Goldschmidt, Leonard, and Bowie Medals, the highest awards of the Geochemical Society, the Meteoritical Society, and the American Geophysical Union, respectively (*5–13*).

References

(*1*) Clayton, R. N. Isotopes: From Earth to the Solar System. *Annu. Rev. Earth Planet. Sci.* **2007**, *35*, 1–19.

(*2*) Marvin, U. B. Oral Histories in Meteoritics and Planetary Science: II. Robert N. Clayton. *Meteorit. Planet. Sci.* **2001**, *36*, A269–A274.

(*3*) Goldsmith, J. R. William Bowie Medal to Robert N. Clayton: Citation. *EOS Trans. Am. Geophys. Union* **1987**, *68*, 624–625.

(*4*) Humayun, M.; O'Neil, J. R. A Special Issue Dedicated to Robert N. Clayton. *Geochim. Cosmochim. Acta* **2003**, *67*, 3097–3099.

(*5*) Clayton, R. N. Oxygen Isotope Fractionation between Calcium Carbonate and Water. *J. Chem. Phys.* **1961**, *34*, 724–726.

(*6*) Northrop, D. A.; Clayton, R. N. Oxygen-Isotope Fractionations in Systems Containing Dolomite. *J. Geol.* **1966**, *74*, 174–196.

(*7*) Clayton, R. N.; Grossman, L.; Mayeda, T. K. A Component of Primitive Nuclear Composition in Carbonaceous Meteorites. *Science* **1973**, *182*, 485–488.

(*8*) Muehlenbachs, K.; Clayton, R. N. Oxygen Isotopic Composition of the Oceanic Crust and its Bearing on Seawater. *J. Geophys. Res.* **1976**, *81*, 4365–4369.

(*9*) Clayton, R. N.; Mayeda, T. K. Correlated Oxygen and Magnesium Isotope Anomalies in Allende Inclusions, I: Oxygen. *Geophys. Res. Lett.* **1977**, *4*, 295–298.

(*10*) Thiemens, M. H.; Clayton, R. N. Ancient Solar Wind in Lunar Microbreccias. *Earth Planet. Sci. Lett.* **1980**, *47*, 34–42.

(*11*) Humayun, M.; Clayton, R. N. Potassium Isotope Cosmochemistry: Genetic Implications of Volatile Element Depletion. *Geochim. Cosmochim. Acta* **1995**, *59*, 2131–2148.
(*12*) Cook, D. L.; Wadhwa, M.; Janney, P. E.; Clayton, R. N.; Davis, A. M. High-Precision Measurements of Nickel Isotopes in Metallic Samples via Multi-Collector ICPMS. *Anal. Chem.* **2006**, *78*, 8477–8484.
(*13*) Clayton, R. N. Oxygen Isotopes in the Early Solar System – A Historical Perspective. *Rev. Mineral Geochem.* **2008**, *68*, 4–14.

Keith A. Nier
Independent Historian, Madison, NJ, USA

Norman David Coggeshall

b. May 15, 1916
d. April 22, 1995

Norman Coggeshall was an American physicist known for his work in mass spectrometry and gas chromatography. He worked in other fields of analytical spectroscopy as well and developed apparatus for determination of particulate surface area and pore size distributions.

Norman David Coggeshall was born on 15 May 1916 in Ridge Farm, IL, USA. He received BA (1937), MS (1938) and PhD (1942) all in physics and all from the University of Illinois. He joined the Gulf Research and Development Company in June 1943 and remained there for his entire scientific career. He married Margaret Josephine Danner while still a graduate student and together they raised two sons and two daughters. After retiring from Gulf research in 1981, he moved to Lynn Haven, FL, where he became a private investor and consultant. He died there in April 1995.

Coggeshall's research spanned issues in instrument design, particularly problems of mass discrimination (*1*), ranged to studies of thermal decomposition and ionization and spectra of hydrocarbons (*2–4*), as might be expected in a petroleum company, and included studies of metastable ion decomposition (*5*).

From his work at Gulf, Coggeshall was granted many patents. He was a Fellow of the American Physical Society. His society memberships included the Massachusetts Spectrometry Society, the Spectroscopy Society Pittsburgh, and the American Chemical Society (ACS). In 1969, he received the ACS Award for Chemical Instrumentation in recognition of his work in modern physics and mass spectrometry, and in 1970 he was the recipient of a Resolution of Appreciation from the American Petroleum Institute.

Note: Norman Coggeshall's younger daughter, M. Gwen (Coggeshall) Calabretta, generously contributed to this article.

References

(*1*) Coggeshall, N. D. Discrimination in Mass Spectrometer Ion Sources. *J. Chem. Phys.* **1944**, *12*, 19–23.

(*2*) Coggeshall, N. D.; Kerr, N. F. Mass Spectrometer Studies of Thermal Decomposition Products from Hydrocarbons. *J. Chem. Phys.* **1949**, *17*, 1016–1021.

(*3*) Coggeshall, N. D. Ionization of *n*–Paraffin Molecules. *J. Chem. Phys.* **1959**, *30*, 595.

(*4*) Coggeshall, N. D. Quantitative Relations in the Mass Spectra of *n*-Paraffins. *J. Chem. Phys.* **1960**, *33*, 1247–1252.

(*5*) Coggeshall, N. D. Studies of Metastable Ion Transitions with a 180° Mass Spectrometer. *Chem. Phys.* **1962**, *37*, 2167–2175.

Alfred L. Yergey
National Institute of Child Health and Human Development, NIH, Bethesda, MD, USA

Melvin Barnet Comisarow

b. May 30, 1941

Mel Comisarow is a Canadian scientist, best known for his part in the development of Fourier Transform Ion Cyclotron Resonance (FT-ICR) Spectroscopy. His break-through realization was that a method to detect excited ICR motion was required, something that conventional ICR theory did not offer. Comisarow's rotating electric monopole theory, worked out in the early 1970s but not published in completeness until 1978, provided the theoretical basis for this ICR detection. Since its development in 1974, the FT-ICR instrument has become recognized as a very high performance and versatile type of mass spectrometer that can be used for the study of ion–molecule chemistry and a wide variety of problems in analytical chemistry, especially the analysis of biological molecules and petroleum.

Comisarow was born on 30 May 1941, in Viking, Alberta, Canada and graduated in Honors Chemistry from the University of Alberta, Edmonton, Alberta. He spent the next two years with Dow Chemical, first in Sarnia, Ontario, Canada and then in Wayland, MA, USA. He obtained his PhD from Case Western Reserve University, studying under George Olah (1994 Nobel Laureate in chemistry). He spent 2 years as a postdoctoral fellow with John Baldeschwieler at Stanford University, where he learned about conventional ion cyclotron resonance spectroscopy, the precursor to FT-ICR. Comisarow then joined the Chemistry Department at the University of British Columbia (UBC), Vancouver, BC, Canada, where he is now Emeritus Professor of Chemistry.

In 1966, Richard Ernst, a nuclear magnetic resonance (NMR) spectroscopist at Varian Associates in Palo Alto, CA, developed the Fourier Transform NMR spectrometer, a faster spectrometer that could, via time averaging, produce NMR spectra with enhanced sensitivity. Although NMR had been a valuable spectroscopic technique to the synthetic chemist, it was notorious for its low sensitivity. The enhanced sensitivity was a major breakthrough and, as soon as FT-NMR instruments were commercialized, they were eagerly adopted by the chemical community. In 1991, Richard Ernst received the Nobel Prize in chemistry for his development of FT-NMR.

In September 1972 Alan Marshall, also at UBC, approached Comisarow, who was at the time building a conventional ICR spectrometer, with the proposal that they develop FT-ICR (see this volume, *Alan G. Marshall*). In March 1973, after reaching agreement with Marshall that the FT experiment was inherently faster not inherently more sensitive, Comisarow agreed to try to develop FT-ICR. After discovering in October 1973 that the pulse excitation method that Marshall proposed adapting from Ernst's FT-NMR was unworkable in ICR but that a frequency sweep excitation method was feasible. Using equipment borrowed from other UBC professors, Comisarow worked toward complete conversion of his conventional ICR spectrometer to an FT-ICR instrument. The first successful FT-ICR experiments were conducted in December 1973, and published early in 1974.

The most significant feature of FT-ICR, first recognized, predicted, demonstrated, and published by Comisarow soon after the first FT-ICR experiments, is its high resolution, the highest of any type of mass spectrometer. The enhancement in performance – many times greater than the resolution available from conventional ICR – has no parallel in other forms of Fourier spectroscopy, e.g., FT-NMR has the same resolution as conventional NMR. This feature was used by Rick Smalley in his famous 1988 'shrink wrap' experiment that demonstrated the hollow nature of C_{60}. The high resolution of ion cyclotron resonance was used by Gerry Gabrielse of Harvard to show that the masses of the proton and the antiproton were the same. This latter experiment is a rare example of a chemist first demonstrating a spectroscopic principle that was later used by a physicist to investigate fundamental properties of matter.

In recognition of his FT-ICR work he received the 1989 Barringer Award of the Canadian Society for Analytical Sciences and Spectroscopy, the 1991 Royal Society of Chemistry (UK) Award in Chemical Analysis and Instrumentation, the 1995 American Chemical Society Field Franklin Award for Outstanding Achievement in Mass Spectrometry (shared with A. G. Marshall), the 1996 Canadian Society for Chemistry Fisher Award for Distinguished Contribution in Analytical Chemistry, the 1996 University of British Columbia Killam Research Prize, the 1999 American Society of Mass Spectrometry Award for Distinguished Contribution to Mass Spectrometry (shared with A. G. Marshall), and the 2010 University of Alberta Chemistry Alumni Award for Outstanding Achievement. In 1995, in recognition of his development of FT-ICR, he was elected to Fellowship in the Royal Society of Canada.

Mel Comisarow married Ruth Wilson, an industrial psychologist, who, until her retirement, had her own firm in Vancouver. They have one son, Jeff, who graduated in science from UBC and then from the MD-MBA program at UCLA. He now runs a biotech hedge fund on Wall Street.

The publications listed below are representative of the author's most significant contributions to the field (*1–9*).

References

(*1*) Comisarow, B.; Marshall, A. G. Fourier Transform Ion Cyclotron Resonance Spectroscopy. *Chem. Phys. Lett.* **1974**, *25*, 282–283.

(*2*) Comisarow, B.; Marshall, A. G. Frequency Sweep Fourier Transform Ion Cyclotron Resonance Spectroscopy. *Chem. Phys. Lett.* **1974**, *26*, 489–490.

(*3*) Comisarow, B.; Marshall, A. G. Resolution-Enhanced Fourier Transform Ion Cyclotron Resonance Spectroscopy. *J. Chem. Phys.* **1975**, *62*, 293–295.

(*4*) Comisarow, M. Signal Modelling for Ion Cyclotron Resonance. *J. Chem. Phys.* **1978**, *69*, 4097–4104.

(*5*) Comisarow, B.; Grassi, V.; Parisod, G. Fourier Transform Ion Cyclotron Double Resonance. *Chem. Phys. Lett.* **1978**, *57*, 413–416.

(*6*) Comisarow, B. Fourier Transform Ion Cyclotron Resonance Spectroscopy. *Adv. Mass Spec.* **1980**, *8*, 1605–1698.

(*7*) Parisod, G.; Comisarow, B. FT-ICR Study of the Ion Molecule Condensation Chemistry of $CpCr(CO)_2NS$ and Related Compounds. In: *Advances in Mass Spectrometry, Volume 8A*; Quayle, A., Ed.; Heyden & Son Ltd.: London, UK, 1980, pp. 212–223.

(*8*) Comisarow, B. Signals, Noise, Sensitivity and Resolution in Ion Cyclotron Resonance Spectroscopy. In: *Ion Cyclotron Resonance Spectroscopy*; Wanczek, K. P., Ed.; Springer Verlag: Berlin, 1982; pp 484–513.

(*9*) Lam, Z.; Comisarow, B.; Dutton, G. S.; Weil, A.; Bjarnason, A. The Structural Elucidation of Bacterial Capsular Polysaccharides Using Laser Desorption Ionization Fourier Transform Ion Cyclotron Resonance Spectroscopy (LDI-FT-ICR). *Rapid Commun. Mass Spec.* **1987**, *1*, 83–86.

Melvin B. Comisarow

University of British Columbia, Vancouver, BC, Canada

William Compston

b. February 19, 1931, Fremantle, Western Australia.

Source: http://www.sciencearchive.org.au/scientists/interviews/c/bc.html

William Compston is an Australian geophysicist who was a pioneer in the development of Rb-Sr geochronology, culminating in the successful dating of lunar samples from Apollo 11. His basic training in Physics enabled him to adapt ion beam transport techniques as developed in nuclear physics, to the design and construction of a sensitive thermal ionization mass spectrometer in the late 1960s. His most significant contribution to mass spectrometer design was the Sensitive High Resolution Ion MicroProbe mass spectrometer (SHRIMP), which eliminated chemical separation of the element undergoing analysis from a sample, before analysis. SHRIMP was particularly well suited to U,Th-Pb geochronology, and zircon proved to be an ideal mineral to give SHRIMP ages with greatly improved accuracy as compared to traditional geochronological methods.

Bill lived with his two brothers and a sister in Fremantle, Western Australia where his father operated a butcher's business. Children in the Fremantle area were evacuated during World War II, and Bill moved to Toodyay, about 90 km east of Perth, where he continued to enjoy a happy childhood. At age seven he commenced learning the piano and he has maintained that interest ever since. He may well have achieved professional status as a pianist, were it not for his greater love of science.

His undergraduate education was at the University of Western Australia (UWA) where he studied physics, mathematics and geology. An interest in geology stemmed from his elder brother, who earned a bachelor's degree in geology from UWA some years earlier. Bill's doctoral studies in Physics were linked to his interest in geology by studying the isotopic composition of carbon in natural materials, using a gas source mass spectrometer. In the early 1950 s mass spectrometers were often located in Physics Departments, because instrumentation available at that time was much less reliable, and required significant technical support for maintenance and operation of the instrument.

The mass spectrometry laboratory at UWA was established by Dr. Peter Jeffery in 1951. The home-made single stage Nier-type mass spectrometer produced valuable results on the isotopic composition of carbon in limestones and coals, and the first publication from the research was published in 1955 (*1* and see also this volume, *Alfred Otto Carl Nier*). The study of carbon isotopes was of geological value in that carbon is easily fractionated in living organisms, and thus variations in carbon isotopes provide information on the biogenic origin of materials. Peter Jeffery provided a high level of supervision in experimental physics, and under his tutelage one learnt about electronics, glass working and vacuum technology in addition to more conventional skills.

Bill continued his postgraduate education at the California Institute of Technology working with Dr Sam Epstein (see this volume, *Samuel Epstein*). There he studied oxygen isotopes that enabled the determination of the temperature at which calcium-bearing shells were formed in the past. These studies concluded that large global temperature changes had occurred in the geological past. Bill's postgraduate education in isotope geochemistry was completed at the Carnegie Institution in Washington DC, where he became acquainted with a variety of geochronological techniques (see this volume, *The Development of Mass Spectrometry in the Earth and Planetary Sciences*).

This overseas experience was an excellent preparation for his return to Perth, where he was appointed as a Lecturer in Physics at UWA, and rejoined Peter Jeffery to work on Rb-Sr geochronology. Together they unravelled the dilemma of why whole rock samples of metamorphosed rocks gave older ages than those determined on micas from the same rock unit (*2*). This research gained international recognition for the UWA group. In 1960, Jeffery decided to terminate geochronology at UWA, so Bill moved to the Geophysics Department in the School of Physical Sciences at the Australian National University (ANU) in Canberra. There Bill continued his growing interest in Rb-Sr geochronology, and the culmination of this research came in 1969 when the ANU laboratory was selected as one of the few in the world to date lunar samples from the Apollo Project. The ANU group were successful in obtaining an age for the lunar basalts of approximately 3,800 Ma, and later showed that the initial primitive lunar crust formed at approximately 4,400 Ma.

One of the reasons for their success was that the ANU group possessed excellent experimental facilities. Of critical importance was a thermal ionization mass spectrometer designed by Bill and his graduate student Steve Clement, that coupled high sensitivity with high resolution. This mass spectrometer was used to great advantage by the group in both Rb-Sr and U-Pb analyses. They also utilised computer methods for both data analysis and mass spectrometer control at an early stage in the development of on-line computers (see this volume, *Reflections on the History of Computers in Mass Spectrometry*).

A problem associated with traditional geochronology was that micrograms of the daughter product of the conventional radiometric decay schemes had to be chemically separated from other components in the rock or mineral. Clean-room procedures had to be rigorously followed if the tiny amount of daughter product was to be efficiently measured. However, even with painstaking chemical separation and optimum mass spectrometric techniques, it became increasingly obvious that there were limits on the accuracy with which a sample could be dated using the Rb-Sr technique. A different approach needed to be taken if geochronology was to fulfil its potential in geology. The ANU group decided to use ion sputtering to efficiently ionize a mineral sample, usually zircon, in-situ. The resultant energy spread of the sputtered ions required a double focusing system to select an appropriate energy for the ions, and a magnetic analyser to separate the resultant ion beam (*3*). The ion microprobe methodology, based on a design by Matsuda (*4* and see this volume, *Hisashi Matsuda*), required both an electrostatic as well as a magnetic analyser in order to measure the isotope ratios, but its great sensitivity and high resolution was ideally suited to measuring the isotopes of lead, and therefore in measuring U,Th-Pb ages.

This instrument was an immediate success. In addition to eliminating the need for time-consuming chemical separation, the ion microprobe technique improved the accuracy of geochronological ages by a significant margin over conventional methods. SHRIMP revolutionized geochronology and once the method was proven, scientists flocked to Australia to use a SHRIMP either located at the ANU or Curtin University of Technology, which had acquired the first commercial version (SHRIMP II) in 1992. Six SHRIMPS now operate in Australia and 14 are installed worldwide. A modified SHRIMP, in which the positions of the electrostatic and magnetic analysers are reversed, is called the Reverse Geometry SHRIMP. One of these machines is located at the ANU, and the other at Stanford University in California. Part of the success of constructing SHRIMP was due to the experimental and leadership skills of Compston, and the strength and support of the ANU, where excellent mechanical and electronic workshops existed. It is doubtful that such an ambitious project could have been undertaken at any other Australian University at that time.

Bill Compston has devoted his professional life to science and his research is characterised by excellence in every respect. A person of great tenacity, the quality of his leadership of the ANU mass spectrometry group is exemplified by the successful construction and operation of SHRIMP. The present international status of the Research School of Earth Sciences owes much to his research work and dedication. Bill has received many honours and awards, including the Flinders Medal, the Mawson Medal, the Centenary Medal and the Clunies-Ross award. He is a Fellow of the Australian Academy of Science, the Australian Academy of Technological Sciences and Engineering and the Royal Society of London (*5*).

References

(*1*) Jeffery, P. M.; Compston, W.; Greenhalgh, D.; De Laeter, J. R. On the Carbon-13 Abundance Of Limestones and Coals. *Geochim. Cosmochim. Acta* **1955**, *7*, 255–286.

(*2*) Compston, W.; Jeffery, P. M. Anomalous Common Strontium in Granite. *Nature* **1959**, *184*, 1792–1793.

(*3*) Compston, W. SHRIMP: Origins. Impact and Continuing Evolution. *J. Royal Soc. W. Aust.* **1996**, *79*, 109–117.

(*4*) Matsuda, H. Double Focusing Mass Spectrometers of Second Order. *Int. J. Mass Spectrom. Ion. Phys.* **1974**, *14*, 219–233.

(*5*) Professor Bill Compston: Isotope Geologist. Salt, D., interviewer. In Interviews with Australian Scientists. Australian Academy of Science. Available at: http://www.sciencearchive.org.au/scientists/interviews/c/bc.html (accessed August 18, 2014).

John De Laeter (d.)
Curtin University, Perth, Australia

R. Graham Cooks

b. July 2, 1941

Graham Cooks is a South African-born American chemist whose interests center on the use of mass spectrometry to understand physical phenomena, the development of new instrumentation, and the study of chemical reactions under vacuum and in microdroplets. With this science he seeks to make contributions to practical problems, particularly in surgery (determination of tumor types and margins), food safety (chemical and biological contaminants), and forensics (airport and other public safety concerns). His work on explosives and toxic agent detection, fingerprint analysis, and DESI imaging for disease diagnosis and pharmaceutical analysis has had significant societal impact. Several inventions have been commercialized and four companies launched from his lab. Early in his career, Cooks worked on energy partitioning during metastable ion fragmentation and contributed to the concept and implementation of tandem mass spectrometry and to desorption ionization, including the first matrix-based methods. His interests in the fundamentals of ion chemistry include reactions of ions at surfaces, acceleration of bimolecular organic reactions in microdroplets, spontaneous chiral resolution in molecular clusters, and the possible role of the amino acid serine in the biochemical origins of life.

Cooks was born in Natal, South Africa on July 2, 1941. He was educated at the University of Natal (now KwaZulu-Natal), where he received a B.S. in 1961 and a PhD in 1965 under Frank L. Warren. He then studied with Peter Sykes at Cambridge University, Great Britain, receiving a second PhD in 1967, staying on at Cambridge for a post-doctoral fellowship at Cambridge with Dudley Williams until 1968 (see this volume, *Dudley Howard Williams*). Cooks was assistant professor chemistry at Kansas State University, Manhattan, KS, from 1968–1971. He became co-director of the Mass Spectrometry Center at Purdue University in 1971 and Director in 1973, a position that he held until 1986. Associate Professor of Chemistry from 1975–1980, he became full professor in 1980 and the Henry Bohn Hass Distinguished Professor of Chemistry at Purdue University, a position that he continues to hold.

Cooks has introduced new concepts that promote chemical analysis. These include ionization of samples in the ambient environment, determination of biomarkers and drugs directly from tissue, ion soft landing as a means of preparative scale mass spectrometry, new tandem MS scan modes, miniature mass spectrometers (as seen for example on the TV program CSI and even in local grocery stores), and a kinetic method to rapidly measure thermochemical quantities.

High impact advances in modern science often follow from the conception and development of novel instrumentation. Working in this tradition, Cooks combined soft ionization with two stages of mass analysis to introduce direct analysis of mixtures by tandem mass spectrometry, made the first use of matrices to facilitate desorption ionization, and built the first hybrid, surface-induced dissociation and hand-held miniature mass spectrometers. Recently, his lab introduced desorption electrospray ionization (DESI), the first of a family of powerful high-throughput MS methods that operate in the ambient environment without sample preparation. DESI is being applied to follow the distribution of drug metabolites in untreated animal tissue and image phospholipid biomarker distributions to characterize tumor margins. This latter application will soon form the basis for intrasurgical diagnostics based on mass spectrometry.

Graham Cooks pioneered the use of mass spectrometry to separate and collect pure materials by ion soft landing. His interest in minimizing sample work-up contributed to the development of the ambient ionization methods, including DESI. These same interests also led to the construction of miniature ion trap mass spectrometers and their application to problems of trace chemical detection. He conceived a kinetic method of estimating thermochemical properties based on cluster ion dissociation kinetics and extended it into a general method of quantitative chiral analysis. The books *Metastable Ions* and *Collision Spectroscopy* helped define the field of molecular ion fragmentation. He has taught or contributed to teaching a course in mass spectrometry for several decades and his lectures have reached many thousands of people, worldwide. Cooks has served as mentor to 120 Ph. D. students. He has helped build an infrastructure for analytical instrumentation development which operates from the Bindley Biosciences

Center at Purdue and has included the development of handheld mass spectrometers. His home institution is a leading center for analytical science.

Graham Cooks is a past President of the American Society for Mass Spectrometry and the International Mass Spectrometry Society and a Life Member of several national mass spectrometry societies. He is a past recipient of the ACS awards in Mass Spectrometry and in Analytical Chemistry, the 2012 recipient of the ACS Cotton Medal a recipient of the Royal Society of Chemistry Centennial Prize, and the Ralph N. Adams Award and the Pittsburgh Spectroscopy Awards of Pittcon. He has also received a national award (Monie Ferst Award, 2011) for graduate education. His undergraduate teaching has included instrumental analysis, a subject taught at the advanced undergraduate level where his rigorous style and energetic pace are well received by students.

The publications listed below are representative of the author's most significant contributions to the field (*1–8*).

References

(*1*) Cooks, R. G.; Warren, F. L.; Williams, D. H. Rhizophoraceae Alkaloids. Part III. Cassipourine. *J. Chem. Soc. C: Organic* **1967**, 286–288.

(*2*) Beynon, J. H.; Bertrand, M.; Jones, E. G.; Cooks, R. G. Kinetic Energy Loss in Collision-Induced Fragmentations: Asymmetric Metastable Peaks in Mass Spectra. *Chem. Commun.* **1972**, 341–342.

(*3*) Beynon, J. H.; Brothers, D. H.; Cooks, R. G. Mass Spectrometric Studies using Compounds Isotopically Labeled at Natural Abundance: A Development from Ion Kinetic Energy Spectrometry. *Anal. Chem.* **1974**, *46*, 1299–1302.

(*4*) Unger, S. E.; Day, R. J.; Cooks, R. G. Positive and Negative Secondary Ion Mass Spectra and Mass-Analyzed Ion-Kinetic Energy Spectra of Some Amides, Amines, and Related Compounds: Mechanisms in Molecular SIMS. *Int. J. Mass Spectrom. Ion Phys.* **1981**, *39*, 231–255.

(*5*) Glish, G. L.; McLuckey, S. A.; Ridley, T. Y.; Cooks, R. G. A New Hybrid Sector/Quadrupole Mass Spectrometer for Mass Spectrometry/Mass Spectrometry. *Int. J. Mass Spectrom. Ion Phys.* **1982**, *41*, 157–177.

(*6*) Brodbelt, J. S.; Wysocki, V. H.; Cooks, R. G. Thermochemical vs. Kinetic Control of Reaction in an Ion Trap Mass Spectrometer. *Org. Mass Spectrom.* **1988**, *23*, 54–56.

(*7*) Schey, K. L.; Kenttämaa, H. I.; Wysocki, V. H.; Cooks, R. G. Low-Energy Collisional Activation of Polyatomic Ions with Different Target Gases. *Int. J. Mass Spectrom. Ion Processes* **1989**, *90*, 71–83.

(*8*) Takáts, Z.; Wiseman, J. M.; Gologan, B.; Cooks, R. G. Mass Spectrometry Sampling Under Ambient Conditions with Desorption Electrospray Ionization. *Science* **2004**, *306*, 471–473.

Alfred L. Yergey
National Institute of Child Health and Human Development, NIH, Bethesda, MD, USA

Robert J. Cotter

b. July 15, 1943
d. November 12, 2012.

Robert J. Cotter, an American chemist and a recognized pioneer in the development of time-of-flight (TOF) mass spectrometers and in particular miniaturized mass spectrometers, was born in Washington, DC, on July 15, 1943. He received his B.S. degree in Chemistry from the College of the Holy Cross (1965) and his Ph.D. degree in Physical Chemistry from the Johns Hopkins University (1972). Cotter raised one son, Bruce Robert Cotter. He was married to fellow chemist and Professor Catherine Fenselau (see this volume, *Catherine Fenselau*).

Cotter's thesis work on fundamental studies on proton attachment reactions, the heats of formation of H_3^+, hydronium ion and other gas phase reactions was carried out under Walter S. Koski. In 1974 he served as Assistant Professor of Analytical Chemistry at Gettysburg College, and in 1978 joined the faculty at the School of Medicine of the Johns Hopkins University, Baltimore, Maryland. He ultimately became Professor of Pharmacology and Molecular Sciences and Director of the Middle Atlantic Mass Spectrometry Laboratory, positions he held at his death on 12 November, 2012.

His early contributions include the development of fast heating and laser desorption methods for the analysis of non-tractable biological molecules, specifically the use of infrared laser desorption for peptides, carbohydrates and industrial polymers. Carried out on a time-of-flight mass spectrometer, this work resulted in several fundamental and instrumentation advances including the first use of delayed extraction methods for laser desorption, combined laser desorption/electron impact ionization for creating radical ions from non-volatile cancer drug intermediates, the *curved-field reflectron* for focusing product ions in a tandem TOF mass spectrometer, and miniaturized TOF mass spectrometers incorporating high order focusing and non-linear reflectrons (see this volume, *The Development of Time-of-Flight (TOF) Mass Spectometry*). Many of these miniaturized instruments were developed for and used in the detection of biological threat agents, while a tandem TOF instrument enabled demonstration of the *impulse collision theory* based upon incremental time shifts in the TOF spectra of collided ions. His instrument development also included several improvements in ion trap performance, and at the time of his death he was involved in the design of a low voltage, low power ion trap mass spectrometer for a 2018 Mars mission sponsored by NASA and the European Space Agency.

In addition to his work in instrumentation, his research interests included the identification of class I antigens of the multiple histocompatibility (MHC) complex, analysis and biosynthesis of lipid-A structures from gram-negative bacteria, structures of the amyloid proteins, epigenetics and modifications to histones including acetylation, methylation, ubiquitylation and SUMOylation, cardiac biomarkers and the so-called *albuminome*, and oxidative stress biomarkers involved in HIV-associated dementia, multiple sclerosis and Parkinson's Disease.

His awards include the American Chemical Society's Analytical Division Award in Chemical Instrumentation (2009) and the Frank H. Field and Joe Louis Franklin Award for Outstanding Achievement in Mass Spectrometry (2011) and The ASMS Distinguished Contributionin Mass Spectrometry Award (2011). He was an active member of the American Society of Mass Spectrometry (ASMS) serving as a board member (1990–1), and President (1998–9). He was also a member of the Board of Directors of the US Human Proteome Organization (USHUPO) where he served as Treasurer (2005–9). He served as a member of the Editorial Boards of Mass Spectrometry Reviews, Rapid Communications in Mass Spectrometry, and an Associate Editor of Clinical Proteomics. He published more than 300 peer-reviewed papers, and is the author of a book on time-of-flight mass spectrometry published by the American Chemical Society. He held 16 US patents and several world patents on time-of-flight and ion trap instrumentation.

The publications cited below are representative of the author's most significant contributions to the field (*1–10*).

Bob Cotter was a congenial colleague and a talented musician. In the evenings at the ASMS annual meetings he could often be found in hospitality suites playing piano. In addition he had a knack for writing humorous lyrics on mass spectrometry themes set to popular tunes. One example of this was a musical review that he wrote and staged at the 1991 ASMS meeting in Nashville, for which he persuaded a number of colleagues from the Baltimore MS Discussion Group to participate in the singing and dancing.

References

(*1*) Cotter, R. J. *Time-of-Flight Mass Spectrometry: Instrumentation and Applications in Biological Research*; American Chemical Society: Washington DC, 1997.

(*2*) Cotter, R. J.; Koski, W. S. Reactions of D_3O^+ with D_2: Proton Affinity of Water. *J. Chem. Phys.* **1973**, *59*, 784.

(*3*) Yuan, Z-M.; Fenselau, C.; Dulik, D. M., *et al.* Laser Desorption Electron Impact: Application to a Study of the Mechanism of Conjugation of Glutathione and Cyclophosphamide. *Anal. Chem.* **1990**, *62*, 868–870.

(*4*) Welch, A. R.; Woods, A. S.; McNally, L. M.; Cotter, R. J.; Gibson, W. A Herpesvirus Maturational Proteinase, Assembling: Identification of Its Gene, Putative Active Site Domain and Cleavage Site at the Carboxyl End of the Assembly Protein Precursor. *Proc. Natl. Acad. Sci.* **1991**, *88*, 10792–10796.

(*5*) Roher, A. E.; Lowenson, J. D.; Clarke, S., *et al.* β-Amyloid (1–42) is a Major Component of Cerebrovascular Amyloid Deposits: Implications for the Pathology of Alzheimer Disease. *Proc. Natl. Acad. Sci.* **1993**, *90*(22), 10836–10840.

(*6*) Prieto, M. C.; Kovtoun, V. V.; Cotter, R. J. Miniaturized Linear Time-of-Flight Mass Spectrometer with Pulsed Extraction. *J. Mass Spectrom.* **2002**, *37*, 1158–1162.

(*7*) English, R. D.; Warscheid, B.; Fenselau, C.; Cotter, R. J. *Bacillus* Spore Identification Using Proteolytic Mapping and a Miniaturized MALDI TOF Mass Spectrometer. *Anal. Chem.* **2003**, *75*, 6886–6893.

(*8*) Cotter, R. J.; Gardner, B.; Iltchenko, S.; English, R. D. Tandem Time-of-Flight Mass Spectrometry with a Curved Field Reflectron. *Anal. Chem.* **2004**, *76*, 1976–1981.

(*9*) Gundry, R. L.; Fu, Q.; Jelinek, C. A.; Van Eyk, J. E.; Cotter, R. J. Investigation of an Albumin-Enriched Fraction of Human Serum and its Albuminome. *Proteomics Clin. Appl.* **2007**, *1*, 73–88.

(*10*) Ilchenko, S.; Cotter, R. J. Collision Energetics in a Tandem time-of-flight (TOF/TOF) Mass Spectrometer with a Curved-Field Reflectron. *Int. J. Mass Spectrom.* **2007**, *265*, 372–381.

Robert J. Cotter (d.)
School of Medicine, The Johns Hopkins University, Baltimore, MD, USA

Willi Dansgaard

b. August 30, 1922
d. January 8, 2011

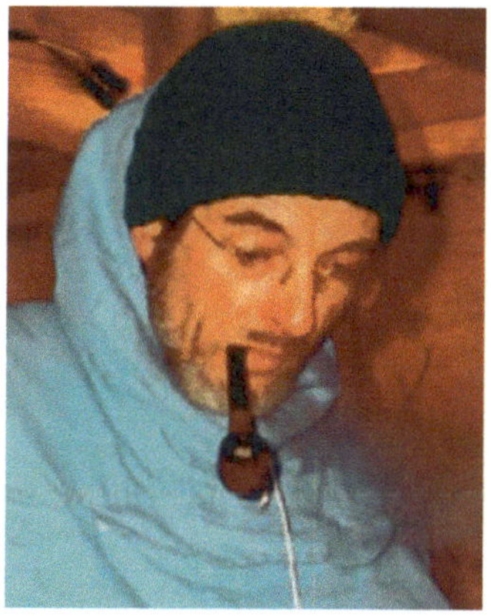

Willi Dansgaard was a Danish geophysicist whose path-breaking studies of cores drilled from ice caps revealed repeated dramatic climate changes in earth's history and helped determine the influence of orbital variations on the earth. Dansgaard discovered that the different temperatures of air masses produce sufficient differences in the isotopic make-up of precipitation that these can be detected by mass spectrometry and that measurements of $^{18}O/^{16}O$ ratios can be used to understand the record of earlier precipitation and climate history preserved in the Greenland and Antarctic ice caps.

Dansgaard was born in 1922 Copenhagen; all his formal education took place there. He studied biology and physics at the University of Copenhagen, doing some work in the Biophysical Laboratory there in the process. Upon finishing his degree in physics in 1947, he joined the Danish Meteorological Institute and was posted for a year to the Geomagnetic Observatory at Godhavn (now Qeqertarsuaq) on Disko Island on the western coast of Greenland. Both he and his wife developed a life-long fascination with that land and its ice. After returning to Copenhagen he got a job in the University's Biophysical Laboratory in 1951, setting up and running a new mass spectrometer for tracer studies in biological research (*1–4*).

Dansgaard's position in the Biophysical Laboratory allowed sufficient time for other research and he began studying the $^{18}O/^{16}O$ ratios in samples of rain from different cloud and front systems in the summer of 1952. He was able to show that once internal processes in the cloud were considered, the colder the air mass, the less ^{18}O in the drops that condensed and fell (*5*). Obtaining a wide range of water samples any way he could – from flying through and above large storm clouds to asking overseas agents of a Danish company to collect and send them – he established that climatic differences were identifiable through isotopic ratios in water. In the publication of his results, he forecast that measurement of $^{18}O/^{16}O$ ratios in distinct layers of old ice would make it possible to determine average temperatures of earlier years going back several centuries (*6*). He also stated he would make such measurements whenever it became possible; he spent most of the rest of his life carrying out that program.

In the course of a fellowship at Northwestern and Chicago Universities in 1954–1955 he collected water samples around the US. He began serious investigation of ice samples with pieces of Norwegian glaciers and Greenland icebergs at the end of the 1950s, initially as part of the International Geophysical Year (*7*). He further expanded his sources of precipitation samples through involvement with a global network set up under the joint aegis of the International Atomic Energy Agency and the World Meteorological Organization to monitor nuclear fallout. All this led to his doctoral thesis in 1961 (*The Isotopic Composition of Natural Waters*), to a professorship in the University of Copenhagen, and to a landmark publication (*8*).

Dansgaard began a series of major international collaborative scientific ventures through an arrangement with the American physicist Chester Langway in 1966. His lab's first effort in this series was a project to analyze isotopically the annual layers in the ice core that was drilled all the way to the bedrock under the northern Greenland ice cap at an American army facility called Camp Century. Over the following quarter century, his collaborations multiplied to involve groups in Switzerland, France, and several other countries, to include many more ice cores, both longer and shorter, from both Greenland and Antarctica, and to incorporate several other lines of analysis, including particulates and greenhouse gases. This research has worked out in remarkable detail the history of climate on a global scale, reaching back almost a million years. It revealed the surprisingly short span of years in which major warming has taken place several times. It established the tight linkage of overall temperatures with the quantities of carbon dioxide and methane. It demonstrated a very close correlation of the isotopic record in ancient ice with that found in oceanic sediments of the same eras. And this research revealed that periods of major cooling and glaciation were indeed linked, though in a complex manner, to those variations in the earth's orbit singled out in Milanković's theory of the origin of ice ages (*9–14*, and see this volume, *The Development of Mass Spectrometry in the Earth and Planetary Sciences*).

Dansgaard retired in 1992 from the directorship of the Geophysical Isotopes Laboratory (since renamed the Center for Ice and Climate of the University's Niels Bohr Institute). In 1995 he was a co-winner of the Crafoord Prize in 1995, which is widely regarded as an equivalent to the Nobel Prize for the fields in which it is given. The geologically-rapid climate fluctuations discovered in his research are now named Dansgaard-Oeschger events, honoring him and his most influential collaborator, the Swiss scientist Hans Oeschger.

As professor emeritus Dansgaard was still active and also found time to write a memoir (*15,16*). He died in 2011, survived by his three children.

References

(*1*) Citterio, M. In Memorium: Prof. Dr. Willi Dansgaard 1922–2011. *Studia UBB Geologia.* **2011**, *56*, 43–44.

(*2*) Langway, C. C. Jr. Willi Dansgaard (1922–2011). *Arctic* **2011**, *64*, 385–387.

(*3*) Sweet, W. *Kicking the Carbon Habit: Global Warming and the Case for Renewable and Nuclear Energy*; Columbia University Press: New York, 2006.

(*4*) Martin, D. Willi Dansgaard Dies at 88; Read Climates in Old Ice. *N. Y. Times*, 28 Jan 2011.

(*5*) Dansgaard, W. The Abundance of ^{18}O in Atmospheric Water and Water Vapour. *Tellus* **1953**, *5*, 461–469.

(*6*) Dansgaard, W. The ^{18}O-Abundance in Fresh Water. *Geochim. Cosmochim. Acta.* **1954**, *6*, 241–260.

(*7*) Dansgaard, W.; Nief, G.; Roth, E. Isotopic Distribution in a Greenland Iceberg. *Nature* **1960**, *185*, 232–235.

(*8*) Dansgaard, W. Stable Isotopes in Precipitation. *Tellus* **1964**, *16*, 436–468.

(*9*) Dansgaard, W.; Johnsen, S. J.; Møller, J.; Langway, C. C. Jr. One Thousand Centuries of Climatic Record from Camp Century on the Greenland Ice Sheet. *Science* **1969**, *166*, 377–380.

(*10*) Dansgaard, W.; Johnsen, S. J.; Clausen, H. B.; Langway, C. C. Jr. Climate Record Revealed by the Camp Century Ice Core. In: *Late Cenozoic Glacial Ages Symposium 1969*; Turekian, K. K., Ed.; Yale University Press: New Haven, CN, USA, 1971; pp 37–56.

(*11*) Johnsen, S. J.; Dansgaard, W.; Clausen, H. B.; Langway, C. C. Jr. Oxygen Isotope Profiles through the Antarctic and Greenland Ice Sheets. *Nature* **1972**, *235*, 429–434.

(*12*) Dansgaard, W.; Johnsen, S. J.; Clausen, H. B.; Hammer, C. U.; Langway, C. C. Jr. Stable Isotope Profile through the Ross Ice Shelf at Little America V, Antarctica; *Proceedings of the Symposium on Isotopes and Impurities in Snow and Ice*; International Association of Hydrological Sciences Press: Wallingford, UK, 1977; pp 322–325.

(*13*) Langway, C. C. Jr., Oeschger, H., Dansgaard, W., Eds. *Geochemistry and the Environment*; American Geophysical Union: Washington, DC, 1985.

(*14*) Shackleton, N. J.; Hall, M. A.; Vincent, E. Phase Relationships Between Millennial-Scale Events 64,000–24,000 Years Ago. *Paleoceanography* **2000**, *15*, 565–569.

(*15*) Dansgaard, W.; Johnsen, S. J.; Clausen, H. B., *et al.* Evidence for General Instability of Past Climate from a 250-kyr Ice-Core Record. *Nature* **1993**, *364*, 218–220.

(*16*) Dansgaard, W. *Frozen Annals: Greenland Ice Cap Research*; Narayana Press: Odder, Denmark, 2005.

Keith A. Nier
Independent Historian, Madison, NJ, USA

Anne Dell

b. November 9, 1950

Anne Dell is a British/Australian mass spectrometrist widely noted for work in the general field of glycobiology. She has focused particularly on the development and application of ultra-high sensitivity mass spectrometric strategies for glycopolymer structure analysis.

Anne Dell, the youngest of seven children, spent her early life in rural Western Australia. She was taught by her mother at home via Correspondence School lessons until the age of 11 and then attended schools in Kalamunda and Midland before taking a Chemistry degree at the University of Western Australia. She graduated in 1971 with a first class honours degree. With support from an 1851 Exhibition Award and a British Council travel grant she went to Cambridge, UK, for her postgraduate training. She obtained a PhD in Biopolymer Mass Spectrometry in 1975 from the University Chemical Laboratory under the supervision of Professor Howard R. Morris FRS (see this volume, *Howard Redfern Morris*). Subsequently she was a post-doctoral fellow in the Biochemistry Department, Imperial College London before taking up a lectureship in the same department in 1979. She was promoted to a Personal Chair in 1991 and was Head of the Department of Biochemistry from 1999 to 2001. She has one daughter who was born in 1984.

Under the supervision of Howard Morris, Anne's PhD research at Cambridge University focused on the development and application of mass spectrometry (MS) for peptide sequencing. During that period, she also had the opportunity to undertake collaborative research with biologists at the Medical Research Council (MRC) Laboratory of Molecular Biology, early exposure to interdisciplinary research that would shape and enrich her career with many subsequent collaborations with scientists around the world. In 1975 she joined Howard Morris' new biopolymer MS laboratory at Imperial College London as a postdoctoral research fellow, where she had access to unique high-field magnet instrumentation that had emerged from Morris' research in high molecular weight mass spectrometry. Her independent research career began in 1979, when she was appointed to a lectureship at Imperial. She has remained at Imperial for her subsequent academic career.

Since the late 1970s Anne has been involved with the development and application of ultra-high sensitivity MS methodologies coupled with microchemical manipulations for the characterisation of complex carbohydrates and glycoconjugates of biological and medical importance. Using the soft ionisation techniques of field desorption and fast atom bombardment to glycopolymers, Dell, in collaboration with Clint Ballou (UC Berkeley) obtained high molecular weight data that not only helped solve a long-standing structural problem in mycobacterial polysaccharide research but was also considered an important breakthrough in carbohydrate analysis. Two collaborations arising from these results were especially important for the development of glycobiology. First, research with Peter Albersheim's group (University of Colorado) yielded the structures of carbohydrates implicated in hormone-like roles in plants (the 'oligosaccharins') and helped to provide impetus for the creation of the Complex Carbohydrate Research Centre at the University of Georgia, now a world-leading centre for glycobiology. Second, a collaboration with Minoru Fukuda (La Jolla Cancer Research Foundation) in the 1980s yielded a series of papers that delineated the complex carbohydrate structures on human erythrocytes and leukocytes. Highlights of this work include the first molecular definition of a human disease (HEMPAS) caused by defects in glycosylation and the identification of sialyl Lewis^{X} as a major carbohydrate epitope on neutrophils. The latter work led to progress in defining possible ligands for the selectins when these molecules were identified towards the end of the 1980s. The leukocyte studies provided information pertaining to the biosynthetic pathways operating in these cells and laid the foundation for on-going immunological and biochemical research into the roles of carbohydrates in the immune system. Anne's characterisation, in the late 1980s, of the glycan architecture on the glycoprotein hormone that controls erythropoiesis (Erythropoietin) provided the first evidence that the recombinant product carried natural glycosylation and could be safely used to treat anaemia.

From the 1990s onwards, with her laboratory now having MALDI and electrospray technologies, Anne devised methodologies that have populated the public glycomic repositories of the NIH Consortium for Functional Glycomics. Amongst the 1990s highlights were the characterisation of schistosomal glycans, meningococcal pilins, mycobacterial cell walls and human reproductive glycoproteins (the glycodelins) that catalysed new avenues of research in parasitology, infectious diseases and developmental biology. Work of the 21st century includes analysis of N-linked glycosylation in *Campylobacter jejuni* and glycomics analysis of novel glycosylation in knockout mouse models and identification of the human sperm receptor as sialyl Lewis^x.

Dell's work has been recognized with a number of awards and honors. She received the Tate and Lyle Medal in 1986, the Whistler Award of the International Carbohydrate Organisation in 2000, the Haworth Memorial Medal of the Royal Society of Chemistry in 2003, and the International Glycoconjugates Organisation Award in 2005. She was elected Fellow of the Royal Society in 2002, Member of the European Academy of Science in 2004, and a Fellow of the Academy of Medical Sciences in 2008. She received an Honorary Doctorate of Science from the University of Western Australia in 2010. In recognition of her services to science, she was made Commander of the British Empire (CBE) in 2009. Some representative papers are listed here (*1–5*).

References

(*1*) Khoo, K. H.; Sarda, S.; Xu, X., *et al.* A Unique Multifucosylated -3GalNAc Beta 1→4GlcNAc Beta 1→3 Gal Alpha 1- Motif Constitutes the Repeating Unit of the Complex O-Glycans Derived from the Cercarial Glycocalyx of Schistosoma mansoni. *J. Biol. Chem.* **1995**, *270*, 17114–17123.

(*2*) Dell, A.; Morris, H. R.; Easton, R. L., *et al.* Structural Analysis of the Oligosaccharides Derived from Glycodelin, a Human Glycoprotein with Potent Immunosuppressive and Contraceptive Activities. *J. Biol. Chem.* **1995**, *270*, 24116–24126.

(*3*) Dell, A.; Morris, H. R. Glycoprotein Structure Determination by Mass Spectrometry. *Science* **2001**, *291*, 2351–2356.

(*4*) Dell, A.; Chalabi, S.; Easton, R. L., *et al.* Murine and Human Zona Pellucida 3 Derived from Mouse Eggs Express Identical O-Glycans. *Proc. Natl. Acad. Sci. USA* **2003**, *100*, 15631–15636.

(*5*) Pang, P. C.; Chiu, P. C.; Lee, C. L., *et al.* Human Sperm Binding is Mediated by the Sialyl-Lewis^x Oligosaccharide on the Zona Pellucida. *Science* **2011**, *333*, 1761–1764.

Anne Dell

Department of Life Sciences, Imperial College London

Arthur Jeffrey Dempster

b. August 14, 1886
d. March 11, 1950

Arthur Dempster was born in Toronto, Canada on August 14, 1886. His family owned a bread company that has since become nationally known in Canada as 'Dempster's Bread,' and in which he worked during summers. He attended the University of Toronto, excelling in several subjects and obtaining the BA in 1909. The following year he obtained the MA. His interest in physics led him to Germany in 1911 to study for the PhD, a destination that was still popular despite the recent development of graduate programs in the United States. Before leaving Toronto, he had submitted for publication 'Mobilities of Ions in Gases at High Pressures,' presaging his lifelong interest in positive ions (*1*). He had been awarded one of the 1851 Exhibition Scholarships, the famous student research awards established with profits from the great London Exhibition of 1851. As was the custom, he studied at more than one university (Munich and Göttingen) before choosing Professor Wien at Würzburg as his thesis advisor and his thesis subject as the study of the *Kanalstrahlen* (channel rays) that had been discovered by Goldstein in 1886. These had subsequently been shown by Wien to comprise a beam of positively charged particles (see this volume, *Wilhelm Carl Werner Otto Fritz Franz Wien*). In 1914, however, at the outbreak of the Great War, he returned to America, having caught the last train out of Würzburg before enemy aliens were interned. Another Canadian student who took the next train was interned for four years. Dempster's research was well advanced but his doctoral thesis had yet to be accepted.

In his three years in Germany, in addition to physics and the language, he had learned much of German rules of behavior, including those governing the student dueling societies. On one occasion, whilst walking with a German friend, he met another German student who later challenged Dempster's friend to a sabre duel because Dempster had *had his hands in his pockets*.

Back in America, he presented his thesis to the University of Chicago, under the title: *Some properties of low energy Canal Rays*, and received his PhD from that institution in 1915 (*2*). He was appointed an assistant in the department and became a naturalized US citizen. In 1917, he enlisted in the Signal Corps, serving until 1919 when he returned to Chicago where, except for leaves of absence, he remained for the rest of his life. In 1917, he constructed one of the first mass spectrometers, which he employed in the next two years to determine the isotopic constitution of lithium, magnesium, potassium, calcium, and zinc (*3* and see this volume, *A History of Magbetic and Electrostatic m/z Analyzers*). Meanwhile, in 1919 at Cambridge University, F.W. Aston (encouraged by J. J. Thomson's 'positive ray parabola apparatus') had constructed a mass spectrograph (see this volume, *Francis William Aston* and *Joseph John Thomson*). With this instrument he confirmed (1920) that neon existed in two isotopic forms and rapidly proceeded to determine the isotopic abundances for neon, nitrogen, chlorine, mercury and the noble gases. Thus, in the period 1918–1920, Dempster and Aston not only demonstrated the existence of isotopes in numerous nonradioactive elements, but thereby explained why some chemical atomic weights were approximately integral (on an oxygen = 16 scale) whilst others were wildly fractional. Perhaps unfairly, the Nobel Prize for Chemistry for 1922 was awarded to Aston alone, possibly because he was closer to the scientific centre of gravity than a young assistant professor in Chicago.

In 1927, Dempster became full professor at the University of Chicago and in the next 25 years directed numerous PhD theses, including that of this author in 1942 (see this volume, *Henry E. Duckworth*). During much of this time it was Chicago's policy that graduate students should publish under their own names, as sole authors. Thus, much research that Dempster suggested, and closely followed, bears no mention of his name, except in a thesis or as acknowledgement at the end of a scientific article.

In the period 1922/1934, papers appeared in his own name on the passage of positive ions through hydrogen and helium, the interference pattern of individual light quanta and the reflection of positive ions from crystal surfaces (*4*).

But in 1935 he turned his full attention again to mass spectroscopy with the completion of a double-focusing mass spectrometer, constructed with the aid of a grant from the Penrose Fund of the American Philosophical Society. In this instrument, the ion source was a high frequency spark between metal electrodes that could deliver singly- and multiply-charged ions of any element present (*5*,*6*). This filled a major gap in the important work that Aston had continued, but which relied primarily on ions from gaseous materials.

In the next five years, with this powerful new instrument, Dempster discovered 33 new isotopes in 17 elements, including the rare uranium isotope at mass 235 which, in 1938, was shown by Otto Hahn to undergo fission upon the absorption of a slow neutron (*7*). He also determined the masses of many isotopes of the heavy elements, providing the shape of the 'Packing Fraction Curve' (roughly equivalent to the Binding Energy per Nucleon Curve, which superseded it) and, thereby, the means to estimate the energy that would be released if very heavy atoms were to be split into lighter ones (*8*,*9* and see this volume, *A History of Mass Spectrometry in Nuclear Science*). It was a period of remarkable research achievement. Further, in the years preceding US entry into the War, Dempster's research students used this instrument to extend his atomic mass determinations (*10*).

It should be noted that, almost contemporaneously with Dempster's, two other double-focusing instruments were constructed, at Harvard (Bainbridge and Jordan) and at the University of Vienna (Mattauch and Herzog) (see this volume, *Kenneth Tompkins Bainbridge, Richard Franz Karl Herzog*, and *Josef Heinrich Elisabeth Mattauch*). These were of different designs and were also used for the precise determination of atomic nasses, but primarily those of the lighter elements, whereas Dempster's instrument lent itself to the study of heavier atoms.

In 1941, when the 'Metallurgical Laboratory' was established at the University of Chicago to ascertain the feasibility of an explosive chain reaction, based on the fission of ^{235}U, Dempster was named Chief Physicist and he and his associates employed mass spectrometers in various ways related to the project and its subsequent developments. Chief amongst this work was the study of abnormal isotopic abundances in samples of cadmium, samarium, and gadolinium that had been exposed to lengthy neutron radiation. In this way, he established that the large neutron capture cross sections of these elements could be attributed to ^{113}Cd, ^{149}Sm, ^{163}Gd, and ^{165}Gd (*11*,*12*).

At the end of the War, when the Metallurgical Laboratory morphed into the Argonne National Laboratory, Dempster assumed a position in the latter as a Division Director, whilst on indefinite leave from the University.

Dempster received many honors during his lifetime. These included the AAAS Prize (1930), Lewis Award and Glasham Gold Medal of the American Philosophical Society (1932), election to the National Academy of Sciences (1937), Presidency of the American Physical Society (1944), and a D.Sc. (honorary) from University of Toronto (1947).

Throughout his career Dempster took an active role in discussions of University policy and the importance of professorial contributions to it. He took his share of lecturing to advanced students, although he was uncomfortably shy in doing so. His students were particularly appreciative of his vast knowledge of the current physics literature.

Dempster had a fine bearing, although he was short in stature. With strangers and most others he was reserved, but he relaxed in the company of close friends. In 1926 he married Germaine Collette, a Belgian mediaeval scholar, who became an authority on Chaucer, to the great pride of her husband. They had no children.

Arthur Dempster died suddenly on March 11, whilst vacationing in Florida from his still very active research activities (*13*).

References

(*1*) Dempster, A. J. On the Mobility of Ions in Air at High Pressures. *Phys. Rev.* **1912**, *24*, 1.

(*2*) Dempster, A. J. Properties of Slow Canal Rays. *Phys. Rev., New Series* **1916**, *8*, 651–663.

(*3*) Dempster, A. J. A New Method of Positive Ray Analysis. *Phys. Rev.* **1918**, *11*, 316–324.

(*4*) Dempster, A. J.; Batho, H. F. Light Quanta and Interference. *Phys. Rev.* **1927**, *30*, 644.

(*5*) Bartky, W.; Dempster, A. J. Paths of Charged Particles in Electric and Magnetic Fields. *Phys. Rev.* **1929**, *33*, 1019–1022.

(*6*) Dempster, A. J. New Methods in Mass Spectroscopy. *Proc. Amer. Phil. Soc.* **1935**, *75*, 755–767.

(*7*) Dempster, A. J. Isotopic Constitution of Uranium. *Nature* **1935**, *136*, 180.

(*8*) Dempster, A. J. The Atomic Masses of the Heavy Elements. *Phys. Rev.* **1938**, *53*, 64–75.

(*9*) Dempster, A. J. The Energy Content of Heavy Nuclei. *Phys. Rev.* **1938**, *53*, 869–874.

(*10*) Wilkins, T. R.; Dempster, A. J. The Radioactive Isotope of Samarium. *Phys. Rev.* **1938**, *54*, 315.

(*11*) Lapp, R. E.; Van Horn, J. R.; Dempster, A. J. The Neutron-Absorbing Isotopes in Gadolinium and Samarium. *Phys. Rev.* **1947**, *71*, 745.

(*12*) Dempster, A. J. Isotopic Changes in Cadmium by Neutron Absorption. *Phys. Rev.* **1947**, *71*, 829.

(*13*) Allison, S. K. Arthur Jeffrey Dempster, 1886–1950. *Biographical Memoirs of the National Academy of Sciences*; National Academy Press: Washington, D.C., 1952; vol. 27, pp 318–333. Available at: http://www.nasonline.org/publications/biographical-memoirs/memoir-pdfs/dempster-arthur.pdf (accessed October 21, 2014).

Henry E. Duckworth (d.)
University of Manitoba, Winnipeg, MB, Canada

M. Bonner Denton

b. June 15, 1944

M. Bonner Denton is an American chemist who has specialized in improving detection technology for analytical chemistry. Denton's research efforts have materially advanced many areas of analytical chemistry, ranging from analysis of difficult matrix samples to revolutionizing low light spectroscopy; new approaches for selective photoionization; unique modes of quadrupole operation; and most recently, significantly improving high stability ion detector arrays. Denton has extended earlier contributions in multiplex optical spectroscopy to mass spectrometry through the development of a new generation of ion detection technologies.

Denton was born in Beaumont, Texas, USA. He received his Bachelor of Science and Bachelor of Arts in 1967 from Lamar State College of Technology in Beaumont. In 1972 he received his PhD from the University of Illinois as a student of Howard Malmstadt, widely known for his emphasis on improving electronic instrumentation for chemistry. Denton joined the faculty of the University of Arizona in 1971, became Professor of Chemistry in 1980, and also Professor of Geosciences in 1993. His research interests in addition to mass spectrometry include analytical instrumentation, optical imaging instrumentation, laboratory automation, ion mobility and chemical forensics. His most recent work has been on improved detection of explosive substances.

Denton first worked on mass spectrometers as a graduate student. In 1985 he did pioneering work in laser vacuum UV single photon selective photoionization from a complex matrix. Since then, vacuum UV photoionization has become an expanding field. Denton and his group also have explored unconventional modes of quadrupole operation, including the notch and bandpass filters.

Denton's group started exploring scientific charge-coupled devices (CCD's) and charge-injection devices (CID's) in the late 1970s. Their work revolutionized many areas of optical spectroscopy by solving the problem of simultaneous detection of large ranges of wavelengths. Denton extended his successes in multiplex optical spectroscopy to mass spectrometry through the development of a new generation of ion detection technologies based on capacitive transimpedance amplifier (CTIA) devices. These custom integrated circuits provide the stability of Faraday cups combined with sensitivities approaching those of ion multipliers, no upper mass limitation, and a dynamic range exceeding both. Additionally, just as CCD and CID devices allowed simultaneous wavelength coverage in optical spectroscopy, these new CMOS-based detectors are readily fabricated into large arrays suitable for simultaneous detection on focal-plane-based mass spectrometers. A new generation of inductively-coupled plasma-based isotope ratio mass spectrometers has been commercially introduced, providing simultaneous observation of all isotopes from lithium through uranium (see this volume, *The Development of Inductively-Coupled Plasma Mass Spectrometry (ICPMS)*). This same CTIA-based detection scheme has been further applied to ion mobility spectrometry where the ability to operate at or near atmospheric pressures is required.

Denton has worked with many companies over the years to develop a variety of new aspects of chemical instrumentation. He served on the editorial boards of numerous journals and as President of The Society of Applied Spectroscopy and Chair of the Analytical Division of the American Chemical Society. Among his honors have been the ACS Division of Analytical Chemistry Award in Chemical Instrumentation, the Spectroscopic Society of Pittsburgh's Spectroscopy Award, and the Eastern Analytical Symposium Award for Outstanding Achievements in the Fields of Analytical Chemistry. A few examples of his publications are listed below (*1–5*).

When not improving and applying analytical instrumentation, Denton enjoys working on setting land speed records for modified sports cars.

References

(*1*) Huth, T. C.; Denton, M. B. Complex Mixture Analysis by Photoionization Mass Spectrometry with a VUV Hydrogen Laser Source. *Analytical Chimica Acta.* **1987**, *192*, 165–173.

(*2*) Zumwalt, M. C.; Denton, M. B. Photoionization Mass Spectrometry of C_{60} and C_{70} with a Vacuum Ultraviolet Hydrogen Laser Source. *Appl. Spectrosc.* **1993**, *47*, 1739–1741.

(*3*) Knight, A.; Sperline, R.; Hieftje, G., *et al.* The Development of a Micro Faraday Array for Ion Detection. In Special Issue "Detectors and the

Measurement of Mass Spectra," Birkinshaw, K., Ed. *Int. J. Mass Spectrom.* **2002**, *215*, 131–139.
(*4*) Barnes, J. H.; Schilling, G. D.; Hieftje, G. M., *et al.* Use of a Novel Array Detector for the Direct Analysis of Solid Samples by Laser Ablation Inductively Coupled Plasma Sector–Field Mass Spectrometry. *J. Am. Soc. Mass Spectrom.* **2004**, *15*, 769–776.
(*5*) Felton, J. A.; Schilling, G. D.; Ray, S. J., *et al.* Evaluation of a Fourth-Generation Focal Plane Camera for Use in Plasma–Source Mass Spectrometry. *J. Anal. At. Spectrom.* **2011**, *26*, 300–304.

M. Bonner Denton
University of Arizona, Tucson, AZ, USA

Vernon Hamilton Dibeler

b. July 20, 1918

Vernon Dibeler is an American chemical physicist who devoted his entire career to research at the National Bureau of Standards (NBS, now National Institute of Standards and Technology, NIST). Mass spectrometry was the primary focus of his work, in particular using photoionization to probe the molecular bond strengths and the electronic structure of ions.

Vernon Dibeler was born in 1918 in Elizabeth, NJ, USA. He earned a BA and MA in chemistry in 1939 and 1940 at Duke University in Durham, NC. He joined the NBS in 1942. Dibeler married his wife Anne in 1943; they would have three children.

While working at NBS he won a National Institutes of Health Pre-doctoral Fellowship which he used at Columbia University, NY, in 1947–1948; he completed his Ph.D. in chemistry from Columbia in 1950. Dibeler rose to become the head of the mass spectrometry section at the NBS by 1961 but preferred to spend his time in research and in 1962 he reverted to his non-administrative status in favor of Henry Rosenstock (*1–2*).

Dibeler pursued his own interests in various aspects of mass spectrometry while carrying out the Bureau's programs and policies. His research was in part oriented toward isotopic questions but more generally focused on molecular ionization (*3–9*). He served as Chariman of the ASTM E-14 Committee on Mass Spectrometry, the first US mass spectrometry scientific organization, for 1961 and 1962. Among other honors he received the Gold Medal Award of the Department of Commerce in 1969.

References

(*1*) Biographical Records for Vernon H. Dibeler. Archives, National Institute of Standards and Technology, Gaithersburg, Maryland, US, 1962.

(*2*) *American Men and Women of Science, 14th Edition,* R. R. Bowker, New Providence, New Jersey, US, 1979, *s.v.* "Dibeler, Vernon Hamilton."

(*3*) Dibeler, V. H.; Mohler, F. L. Analysis by the Mass Spectrometer of a Liquified Hydrocarbon Mixture Containing C_3–C_5 Paraffins and Olefins. *J. Res. Natl. Bur. Stand.* **1947**, *39*, 149–154.

(*4*) Dibeler, V. H. The Isotope Reference Sample Program at the National Bureau of Standards. In: *Nuclear Processes in Geologic Settings*; Hurley, P. M., *et al.*, Eds.; National Academy of Sciences–National Research Council: Washington, DC, 1956; pp 55–61.

(*5*) Shields, W. R.; Craig, D. N.; Diebler, V. H. Absolute Mass Spectrometric Determination of the Atomic Weight of Silver. In: *Proceedings of the International Conference on Nuclear Masses*; Duckworth, H. E., Ed.; Univ. of Toronto Pr.: Toronto, Canada, 1960; pp 519–531.

(*6*) Dibeler, V. H.; Reese, R. M.; Franklin, J. L. Mass Spectrometric Study of Cyanogen and Cyanoacetylenes. *J. Am. Chem. Soc.* **1961**, *83*, 1813–1818.

(*7*) Dibeler, V. H.; Robert, M.; Reese, R. M. Mass Spectrometric Study of Photoionization. I. Apparatus and Initial Observations on Acetylene, Acetylene-d_2, Benzene, and Benzene-d_6. *J. Res. Nat. Bur. Stand. Sec. A: Phys. Ch.* **1964**, *68A*, 409–417.

(*8*) Dibeler, V. H.; Walker, J. A.; Rosenstock, H. M. Mass Spectrometric Study of Photoionization. V. Water and Ammonia. *J. Res. Nat. Bur. Stand. Sec. A: Phys. Chem.* **1966**, *70A*, 459–463.

(*9*) Dibeler, V. H.; Walker, J. A. Mass Spectrometric Study of Photoionization. XIV. Nitrogen Trifluoride and Trifluoramine Oxide. *Inorg. Chem.* **1969**, *8*, 1728–1733.

Keith A. Nier

Independent Historian, Madison, NJ, USA

Carl Djerassi

b. October 29, 1923
d. January 30, 2015

Carl Djerassi was an American chemist and writer most famous for research on the hormones that were used in creating the birth control pill. He was one of the first natural product chemists to extensively develop and use mass spectrometry for his work. He participated in laying the ground rules for interpreting mass spectra of organic compounds and co-authored several textbooks. Moreover, he was one of the key actors in the effort to develop artificial intelligence with respect to mass spectrometry.

Djerassi was born in Vienna, Austria in 1923; his parents were Jewish. He and his mother fled after the German annexation of Austria in 1938 and eventually reached the United States. Djerassi studied chemistry and biology at Kenyon College, graduating with his A.B. in 1942. He received his Ph.D. in organic chemistry at the University of Wisconsin in 1945. He joined CIBA Pharmaceuticals Co., Summit, New Jersey in the same year, and in 1949 became associate director of chemical research of the Syntex Corporation in Mexico City. In 1952, he moved to Wayne State University, where he was associate professor from 1952–54 and professor from 1954–59. In 1959 he joined the chemistry faculty of Stanford University, where he remained until his retirement in 2002. Along with his university duties, he continued to hold industry positions at Syntex Co. of Mexico City (1957–60), Syntex Co. of Palo Alto (1960–72), and Zoecon Co., Palo Alto (1968–88). He was been married three times. After becoming Emeritus Professor of Chemistry at Stanford, he concentrated on his literary work and on supporting the arts.

Carl Djerassi is, together with Klaus Biemann, Fred W. McLafferty, Einar Stenhagen, and a few others, one of the pioneers of organic mass spectrometry (see this volume, *John Herbert Beynon, Klaus Biemann, Fred W. McLafferty, Ragnar Erik Ryhage*, and *Einar August Stenhagen*). In the 1940s, when commercial instrumentation became available, mass spectrometry was used extensively in the chemical and the petroleum industries. For research into the structures of organic compounds, however, mass spectrometry faced formidable hurdles because the spectra of the molecules were very difficult to interpret. In the late 1950s and early 1960s, reaction mechanisms conceived in physical organic chemistry were applied in an effort to make sense of the spectra. Djerassi in the 1960s worked extensively to lay the foundations for the interpretation of mass spectra of organic compounds in the terminology of physical organic chemistry. Together with Dudley Williams and Herbert Budzikiewicz, he wrote influential textbooks and developed the language for analyzing the phenomena at hand (see this volume, *Dudley Howard Williams*). By the 1960s, Djerassi was already a very experienced research chemist, famous for his contributions to the partial synthesis of cortisone and his research into the steroid hormones that were used in the contraceptive pill. In addition, he had made major forays into the field of optical rotatory dispersion (ORD) for structural elucidation of organic molecules. When he moved to Stanford in 1959, Djerassi immediately set out to advance the uses of mass spectrometry in the fields of alkaloid and steroid chemistry, later also the chemistry of sterol compounds. He was not so much concerned with the development of new instrumentation but set out instead on a program to use isotopically-labeled compounds, systematically studying the effects that chemical alterations of the elemental composition of sample molecules had on their mass spectra. In doing so, he crucially contributed to the foundation of what can be called the chemistry of mass spectrometry (*1*, and see this volume, *The Beginnings of Organic Mass Spectrometry*). Djerassi's mass spectrometry research resulted in hundreds of research papers plus reference books and widely-influential texts, a sample of which are listed below (*2–6*).

In the 1960s and 1970s, together with biologist Joshua Lederberg and computer scientist Edward Feigenbaum, Djerassi started the development of methods of artificial intelligence for solving problems in organic mass spectrometry. Essentially, they tried to let the computer algorithm DENDRAL establish all possible structural formulae, given a certain mass spectrum. In an advanced version, called META-DENDRAL, the program even developed new rules for interpreting mass spectra. With these projects, Djerassi found himself in the middle of efforts of the Stanford academic community to apply computerized methods in the areas of exobiology, medicine, and of course artificial intelligence itself. Although the project was moderately successful, in the long term it met with difficulties, due to the large amount of manpower needed to do the programming.

Djerassi has engaged in political discussions, first focused on the social impact of the pill. And in the 1980s, based on his autobiographical writing, Djerassi started a new career as an author of novels and (since 1997) also theater plays (*7–9*). In writing about the working modes of modern science, he established a new genre that he aptly named science in fiction. Having been an ardent art collector for many years, he also started a program for the support of artists.

Carl Djerassi received far more honors than can be listed in this space, including dozens of honorary doctorates. A few of those of particular note in this context are elections to National Academy of Sciences of the US, to its Institute of Medicine, and to the Royal Society of London, and receipt of the Thomson Gold Medal of the International Mass Spectrometry Foundation, and the American Chemical Society's Priestley Medal. An oral history interview with Djerassi conducted by Jeffrey L. Sturchio and Arnold Thackray on July 31, 1985, has been published on-line by the Chemical Heritage Foundation. (http://www.chemheritage.org/discover/collections/oral-histories/search.aspx?q=&ohtopic=&ohinterviewee=djerassi&ohproject=&ohtranscripts=true).

References

(*1*) Reinhardt, C. *Shifting and Rearranging: Physical Methods and the Transformation of Modern Chemistry*; Science History Publications: Sagamore Beach, MA, USA, 2006.

(*2*) Djerassi, C. Mass Spectrometric Investigations in the Steroid, Terpenoid and Alkaloid Fields. *Pure Appl. Chem.* **1963**, *6*, 575–599.

(*3*) Djerassi, C.; Mutzenbecher, G. v.; Fajkos, J.; Williams, D. H.; Budzikiewicz, H. Mass Spectrometry in Structural and Stereochemical Problems. LXV. *J. Am. Chem. Soc.* **1965**, *87*, 817–826.

(*4*) Budzikiewicz, H.; Djerassi, C.; Williams, D. H. *Mass Spectrometry of Organic Compounds*; Holden-Day: San Francisco, 1967.

(*5*) Djerassi, C. Recent Advances in the Mass Spectrometry of Steroids. *Pure Appl. Chem.* **1978**, *50*, 171–184.

(*6*) Djerassi, C., *et al.* The Dendral Project: Computational Aids to Natural Products Structure Elucidation. *Pure Appl. Chem.* **1982**, *54*, 2425–2442.

(*7*) Djerassi, C.; *Steroids Made it Possible*; American Chemical Society: Washington, DC, 1990.

(*8*) Djerassi, C.; *The Pill, Pygmy Chimps, and Degas' Horse*; Basic Books: New York, 1992.

(*9*) Djerassi, C. *This Man's Pill: Reflections on the 50th Birthday of the Pill*; Oxford University Pr.: New York, 2004.

Photo credit: UW-Madison, University Communications.

Carsten Reinhardt
Chemical Heritage Foundation, Philadelphia, PA, USA

Henry Edmison Duckworth

b. November 1, 1915
d. December 18, 2008

Henry Duckworth was a Canadian physicist noted for design and construction of high-resolution mass spectrometers. The precise measurements of the masses of stable nuclides he made with these instruments provided the backbone of the atomic mass table from titanium to uranium.

Henry Duckworth was born on November 1, 1915, in Brandon, Manitoba, Canada (*1*). His father was a Presbyterian minister who joined the United Church of Canada in 1925, and his mother was a minister's daughter. He attended public schools in Transcona and St. James, Manitoba (now parts of Winnipeg), and United College (then part of the University of Manitoba), where he obtained the BA in 1935. He followed this with a B.Sc. from the University in 1936. He taught high school, and then physics at United College before going to the University of Chicago in 1939, where he completed his PhD in 1942 under the direction of A. J. Dempster (*2*, and see this volume, *Arthur Jeffrey Dempster*). Canada then being at War, he enlisted in the Signals Corps of the Canadian Army but was soon transferred to war research at the National Research Council in Ottawa, where he worked on radar antenna design. In 1944 he was sent to McMaster University to join H. G. Thode's mass spectroscopy group (see this volume, *Henry George Thode*). Thode's group was then a satellite of the Montreal Laboratory and represented Canada's contribution to the development of the atom bomb. Thode was using mass spectroscopy to analyze gaseous fission products, and Duckworth (aided with plans and material provided by Dempster) built a new instrument for the analysis of solids, that was just completed as the War came to an end.

After the war, Duckworth took a post as Assistant Professor of Physics at the University of Manitoba, where he taught for one year; one of his students there was K. G. Standing (see this volume, *Kenneth Graham Standing*). He had been permitted to bring the new mass spectrograph he had built from McMaster to Manitoba. One measurement with this instrument, that of the abundances of the copper isotopes, was completed with the assistance of an undergraduate, B. G. Hogg, who would later be an important collaborator and friend (*3*). The teaching load at Manitoba was very heavy, and Duckworth soon realized that he would need to move elsewhere if he were to establish a research program. In 1946, following a recommendation from Dempster, he was recruited by Wesleyan University, in Middletown, Connecticut.

At that small but celebrated school, the Physics Department consisted of only three faculty members but the students were bright, an M.S. degree was offered, and research was encouraged. Funded by a $10,000 grant from Wesleyan, and later by a grant from the U.S. Atomic Energy Commission, Duckworth built a new double-focusing mass spectrograph, using a Dempster-type spark source and photographic plate detection (*4,5*). This instrument came into operation early in 1949 and was immediately used to search for four unknown but possibly stable isotopes. One of the four, ^{190}Pt, was demonstrated, at an abundance of 0.012% – this turned out to be the last remaining stable isotope to be found – while the other three hypothetical isotopes, ^{100}Pd, ^{178}W, and ^{202}Pb were shown not to exist (*6,7*).

In 1949, M. Goeppert-Mayer and J. H. D. Jensen predicted that nuclear configurations containing 28, 50, 82 and 126 protons or neutrons would be particularly stable. The Duckworth group set to work to measure the masses of isotopes in the regions around 28, 50 and 82 protons, and 50, 82 and 126 neutrons. In a series of papers Duckworth and his students provided experimental evidence for the importance of these 'magic numbers' in conferring stability upon isotopes (*8*).

Realizing that the next stage of his research career would require access to PhD students, Duckworth reluctantly left Wesleyan University in 1951 and returned to Canada to an appointment as Professor of Physics at McMaster University. The last mass measurement made at Wesleyan before the mass spectrograph was dismantled and moved to McMaster was that of the rare isotope ^{234}U, using a sample that had been enriched for ^{235}U. This sample had been obtained from the Atomic Energy Commission by Glenn Seaborg (who was about to receive the Nobel Prize for his work on the transuranic elements), so that Duckworth could measure directly the mass differences between uranium, thorium, and lead isotopes, as a check of the accuracy of the multidecay schemes connecting those elements. The full set of measurements for ^{234}U, ^{235}U, ^{238}U, ^{232}Th, and ^{208}Pb, was completed after the instrument was reassembled at McMaster (*9*). Soon after, samples of the recently purified rare earth elements (lanthanides) were obtained and studied, revealing a new region of nuclear stability, unexplained by then current theory (*10*).

While the mass spectrograph built at Wesleyan was still in use, work was started on a much larger double-focusing instrument, with a similar overall geometry but scaled up by a factor of twelve and with electronic ion detection in place of photographic plates (see this volume, *A History of Mass Spectrometry in Nuclear Science*). The first measurements involving the full instrument were reported in 1960, seven years after the start of the project (*11*). The instrument was used to make high-precision measurements of double-neutron separation energies for many nuclei. Among many other results, this work indicated an abrupt change in the shape (or deformation) of nuclei beyond N=90 (*12*).

In 1956, Duckworth attended the Conference on Atomic Masses organized at Mainz, Germany, by H. Hinterberger, to mark the sixtieth birthday of Josef Mattauch, Director of the Max Planck Institute there (*13*, see also this volume, *Heinrich Hintenberger and Josef Heinrich Elisabeth Mattauch*). This conference was so successful in bringing together the practitioners of nuclidic mass measurement (then a small number of individuals), that it was decided to continue the gathering as occasional Mass Conferences. Duckworth was an enthusiastic promoter of these, organizing the first at McMaster in 1960 and the third at Manitoba in 1967 (*14*). At the ninth Mass Conference, at Bernkastel, Germany, in 1992, he was awarded the SUN-AMCO Medal of the International Union of Pure and Applied Physics (IUPAP) for contributions to the precise determination of atomic masses.

In 1958, Cambridge University Press published Duckworth's *Mass Spectroscopy,* one of the earliest books on this subject in English. A second edition, with R. C. Barber and V. S. Venkatasubramanian, appeared in 1986 (*15,16*). Also in 1958, Duckworth proposed that IUPAP establish a Commission on Atomic Masses to, among other things, address the pressing problem of unifying the chemist's standard for atomic weight with the physicist's standard for atomic mass. The Commission was established in 1959, with Josef Mattauch as president, Duckworth as secretary, and an international membership including A. O. Nier, one of the first to propose that a unified standard should be based on the isotope ^{12}C (see this volume, *Alfred Otto Carl Nier*). This proposal passed the General Assemblies of the International Union of Pure and Applied Chemistry and of IUPAP in 1959 and 1960 (*17*).

Duckworth moved to the University of Manitoba in 1965, with an appointment as Vice-President (Development), changed to that of Vice-President (Academic) a year later. He was accompanied by R. C. Barber, who had been a PhD student and then a Postdoctoral Fellow with Duckworth at McMaster and was now appointed Assistant Professor of Physics at Manitoba. The large-radius mass spectrometer came to Manitoba a year later and was incorporated into a collaborative research program by Duckworth and Barber. Duckworth became President of the University of Winnipeg in 1971 but for some years provided a significant part of the funding for the University of Manitoba group and continued to supervise graduate students and postdoctoral fellows until the late 1970s. The last scientific publication bearing his name appeared in 1995 (*18*).

During his career, Duckworth served on all the major scientific councils in Canada – the National Research Council, the Science Council, the Defense Research Board, and the Natural Sciences and Engineering Research Council. A charter member of the Canadian Association of Physicists (CAP) in 1945, he served as President in 1960. In 1954 he became a Fellow of the Royal Society of Canada (RSC), President in 1971, and a Fellow of the American Physical Society. He received the CAP's Medal of Achievement in 1964 and the Tory Medal of the RSC in 1965. He was appointed an Officer of the Order of Canada in 1976. He was awarded ten honorary degrees.

Duckworth retired from the University of Winnipeg presidency in 1981 but became Chancellor of the University of Manitoba in 1986 and retired from that post in 1992. One of his activities in retirement was writing an autobiography that includes many details of his research career while still meant for a general reader (*1*). He died in Winnipeg in 2008.

References

(*1*) Duckworth, H. E. *One Version of the Facts: My Life in the Ivory Tower*; University of Manitoba Press: Winnipeg, Canada, 2000.

(*2*) Duckworth, H. E. New Packing Fractions and the Packing Fraction Curve. *Phys. Rev.* **1942**, *62*, 19–28.

(*3*) Duckworth, H. E.; Hogg, B. G. Relative Abundance of the Copper Isotopes and the Suitability of the Photometric Method for Detecting Small Variations in Isotopic Abundance. *Phys. Rev.* **1947**, *71*, 212.

(*4*) Duckworth, H. E. Design of Double-Focusing Mass Spectrographs. *Phys. Rev.* **1948**, *73*, 543.

(*5*) Duckworth, H. E. A Large Dempster-Type Double-Focusing Mass Spectrograph. *Rev. Sci. Instrs.* **1950**, *21*, 54–59.

(*6*) Duckworth, H. E.; Black, R. F.; Woodcock, R. F. A New Stable Isotope of Platinum. *Phys. Rev.* **1949**, *75*, 1438–1439.

(*7*) Duckworth, H. E.; Black, R. F.; Woodcock, R. F. Search for Stable Pd^{100}, W^{178}, and Pb^{202}. *Phys. Rev.* **1949**, *75*, 1616–1617.

(8) Duckworth, H. E.; Preston, R. S. Some New Atomic Mass Measurements and Remarks on the Mass Evidence for Magic Numbers. *Phys. Rev.* **1951**, *82*, 468–470.
(9) Stanford, G. S.; Duckworth, H. E.; Hogg, B. G.; Geiger, J. S. Masses of Pb^{208}, Th^{232}, U^{234}, U^{235}, and U^{238}. *Phys. Rev.* **1952**, *85*, 1039.
(10) Hogg, B. G.; Duckworth, H. E. Evidence for a Region of Extra Nuclear Stability between the 82- and 126-Neutron Shells. *Phys. Rev.* **1953**, *91*, 1289–1290.
(11) Isenor, N. R.; Barber, R. C.; Duckworth, H. E. First Measurements with the McMaster Double-focusing Mass Spectrometer. In: Duckworth, H. E., Ed.; *Proceedings of the International Conference on Nuclidic Masses, McMaster University, Hamilton, Spetember 12–16, 1960.* University of Toronto Press: Toronto, 1960, pp 439–444.
(12) Barber, R. C.; Duckworth, H. E.; Hogg, B. G.; Macdougall, J. D.; McLatchie, W.; Van Rookhuyzen, P. The Mass Effect Corresponding to the Onset of Nuclear Deformation in the Region $N \sim 90$. *Phys. Rev. Lett.* **1964**, *12*, 597–600.
(13) Hintenberger, H., Ed.; *Nuclear Masses and their Determination*; Pergamon Press: New York, 1957.
(14) Duckworth, H. E., Ed.; *Proceedings of the International Conference on Nuclidic Masses, McMaster University, Hamilton, September 12–16, 1960*. University of Toronto Press: Toronto, 1960.
(15) Duckworth, H. E. *Mass Spectroscopy*; Cambridge University Press: Cambridge, 1958.
(16) Duckworth, H. E.; Barber, R. C.; Venkatasubramanian, V. S. *Mass Spectroscopy*, 2nd ed; Cambridge University Press: Cambridge, 1986.
(17) Duckworth, H. E.; Nier, A. O. C. Evolution of the Unified Scale of Atomic Mass, Carbon 12=12u". *Int. J. Mass Spectrom. Ion Procs.* **1988**, *86*, 1–19.
(18) Williams, P.; Barber, R. C.; Sharma, K. S.; Duckworth, H. E. Unsuitability of $^{12}C_{60}$ as a Standard of Atomic Weight. *J. Am. Soc. Mass Spectrom.* **1995**, *6*, 1243–1246.

Harry W. Duckworth
Department of Chemistry, University of Manitoba, Winnipeg, MB, Canada

Geoffrey Eglinton

b. January 11, 1927

Geoffrey Eglinton is a British chemist, trained in synthetic organic chemistry, who has spent much of his career developing and exploiting chromatographic and spectrometric (particularly infrared and mass spectrometry) techniques in biogeochemical studies of natural environments and ancient sediments. His major work has been on the development of the molecular biomarker approach to biogeochemistry through the intensive application of gas chromatography/ mass spectrometry (GC/MS) and of compound specific, isotope ratio CG/MS.

Eglinton was born in Cardiff, Wales, UK, in 1927. He earned a BSc from Manchester University and he completed his PhD in organic chemistry in 1951, also from Manchester, under the direction of Mark Whiting. He held positions as Public Health Fellow in Columbus, Ohio, US (1951–52) and Imperial Chemical Industries Fellow, Liverpool University, UK (1952–54) before going to Glasgow University as Lecturer, then Reader, in Chemistry (1964–68). From 1968 to 1993 he was part of the faculty at the University of Bristol, UK, first as Reader, then as Professor of Organic Geochemistry. He has been Professor Emeritus in the Earth Sciences Department at Bristol from 1993 onwards. He and his wife had three children.

Eglinton's interest in mass spectrometry was initiated by Rowland Reed at Glasgow University and further developed by working with Alma Burlingame at the University of California, Berkeley, US (see this volume, *Rowland Ivor Reed*). Then, in 1964, Eglinton and C.J.W. Brooks received UK Research Council funding for purchase of the first fully integrated GC/MS instrument (from LKB, Stockholm; now lodged with The Science Museum) to be sited in Glasgow University. Building on previous experience with improved resolution gas chromatograph columns, Eglinton and Brooks rapidly developed biomedical, environmental and organic geochemical applications for this new, fast and highly sensitive analytical tool. These studies led to the development of molecular stratigraphy and molecular mapping in palaeo-oceanography and other fields and to the initiation of the Organic Geochemistry Unit and the Biogeochemistry Research Centre at Bristol. Molecular biomarker analyses based on coupled chromatographic–mass spectrometric techniques are now applied worldwide in many areas of science. Some representative publications are listed below (*1–7*).

Eglinton has received medals from NASA, and from the Chemical, Geochemical, Geological, European Geochemical, and Chromatographic Societies, as well as the Royal Society. He was elected to the Royal Society in 1976. In 2008, he was awarded the Dan David Prize in the field of Geosciences.

References

(*1*) Eglinton, G.; Scott, P. M.; Belsky, T.; Burlingame, A. L.; Calvin, M. Hydrocarbons of Biological Origin from a One-Billion-Year-Old Sediment. *Science* **1964**, *145*, 263–264.

(*2*) Eglinton, G.; Calvin, M. Chemical Fossils. *Sci Am.* **1967**, *216*, 32–43.

(*3*) Henderson, W.; Wollrab, V.; Eglinton, G. Identification of Steroids and Triterpenes from a Geological Source by Capillary Gas-Liquid Chromatography and Mass Spectrometry. *Chem. Commun.* **1968**, *13*, 710–712.

(*4*) Eglinton, G., Murphy, Sister M. T. J., Eds. *Organic Geochemistry: Methods and Results*; Springer: Heidelberg, New York, 1969.

(*5*) Gray, N. A. B.; Zoro, J. A.; Gronneberg, T. O.; Gaskell, S. J.; Cardoso, J. N.; Eglinton, G. Automatic Classification of Mass Spectra by a Laboratory Computer System. *Anal. Letts.* **1975**, *8*, 461–477.

(*6*) Eglinton, T. I.; Eglinton, G. Molecular Proxies for Palaeoclimatology. *Earth Planetary Sci. Lett.* **2008**, *275*, 1–16.

(*7*) Gaines, S. M.; Eglinton, G.; Rullkoetter, J. *Echoes of Life: What Fossil Molecules Reveal about Earth History*; Oxford University Press: New York, etc., 2009.

Geoffrey Eglinton
University of Bristol, Bristol, UK

Christie George Enke

b. July 8, 1933

Christie George (Chris) Enke is an American analytical chemist known for his work in fundamental electrochemistry, electronics and computer control of chemical instrumentation, tandem mass spectrometry, distance-of-flight mass spectrometry (DOF), and fundamental studies in electrospray ionization (see this volume, *A Perspective on the Development of Tandem Mass Spectrometry* and Chapter 3 (Volume 8): *Electrospray Ionization: How and When It Works*).

Enke was born in Minneapolis, MN, on July 8, 1933. He received his bachelor's degree from Principia College, Elsah, IL and went on to study the electrochemical formation of surface oxides on Pt with Herbert Laitinen at the University of Illinois Urbana/Champaign where he earned Ph.D. He taught at Princeton University as Instructor and Assistant Professor from 1959–1966. Thereafter, he joined the Michigan State University faculty first as Associate Professor, then Professor, and in 1994 Emeritus Professor. In this year he became Professor at University of New Mexico, a position he maintained until 2006, when he retired from full-time duties and became Emeritus Professor from UNM as well.

His principal contributions to mass spectrometry have been the discovery that low-energy collision-induced dissociation could be accomplished efficiently in an RF quadrupole and that incorporation of this device between two quadrupole mass filters could be used to create a tandem quadrupole mass spectrometer, the demonstration of the great analytical power of tandem mass spectrometers with unit mass resolution in both stages, the introduction of computer control in mass spectrometric instrumentation, the development of the equilibrium partition theory for electrospray ionization, the invention of distance of flight mass spectrometry (DOF-MS), and the training of a large number of mass spectrometrists now working in academics and industry (*1–6*). The triple quadrupole mass spectrometer is still one of the most widely used forms of tandem mass spectrometer and is the instrument of choice for quantitation and sensitivity in targeted analysis. The equilibrium partition theory, which explains electrospray selectivity and matrix effects for many analytes, has been widely cited and applied in a variety of studies (*7–11*). DOF-MS provides physical separation of mass-selected ions, as well as the potential for scanless MS/MS (*12*).

In 1962, Howard Malmstadt and he wrote the pioneering book, *Electronics for Scientists*. Then Malmstadt, Stan Crouch and he went on to write several more texts and laboratory manuals (thirteen in all) in the electronics of laboratory instrumentation. This same team developed and presented the hands-on ACS short course, 'Electronics for Laboratory Instrumentation' scores of times beginning in 1979. Chris's interest in the electronics of chemical instrumentation and the advent of the laboratory computer led him from electrochemical research into development of computer-controlled optical spectroscopic instrumentation. While conceptualizing an advanced automated analytical instrument he and his student Rick Yost hypothesized a tandem quadrupole mass spectrometer to accomplish sequential separation and identification in a single, computer-controlled instrument (see this volume, *Richard A. Yost*). The resulting triple quadrupole mass spectrometer brought him finally (and happily!) into research in mass spectrometry instrumentation and related phenomena.

Throughout his career, Chris has remained active in both fundamental research and the development of new teaching materials and methods. His research interests began with electroanalytical chemistry. His Ph.D. thesis was the first publication of an instrumental application of operational amplifiers. While at Princeton, his research focused on high-speed electrochemical charge transfer kinetic studies. These carried over to his first years at Michigan State University where the laboratory computer as a potential automation and control device caught his attention. In 1970, he invented the bipolar pulse method for measuring electrolytic conductance (now universally employed), and a year later wrote his "data domains" analysis of the modes by which electronic signals could encode data. He developed computer-controlled stopped-flow kinetics instruments and, in 1976, an array detector spectrometer (one of the first vidicon applications). The first triple-quadrupole mass spectrometer paper was published in 1978, and when microprocessors became available in the early 80's, his group worked on a distributed microprocessor control system for that instrument. Chris's work on computer-controlled instrumentation focused on using the computer to create versatility in application and ease of function (*13*). His early concept of placing the functions of user interface and real-time instrument control into two separate processors continues to be the practice in modern instrumentation.

Fascinated by the increased information available from the tandem MS instrument, he worked for several years on artificial intelligence for the interpretation of MS/MS spectra. A collaborative paper often cited as the first instance of using an RF-only quadrupole as a linear ion trap was published in 1988. By 1993, his group had demonstrated a tandem time-of-flight instrument that employed photodissociation. This was virtually coincident with Cotter's tandem TOF that used collisional dissociation (*14*). His use of the Bradbury-Nielson ion gate in his tandem time-of-flight instrument, in addition to the analysis of its design, brought this device into common use in TOF and tandem TOF instruments (*15*). In 1995, an inorganically inclined graduate student studying crown complexes of silver ion by electrospray showed dramatic selectivity for these species among the ions in solution. Curiosity regarding the selectivity mechanism led Chris to develop the equilibrium partition theory of electrospray ionization (*5*). This concept, which faced early opposition, has been shown to apply to ions that are created by the ion evaporation (as opposed to the charge residue) model. It enables prediction of relative sensitivity from hydrophobicity information or relative retention times in reverse-phase chromatography. Most recently, his work on distance-of-flight mass spectrometry has revealed the means by which the initial energy dispersion of an ion batch can be brought to focus at the same flight time at m/z-dependent distances. This is an entirely new type of mass analyzer that offers the advantages of spatial dispersion and array detection in a simple format (*6*). Even though Chris' work appears to span a number of unrelated areas of analytical chemistry, what he learned in each area became a valuable component in the next, enabling him to bring a fresh perspective to each new area of focus.

Along with his scientific accomplishments, Chris has always valued his opportunity to teach and mentor many fine new scholars and professionals. Independently and with his students, he has published over 130 papers, 16 book chapters, and obtained 12 patents. He has also written an introductory analytical chemistry text called, *The Art and Science of Chemical Analysis* that was published by Wiley in December, 2000.

He has won the 1974 ACS award in Chemical Instrumentation, the 1989 Computers in Chemistry Award, the 2003 Award for Excellence in the Teaching of Analytical Chemistry, and the 2011 ACS award in Analytical Chemistry. He and Rick Yost share the 1993 ASMS award for Distinguished Contribution in Mass Spectrometry. He served as Chair of the Physical Electrochemistry Division of the Electrochemical Society (1969–71), Chair of the Division of Computers in Chemistry Division of ACS (1981), President of ASMS (1994–96) and as Chair of the Analytical Division of ACS (2005–7). He has also served on a number of journal editorial boards, advisory boards, and review panels.

Chris lives near Albuquerque, NM with his wife Bea Reed. He has two children (David and Anne Enke from his first marriage) and a step child (Gillian Reed) who have collectively produced four grandchildren. Failing to understand the concept of 'full' retirement, he is currently Adjunct Professor at Indiana University, collaborating with Professor Gary Hieftje on DOF-MS. He is also developing a 'theory of complex mixtures' by which one could determine the measurement qualities required to detect any given fraction of the total number of components in the mixture (*16*).

References

(*1*) Yost, R. A.; Enke, C. G. Selected Ion Fragmentation with a Tandem Quadrupole Mass Spectrometer. *J. Am. Chem. Soc.* **1978**, *100*, 2274–2275.

(*2*) Yost, R. A.; Enke, C. G. Triple Quadrupole Mass Spectrometry for Direct Mixture Analysis and Structure Elucidation. *Anal. Chem.* **1979**, *51*, 1251A–1256A.

(*3*) Enke, C. G. Computers in Scientific Instrumentation. *Science* **1982**, *215*, 785–791.

(*4*) Kristo, M. J.; Enke, C. G. System for Simultaneous Count/Current Measurement with a Dual-Mode Photon/Particle Detector. *Rev. Sci. Instrum.* **1988**, *59*, 438–442.

(*5*) Enke, C. G. A Predictive Model for Matrix and Analyte Effects in Electrospray Ionization of Singly-Charged Analytes. *Anal. Chem.* **1997**, *69*, 4885–4893.

(*6*) Enke, C. G.; Dobson, G. S. Achievement of Energy Focus for Distance-of-Flight Mass Spectrometry with Constant Momentum Acceleration and an Ion Mirror. *Anal. Chem.* **2007**, *79*, 8650–8661.

(*7*) Schwartz, J. C.; Wade, A. P.; Enke, C. G.; Cooks, R. G. Systematic Delineation of Scan Modes in Multidimensional Mass Spectrometry. *Anal. Chem.* **1990**, *62*, 1809–1818.

(*8*) Cole, M. J.; Enke, C. G. Direct Determination of Phospholipid Structures in Bacteria by Fast Atom Bombardment/Tandem MS. *Anal. Chem.* **1991**, *63*, 1032–1038.

(*9*) Cech, N.; Enke, C. G. Relating Electrospray Ionization Response to Nonpolar Character of Small Peptides. *Anal. Chem.* **2000**, *72*, 2717–2723.

(*10*) Amad, M.; Cech, N.; Jackson, G.; Enke, C. G. Importance of Gas-Phase Proton Affinities in Determining the Electrospray Ionization Response for Analytes and Solvents. *J. Mass Spectrom.* **2000**, *35*, 784–789.

(*11*) Cech, N.; Krone, J. R.; Enke, C. G. Predicting Electrospray Response from Chromatographic Retention Time. *Anal. Chem.* **2001**, *73*, 208–213.

(*12*) Graham, A. W.; Ray, S. J., *et al.* First Distance-of-Flight Instrument: Opening a New Paradigm in Mass

Spectrometry. *J. Am. Soc. Mass Spectrom.* **2011**, *22*, 110–117.

(*13*) Myerholtz, C. A.; Schubert, A. J.; Kristo, M. J.; Enke, C. G. A Versatile Software Control System for a Triple Quadrupole Mass Spectrometer. *Instruments and Computers* **1985**, *3*, 11–12.

(*14*) Seeterlin, M. A.; Vlasak, P. R.; Beussman, D. J.; McLane, R. D.; Enke, C. G. High Efficiency Photo-Induced Dissociation of Precursor Ions in a Tandem Time-of-Flight Mass Spectrometer. *J. Am. Soc. Mass Spectrom.* **1993**, *4*, 751–754.

(*15*) Vlasak, P. R.; Beussman, D. J.; Davenport, M. R.; Enke, C. G. An Interleaved Comb Ion Deflection Gate for M/Z Selection in Time-of-Flight Mass Spectrometry. *Rev. Sci. Instrum.* **1996**, *67*, 1–5.

(*16*) Enke, C. G.; Nagels, L. J. Undetected Components in Natural Mixtures: How Many? What Concentrations? Do They Account for Chemical Noiseical Noise? What Is Needed to Detect Them? *Anal. Chem.* **2011**, *83*, 2539–2546.

Christie G. Enke
Department of Chemistry and Chemical Biology, University of New Mexico, Albuquerque, NM, USA, and Department of Chemistry, Michigan State University, East Lansing, MI, USA

Samuel Epstein

b. December 9, 1919
d. September 17, 2001

Samuel Epstein was a Canadian-American geochemist who helped establish stable isotope geochemistry as a major approach in the earth and planetary sciences. He was instrumental in opening the field of paleoclimatology. In these ways, he played a major role in spread of mass spectrometry throughout the earth and planetary sciences, starting shortly after the Second World War at the University of Chicago and then for almost four decades at the California Institute of Technology.

Samuel Epstein was born in 1919 in Kobryn, which had just recently become part of Poland and is now in Belarus (*1–4*). His immediate family emigrated severally to Canada in 1926 and 1927. He graduated from high school in Winnipeg, Manitoba, and enrolled in the University of Manitoba in 1937. He completed his undergraduate work with a degree in both geology and chemistry in 1941 and earned a master's degree in chemistry in 1942. He earned his Ph.D. in chemistry in 1944 at McGill University in Montreal with Carl Winkler as his advisor, working on classified studies of high explosives.

Epstein was assigned to the Canadian part of the Allied work developing atomic weapons during the remainder of World War II. As part of that work he became acquainted with Harry Thode of McMaster University and with the capabilities of Thode's mass spectrometer – the only one in Canada at the time (see this volume, *Henry George Thode*). Epstein also met Diane Vool in Montreal, and they married in 1946. Through continued work with Thode, Epstein obtained a post-doctoral position with Harold Urey at the University of Chicago, where he worked from 1947 to 1952 and where his attention turned again to geological topics (*5*, see this volume, *Harold Clayton Urey*). Epstein became part of a group of young scientists who, influenced by the concentration of highly distinguished established researchers and the open interdisciplinary atmosphere in Chicago's Institute for Nuclear Studies (later the Fermi Institute), would reshape the fields of geology, geophysics, geochemistry, planetary astronomy, meteoritics, and more during the second half of the twentieth century (see this volume, *The Development of Mass Spectrometry in the Earth and Planetary Sciences*).

Urey, Jacob Bigeleisen, and Maria Mayer had calculated that isotopic fractionation would occur during various physical and chemical conditions, and Urey had realized that variations in the ratio of oxygen isotopes in nature due to ocean temperature differences would be recorded in shells. He conjectured that if the ratios in fossil shells could be measured with sufficient precision and reliability, they could provide a record of climates in different geological periods. Urey also realized it would take a team of people with different skills and insights to carry this general conception through to definitive results and practical methods. Samuel Epstein became a key member of the team that not only brought the project to fruition but developed it to a degree of precision far beyond what Urey had envisioned. The results were manifold: increased sensitivity and reliability in isotope-ratio mass spectrometry, proof-of-concept for learning about conditions in different geologic eras through isotopic analysis, development of techniques for practically carrying out this type of research with different kinds of materials, and establishment of the $^{18}O/^{16}O$ ratio carbonate-water temperature scale (*6–8*). The carbonate-water method remains among the most widely used ways to determine paleoclimates.

Harrison Brown, one of the major figures in the Chicago group, moved to the California Institute of Technology in Pasadena in 1951 to create a new center of geochemical research (see this volume, *Harrison Scott Brown*). He invited several others to move at the same time, Epstein among them, and helped induce others (including Urey) to make the same or a similar move to a southern California institution in the next few years. Epstein quickly became part of a growing program with international stature.

Epstein stayed at Cal Tech for rest of his life. He retired from teaching, becoming Professor Emeritus, in 1990 but continued daily research work to the end. He died in 2001, survived by his wife, children, and grandchildren.

In the intervening decades, he and his students continued to use the first mass spectrometer he built there along with newer technology to study the variations in isotopic abundances of hydrogen, carbon, oxygen, and silicon and their implications on questions in fields as diverse as astronomy, geology, climatology, biochemistry, and archeology.

He investigated meteorites and lunar samples, ancient sediments, igneous rocks, mantle materials, ice, snow, and water, plant and animal tissues, fossil bones, and tree rings. A small sample of his papers are listed below (*9–17*). It is only a modest exaggeration to take the title of the 1991 publication of the papers from the 1989 symposium celebrating his career as a plain factual declaration: *Stable Isotope Geochemistry – A Tribute to Samuel Epstein*. What the field has become reflects his influence in myriads ways. The European Association of Geochemistry has named its award for work in isotope geochemistry in his honor.

During his life Samuel Epstein received many awards and honors in recognition of his accomplishments. These include his election to the National Academy of Sciences and to the Royal Society of Canada, as well as receipt of the Day and Wollaston Medals from the US and UK's Geological Societies and the Goldschmidt and Urey awards from the major Geochemical organizations.

References

(*1*) Tayor, H. P. Jr.; Clayton, R. N. *Samuel Epstein: 1919–2001*; National Academy of Sciences Press: Washington, DC, 2008.

(*2*) Epstein, S. The Role of Stable Isotopes in Geochemistries of All Kinds. *Annu. Rev. Earth Planet. Sci.* **1997**, *25*, 1–21. see also http://www.annualreviews.org/doi/pdf/10.1146/annurev.earth.25.1.1.

(*3*) Tayor, H. P. Jr.; O'Neill, J. R.; Kaplan, I. R., Eds. *Stable Isotope Geochemistry: A Tribute to Samuel Epstein: Special Publication Number 3*; Geochemical Society: St. Louis, MO, US, 1991.

(*4*) Epstein, S. Interview by Carol Bugé, Pasadena, California, December 19 and 26, 1985, and January 10, 1986; Oral History Project. California Institute of Technology Archives: Pasadena, CA, 2011. see also http://oralhistories.library.caltech.edu/197/1/Epstein,_S._OHO.pdf.

(*5*) Lounsbury, M.; Epstein, S.; Thode, H. G. The Isotopic Composition of Normal Krypton and Xenon. *Phys. Rev.* **1947**, *72*, 517–518.

(*6*) McKinney, C. R.; McCrea, J. M.; Epstein, S.; Allen, H. A.; Urey, H. C. Improvements in Mass Spectrometers for the Measurement of Small Differences in Isotope Abundance Ratios. *Rev. Sci. Instr.* **1950**, *21*, 724–730.

(*7*) Urey, H. C.; Lowenstam, H. A.; Epstein, S.; McKinney, C. R. Measurement of Paleotemperatures and Temperatures of the Upper Cretaceous of England, Denmark, and the Southeastern United States. *Bull. Geol. Soc. Amer.* **1951**, *62*, 399–416.

(*8*) Epstein, S.; Buchsbaum, R.; Lowenstam, H. A.; Urey, H. C. Carbonate Water Isotopic Temperature Scale. *Bull. Geol. Soc. Amer.* **1951**, *62*, 417–426.

(*9*) Epstein, S.; Mayeda, T. K. Variation of ^{18}O Content of Waters from Natural Sources. *Geochim. Cosmochim. Acta.* **1953**, *4*, 213–224.

(*10*) Clayton, R. N.; Epstein, S. The Relationship between $^{18}O/^{16}O$ Ratios in Coexisting Quartz, Carbonates and Iron Oxides from Various Geological Deposits. *J. Geol.* **1958**, *66*, 352–373.

(*11*) O'Neil, J. R.; Epstein, S. Oxygen Isotope Fractionation in the System Dolomite-Calcite-Carbon Dioxide. *Science* **1966**, *152*, 198–201.

(*12*) Epstein, S.; Taylor, H. P. Jr. $^{18}O/^{16}O$, $^{30}Si/^{28}Si$, D/H, and $^{13}C/^{12}C$ Studies of Lunar Rocks and Minerals. *Science* **1970**, *167*, 533–535.

(*13*) Savin, S. M.; Epstein, S. The Oxygen and Hydrogen Isotope Geochemistry of Clay Minerals. *Geochim. Cosmochim. Acta.* **1970**, *34*, 25–42.

(*14*) Smith, B. N.; Epstein, S. Two Categories of $^{13}C/^{12}C$ Ratios for Higher Plants. *Plant Physiol.* **1971**, *47*, 380–384.

(*15*) Yang, J.; Epstein, S. Relic Interstellar Grains in Murchison Meteorite. *Nature* **1984**, *311*, 544–547.

(*16*) Feng, X.; Epstein, S. Climatic Implications of an 8000-Year Hydrogen Isotope Time Series from Bristlecone Pine Trees. *Science* **1994**, *265*, 1079–1081.

(*17*) Leshin, L.; Epstein, S.; Stolper, E. M. Hydrogen Isotopic Geochemistry of SNC Meteorites. *Geochim. Cosmochim. Acta.* **1996**, *60*, 2635–2650.

Keith A. Nier
Independent Historian, Madison, NJ, USA

Heinz Ewald

b. June 16, 1914, Stolzenau an der Wesser, Germany
d. February 5, 1992, Giessen, Germany

Source: www.staff.uni-mainz.de/bpfeiffe/mpi09.pptx

Heinz Ewald was a German physicist who greatly helped promote mass spectrometry in Germany in the middle of the twentieth century (*1*).

Ewald was born in Germany in 1914 and lived there until his death in 1992. After studying at Jena and Freiburg, he completed a dissertation at the University of Göttingen in 1938 under the direction of Georg Joos (*2*). In 1939, he joined Josef Mattauch's group at the Kaiser Wilhelm-Institut für Chemie in Berlin-Dalhem and concentrated on the development of high resolution mass spectrometry and the determination of nuclear binding energies (see this volume, *Josef Heinrich Elisabeth Mattauch*). He was involved with the group called the *Uranverein,* working to develop atomic energy. He proposed a design for a magnetic isotope separator for concentrating uranium 235, in some ways comparable to the calutrons in the Manhattan Project in the USA. He constructed and in 1946 he published the design for a new version of a Mattauch–Herzog-type double-focusing mass spectrograph (*3–5*, see this volume, *Magnetic and Electrostatic Analyzers before 1960*). He kept this machine with him through the rest of his career; it ended up on display at the II Physikalisches Institut at the University of Giessen, and on the internet.

Ewald remained with the Institute as it moved to Tailfingen (present-day Albstadt), Germany, late in World War II and through the immediate postwar years as it was renamed to become the Max-Planck-Institut für Chemie and moved to the Johannes-Gutenberg-University in Mainz in 1949 (*6*). In addition to his own work in areas such as instrument design and ion dissociation, Ewald and Heinrich Hintenberger (also at the MPI für Chemie) wrote the first German language general textbook on mass spectrometry, *Methoden und Anwendungen der Massenspektroscopie*, published in 1953 – the first of the flurry of general texts that were produced in the middle of the 1950s (*7* and see this volume, *Heinrich Hintenberger*).

In 1952, Ewald took a position at the Physikalische Institut at the Technische Hochscule in Munich. In 1963 he was called to the Justus-Liebig-University in Giessen to become the first director of the II Physikalishes Institut, where he remained until retirement in 1978. Heinz Ewald was married and had several children.

Ewald's work in his later career emphasized separation and precise determination of the masses and related properties of the products of nuclear reactions (*8–10*). Ewald was involved in the creation of the LOHENGRIN fission products spectrometer at the Institute Laue-Langevin in Grenoble, France, and the SHIP heavy ion reaction product separator at the Gesellschaft für Schwerionenforschung (now the GSI Helmholtz Center for Heavy Ion Research in Darmstadt).

References

(*1*) Wollnik, H.; Metag, V. Nachruf: Heinz Ewald (1914–1992). *Phys. Bl.* **1992**, *48*, 610.
(*2*) Ewald, H. Die Analyse und Deutung der Neodymsalzspektren. *Annalen d. Physik*, 1939, Folge 5, Bd *34,* H. 3. Göttingen, Math.-naturwiss. Diss.
(*3*) Ewald, H. Eine Neukonstruktion des Mattauch-Herzog'schen doppelfokussierenden Massenspektrograhen: Die Massen von ^{13}C und ^{15}N. *Z. Naturforschg.* **1946**, *1*, 131–136.
(*4*) Ewald, H. Accuracy of Mass Spectrographic Isotope Mass Measurements. In: Mass Spectroscopy in Physics Research; Circular 522. Mass Spectroscopy Symposium Committee, J. A. Hipple, chair, Ed.; National Bureau of Standards: Washington, D.C., 1953, pp. 37–46.
(*5*) Pfeiffer, B. Atomic Mass Evaluations: Contributions of the Kaiser-Wilhelm-/Max-Planck-Institut für Chemie. Seminar für Kern- und Kosmochemie ("K & K Seminar"). MPI für Chemie, Mainz, 10/2/2010. Powerpoint Slides Available On The Internet at: www.staff.uni-mainz.de/bpfeiffe/mpi09.pptx (accessed June 22, 2014).
(*6*) Elbert, W. History of the Institute. 2007. Max-Planck-Institut für Chemie. Home/ The Institute/ Institutés history. http://www.mpic.de/Institute-s-history.25.0.html?&L=2 (accessed June 22, 2014).
(*7*) Ewald, H.; Hintenberger, H. *Methoden und Anwendungen der Massenspektroscopie*; Ver. Chemie: Weinheim, 1953.
(*8*) Neumann, S.; Ewald, H. Fokussierende Parabelspektrographen. *Zeitschrift für Physik A: Hadrons and Nuclei* **1962**, *169*, 224–238.
(*9*) Ewald, H.; Konecny, E.; Opower, H.; Rösler, H. Ein Massenspektrograph zur Analyse von Spaltproduktstrahlen. *Z. Naturforschg* **1964**, *19a*, 194–200.
(*10*) Armbruster, P.; Asghar, M.; Bocquet, J. P., *et al.* The Recoil Separator Lohengrin: Performance and Special Features for Experiments. *Nucl. Instrum. Meth.* **1976**, *139*, 213–222.

Keith A. Nier
Independent Historian, Madison, NJ, USA

Henry Marshall Fales

b. February 12, 1927
d. October 28, 2010

Henry M. (Hank) Fales was an American organic chemist known for his applications of mass spectrometry to the structural analysis of organic compounds and for his willing adoption of novel techniques to perform these analyses. He was also widely known for his collegiality and for introducing a large number of research scientists to mass spectrometric techniques.

Hank was born in New York City on February 12, 1927 and after graduation from high school in 1944 enlisted in the Navy Air Corp. After WWII, he obtained a B.S. in 1948, and a Ph.D. (with R. A. Barnes) in 1952 from Rutgers University. In 1954, he accepted a research chemist position with W. C. Wildman at the Laboratory of Chemistry at the National Heart, Lung and Blood Institute of the National Institutes of Health (NIH), in Bethesda, Maryland, the institution that would be his professional home until his death on October 28, 2010.

In a research career spanning more than fifty years, Hank introduced many biomedical research scientists to the power of mass spectrometry for organic structure determination. His lab was often the stimulus for the construction and adaptation of new technologies in biomedical applications, including chemical ionization, plasma desorption, and electrospray ionization. The resultant body of science comprises more than 300 publications dealing with natural products, biochemistry, insect pheromones, and protein structures – often the result of collaborations with colleagues from most of the NIH institutes and many other government and academic laboratories. Hank functioned as a catalyst to scientists first in the Washington DC area, when he co-founded a Washington area mass spectrometry discussion group with Fred Saalfeld at the Naval Research Laboratory, Henry Rosenstock of the National Bureau of Standards (now NIST), and Joseph D'Amico of the Food and Drug Administration (see this volume, *Henry Meyer Rosenstock*). Through these meetings and presentations by area scientists, Hank convinced Marvin Vestal to build and install the first commercial chemical ionization (CI) ion source for the MS-902 at NIH in 1968, which led to a series of paradigm-shifting papers on the application of chemical ionization to compounds of biomedical interest (see this volume, *Marvin L. Vestal*).

Hank Fales' interest in applying mass spectrometry to organic structure analysis began when the combustion analyses for the alkaloid compound amaryllisine produced results the explanation for which required one carbon less than usual for related alkaloids. Because Klaus Biemann had recently delivered a lecture series on the interpretation of mass spectra at NIH, Hank and his colleague Robert (Bob) J. Highet sent amaryllisine to Alma (Al) Burlingame for examination in the Biemann laboratory (*1*, see this volume, *Klaus Biemann*). The resulting mass spectral data clarified the structure and convinced Hank that a mass spectrometer was needed at NIH.

Hank and Bob selected the Nier-Johnson geometry AEI- MS902 based on its purported high resolution, demonstrated by the British firm AEI by the separation of $C_2^{13}CH_7$ at m/z 44.058130 from $C_3H_6{}^2H$ at m/z 44.061052, a feat that required a resolution of 15,000 (see this volume, *Magnetic and Electrostatic Analyzers before 1960 and Mass Spectrometry Manufacturers and Markets from the Beginning to 1960*). The purchase was approved by Daniel Steinberg, a noted lipid disease clinical researcher at NIH, because of the applicability of the instrument to understanding the clinical biochemistry of a phytanic acid metabolic disorder (Refsum's disease) (*2*). Fales' and Highet's choice was at odds with conventional wisdom of the time: during the 1960s, the pioneering labs in organic high resolution mass spectrometry in the U.S. (see this volume, *Fred W. McLafferty*) used the Mattauch-Herzog CEC-21–110. In fact, because one famous reviewer remained skeptical of the claimed resolution, Fales' and Highet's early demonstration of deuterium incorporation into cholesterol remained unpublished.

The National Heart Institute Laboratory of Chemistry directed by Fales was a center of activity for structural chemistry in the intramural research program at NIH and the Washington area. The lab under its previous director Evan Horning had developed a reputation as a premier gas chromatography group, and the Fales' section was especially skilled at vapor phase purification of alkaloids on deactivated silanized supports (see this volume, *Evan Charles Horning* and *Marjorie Groothuis Horning*). As the power of gas chromatography and mass spectrometry in biomedical applications became more apparent, scientists from the area met regularly with Fales in the Washington mass spectrometry discussion group to keep each other informally updated on new developments. In 1966 Fales used high resolution mass spectrometry to describe the chemistry of a compound with juvenile hormone activity from balsam fir (*3*). That same year he obtained a mass spectrum for a highly fluorinated compound of molecular weight 3628 given to him by Ian Brown during a trip to the CSIRO in Australia. This demonstration of the potential application of mass spectrometry to the analysis of high molecular weight biologicals generated great interest, since at the time the accepted limits of organic mass spectrometry were below m/z 1000 (*4*).

G.W. (Bill) A. Milne joined the Fales' lab and, when Marvin Vestal's CI source was installed on the MS-902, Hank and Bill produced a series of studies on the chemical ionization spectra of complex molecules (alkaloids, amino acids, peptides, drugs) that heightened the interest of the scientific world in this technique (*5*). The popular appeal of chemical ionization prompted instrument manufacturers to implement CI on quadrupole instruments, which had the advantage of avoiding the discharge problems that high voltages coupled with high CI gas pressures caused on the NIH prototype.

The Fales lab obtained an early version of Vestal's CI – quadrupole mass spectrometer from SRI in Baltimore in 1970 with the intent to demonstrate its utility as an emergency tool for the analysis of fluids (blood, urine, gastric lavage) from drug overdose victims. In collaboration with his physician friend and neighbor Dr. Norman Law, Fales used the instrument at all times of the day and night for rapid characterization of the drugs present (*6*). Recognizing that consistent mass spectrometric characterization of many repeated components suggested the need for a computer-searchable reference library of mass spectra, Fales and Milne began a major effort that Milne and Steve Heller pursued for the collection of additional reference spectra and the development of search-and-retrieval software programs to analyze data first at NIH, and later in collaboration with EPA (*7*). Eventually, the NIH-EPA effort was transferred to NBS (now NIST), where it has grown into an international resource (see this volume, *A History of the Development of Mass Spectral Databases*).

Over the long years of his career, Hank Fales used a combination of organic synthesis, gas or liquid chromatography, nuclear magnetic resonance, IR/UV and x-ray spectroscopy with mass spectrometric methods to elucidate structures of many materials brought to his lab by collaborators. Notably, his collaborations with Murray Blum on insect defensive secretions and caste- or sex-specific pheromones spanned many publications. In the mid-1980s, he convinced Ron McFarlane to duplicate his ^{252}Cf–plasma desorption time-of-flight instrument at NIH, and Fales began a series of studies of high molecular weight compounds of particular interest at NIH (see this volume, *A History of ^{252}Cf–Plasma Desorption Mass Spectrometry*). These analyses were transferred to electrospray ionization, a technique that Hank investigated and utilized in multiple protein structure analyses (*8*).

Hank's international reputation as an innovative scientist was recognized by his peers with his election three times as an officer of the American Society for Mass Spectrometry in 1976 (member at large), 1978 (secretary), and 1992–94 (president). He served on the editorial boards for the Journal of Mass Spectrometry, the Journal of the American Society for Mass Spectrometry, Analytical Chemistry, and Bioanalytical Chemistry and on the advisory panel on Chemistry of the National Science Foundation.

Fales received a number of awards including Chromatographer of the Year of the Washington Chromatography Discussion Group (1973), DHEW Superior Service Awards (1973, 1986), the Hillebrand Prize from the Chemical Society of Washington in 1977, and a Professional Service Award of Alpha Chi Sigma, Washington, D.C. (1988). His bibliography up to the year 2005 was published in the Journal of Biomacromolecular Mass Spectrometry.

References

(*1*) Burlingame, A. L.; Fales, H. M.; Highet, R. J. Structure of Amaryllisine. *J. Am. Chem. Soc.* **1964**, *86*(22), 4976–4979.

(*2*) Steinberg, D.; Mize, C. E.; Avigan, J., *et al.* Studies on the Metabolic Error in Refsum's Disease. *J. Clin. Invest.* **1967**, *46*(3), 313–322.

(*3*) Bowers, W. S.; Fales, H. M.; Thompson, M. J.; Uebel, E. C. Juvenile Hormone: Identification of an Active Compound from Balsam Fir. *Science* **1966**, *154*(752), 1020–1021.

(*4*) Fales, H. M. The Mass Spectrum of a Compound of Formula C72H24O8F128N4P4 and Molecular Weight 3628. *Anal. Chem.* **1966**, *38* (8), 1058–1059.

(*5*) Fales, H. M.; Milne, G. W.; Vestal, M. L. Chemical Ionization Mass Spectrometry of Complex Molecules. *J. Am. Chem. Soc.* **1969**, *91*(13), 3682–3685.

(*6*) Law, N. C.; Aandahl, V.; Fales, H. M.; Milne, G. W. Identification of Dangerous Drugs by Mass Spectrometry. *Clin. Chim. Acta* **1971**, *32*(2), 221–228.

(*7*) Heller, S. R.; Fales, H. M.; Milne, G. W.; Heller, R. S.; McCormick, A.; Maxwell, D. C. Mass Spectral Search System. *Biomed. Mass Spectrom.* **1974**, *1*(3), 207–208.

(*8*) Davis, D. A.; Singer, K. E.; De La Luz Sierra, M., *et al.* Identification of Carboxypeptidase N as an Enzyme Responsible for C-terminal Cleavage of Stromal Cell-Derived Factor-1Alpha in the Circulation. *Blood* **2005**, *105*(12), 4561–4568.

Sanford P. Markey

Laboratory of Neurotoxicology, National Institute of Mental Health, Bethesda, MD, USA

John Bennett Fenn

b. June 15, 1917
d. December 10, 2010

Figure 1
John Fenn at the Nobel Awards ceremony in Stockholm, Sweden, December, 2002.

John Fenn has the unique distinction of being the pioneer of two important scientific fields: supersonic molecular beams and electrospray mass spectrometry, a technique for which he would be awarded the 2002 Nobel Prize in Chemistry (Figs. 1 and 2). Although a chemist by degree, John spent most of his academic life as a professor of Mechanical Engineering at Princeton (1959–67) and of Chemical Engineering (with a joint appointment in Chemistry) at Yale (1967–1993). It was only after his mandatory retirement from Yale that he became solely a Professor of Chemistry at Virginia Commonwealth University (1993), a position held until he passed away December 10, 2010 (*1–3*).

John was born in New York City on June 15, 1917. His father, Herbert, had an Electrical Engineering degree from Rutgers and his mother, Jeanette, was a graduate of Columbia Teacher's College. They met in Sitka, Alaska while serving as missionaries. The family originally lived in Hackensack, NJ, where John attended the State Street Elementary School. One of his fondest memories from this time was sitting in the cockpit and playing with the controls of the Spirit of St. Louis when it was stored in a secret hanger at nearby Teterboro airport shortly after Charles Lindbergh's historic trans Atlantic flight (*1*).

The Fenn family 'lost everything' on the eve of the Great Depression and moved to Berea, KY, in the foothills of the Cumberland Mountains in1928. In retrospect, the move to Kentucky was fortunate because it gave John the opportunity to attend a remarkable institute: Berea College and Allied Schools, founded in 1855 with the mission to educate Appalachian students of limited means, regardless of race or gender, fulfilling the school's motto: "God has made of one blood all the people of the earth" (*4*). Berea was, therefore, integrated and coeducational long before its Ivy League cousins. John entered at the eighth grade of the Training School and continued his education through the College, graduating with a degree in Chemistry (1937). It is difficult to overstate the profound influence Berea had on John: the lectures of an inspirational chemistry professor, Julian Capps (his stories made 'Chemistry dance'); the convocations organized by President William Hutchins (featuring such luminaries as Robert Millikan); the life-long friendships he made; the poems he memorized and would recite verbatim in later years; and simply the time he had as a young man to explore his interests. It was while a student, that John first met Margaret ('Magee') Wilson, who would be his 'bride' when they married in 1939 until her untimely passing in 1992. Of course, the Berea experience was not perfectly idyllic. After a less than stellar performance on a ninth grade Algebra exam, John's teacher wrote in bold red letters; "Never attempt to be a scientist or engineer" (*1*). John later admitted this admonition was on target, but being 'mathematically challenged' may have been a blessing because instead of getting lost among mathematical trees, John was able to envision and discover entirely new forests.

After Berea, John went to Yale for his graduate work in the Chemistry Department under Gusta Akerlof. John was extremely fond of 'Gus,' but the research experience, consisting of thousands of emf (electromotive force) measurements of electrolytic solutions, was less than satisfying: "My dissertation attested to the sterility of that project, consisting as it did of 45 pages of tables with only three pages of text!... So narrow was the focus that I left Yale halfway believing 'Physical Chemistry was the study of slightly contaminated water'" (*1*). The Yale experience, however, was "happy and rewarding" because "my perspectives were stretched by memorable if abstruse encounters with great minds, including lectures on resonance by Linus Pauling and two required courses in statistical mechanics under Lars Onsager (known to students as Norwegian I and II)" (*1*).

After graduating from Yale with a PhD in 1940 John worked as a chemist for Monsanto (Anniston, AL) and Sharples Chemical (Wyandotte, MI). In 1945, he again moved his family (by then including two daughters and a soon-to-arrive son) to Richmond, VA where he was one of the original employees of Experiment Inc., an R&D company founded by his friend James Mullen, to investigate the new field of jet propulsive systems. It was while at Experiment, Inc. John returned to research and published his first paper in 1949 (*5*), when he was 32. After a few more publications, this late bloomer began to gain "some recognition in the combustion community," leading to an appointment as director of Project SQUID, financed by the Office of Naval Research (ONR), and

Figure 2
John Fenn at the Chemical Heritage Foundation with his molecular beam machine on which electrospray ionization was developed.

based at Princeton University. As director (with the rank of professor), John was responsible for administering grants to various combustion and jet propulsion research labs. In 1955, he accepted an offer from the ONR for a one-year appointment as its Liaison Officer in London. In this capacity, he made "many, many friends" in European labs. During this London tenure, he began to "fantasize about using micro-rockets in vacuum as molecular beam sources" to investigate the kinetics and mechanisms of high temperature reactions and, thus, setting "the stage for the rest of my scientific life" (*1*). Upon returning to Princeton, where a sudden vacancy had opened a position, John was formally invited to join the Mechanical Engineering faculty, teaching a course in thermodynamics. An NSF grant (1959), allowed his London 'fantasies' to be turned into reality. The pumping speed of the large (32") diffusion pumps purchased with the grant overcame the limitations of previous researchers and enabled him to convince the skeptics attending a 1965 Gordon Conference to give up their effusive sources for his supersonic beams (*6–8*). The 1986 Nobel Laureate in Chemistry, Dudley Herschbach, acknowledged John's contribution to his research: "Like many others since, our adoption of supersonic nozzles was spurred by John Fenn and Jim Anderson, ardent evangelists among the chemical engineers exploring fluid flow in nozzles" (*9*).

John moved his laboratory to the Engineering and Applied Science Department at Yale in 1967. The uniqueness of the Yale environment allowed him to collaborate with researchers in other disciplines. Among these was Seymour ('Sandy') Lipsky from the Yale Medical School who first made John aware of Malcolm Dole's work using electrospray ionization (ESI) to produce gas phase ions of molecules too large to be 'vaporized' by conventional means (*10–11*). John's ESI research would ultimately provide mass spectrometry with a way to analyze biomolecules, work for which he would share the Nobel Prize in Chemistry in 2002 (see this volume, *The Background to Electrospray* and *The Development of Electrospray Ionization (ESI) for Mass Spectrometry*).

John was a gifted writer and educator. He had an uncanny ability to strip down complicated problems to their most basic elements and explain them in an elegant, almost poetic, manner as attested to by his autobiographies and thermodynamics book 'Engines, Energy and Entropy' (*1–3,12*). It was, however, John's wit, charm, and concern for others that endeared him to his myriad students, colleagues, and friends. Behind his back, his students referred to him as 'Papa' Fenn because when a student had a problem, whether academic, social, or financial, he/she went to 'Papa' for help and advice. His door was always open. He arranged course schedules, visas, accommodations, first jobs, and first dates with future wives. When his grants were frozen due to a dispute with the administration, John paid his graduate students out of his own pocket. He was humble, kind, and generous, yet stubborn and uncompromising in matters of principle, a trait that did not win him many friends among bureaucrats over the years.

John came to Yale, in part, to help re-build the engineering program. While he often ruffled feathers, his early relationship with the Yale administration is perhaps best summarized in a letter the then President Bart Giamatti sent to John on the occasion of the 'Fennfest' held at Yale to celebrate John's 65th birthday in 1983. "After much searching of the literature," Giamatti wrote, "I identified a single source which contains a list of personal characteristics which seem to have some application in talking about John Fenn. I refer you to the Boy Scout Handbook, where a true Boy Scout is described as being trustworthy, loyal, helpful, friendly, courteous, kind, obedient, cheerful, thrifty, brave, clean, and reverent. Now I am not saying that John is all of these things since he is not, for example, notably obedient. I do observe, however, that nine out of twelve is not bad." Giamatti then went on to thank John for "the brilliance, dedication, and citizenship – we are all in your debt" (*9*). Unfortunately, by 1993 bureaucratic disputes, eventually leading to litigation, soured the Yale/Fenn relationship (*9,13*). For this and more personal reasons John accepted an invitation to join the faculty at Virginia Commonwealth University and moved his lab to Richmond, where he married Freda Mullen. John was blessed with two wonderful marriages.

John was the author of more than 100 research articles and held some two dozen patents; he recorded his own views concerning the importance of many of his papers (*14*). He was a member of the American Chemical Society, the American Society for Mass Spectrometry, the American Association of University Professors, and the Alexander von Humboldt Association of America. He was honored by the American Society for Mass Spectrometry with the Award for Distinguished Contributions in Mass Spectrometry in 1992. He became a fellow of the American Academy of Sciences in 2000 and was elected to the National Academy of Sciences in 2003. In 2002, he shared the Nobel Prize in Chemistry for "development of soft desorption ionization methods for mass spectrometric analyses of biological macromolecules" (*15* and see this volume, *Koichi Tanaka*).

Other awards included the Yale Science and Engineering Award for Distinguished Contributions to Basic and Applied Science in 1999; the Award for Advancements in Chemical Instrumentation from the American Chemical Society in 2000; the Thomson Medal from the International Society of Mass Spectrometry in 2000; the Award for Outstanding Contributions to Biomolecular Technologies from the Association of Biomolecular Resource Facilities in 2002. and the Wilbur Cross Medal from the Yale Graduate School Alumni Association in 2003.

References

(*1*) Fenn, J. B. Research in Retrospect: Some Biograffiti of a Journeyman Chemist. *Annu. Rev. Phys. Chem.* **1996**, *47*, 1–41.

(*2*) Fenn, J. B. Autobiography. In: *Les Prix Nobel. Nobel Prizes 2002*; Frängsmyr, T., Ed.; Nobel Foundation: Stockholm, 2003.

(*3*) John, B. Fenn – Nobel Lecture: Electrospray Wings for Molecular Elephants. Nobelprize.org. http://www.nobelprize.org/nobel_prizes/chemistry/laureates/2002/fenn-lecture.html (accessed 30 November 2012).

(*4*) Berea College website: www.berea.edu (accessed May 14, 2013).

(*5*) Mullen, J. W. II; Fenn, J. B.; Irby, M. R. The Ignition of High-Velocity Streams of Combustible Gases by Heated Cylindrical Rods. In: *Third Symposium on Combustion, Flame, and Explosion Phenomena*; Lewis, B., Hottel, H. C., Nerad, A. I., Eds.; Williams & Wilkins Co: Baltimore, 1949; pp 317–329.

(*6*) Deckers, J.; Fenn, J. B. High-Intensity Molecular Beam Apparatus. *Rev. Sci. Instrum.* **1963**, *34*, 96–100.

(*7*) Anderson, J. B.; Fenn, J. B. Velocity Distribution in Molecular Beams from Nozzle Sources. *Phys. Fluids* **1965**, *8*, 780–787.

(*8*) Abuaf, N.; Anderson, J. B.; Andres, R. P.; Fenn, J. B.; Marsden, D. G. H. Molecular Beams with Energies Above One Electron Volt. *Science (Washington, DC)* **1967**, *155*, 997–999.

(*9*) A Memorial Tribute to John Fenn (1917–2010). Yale School of Engineering and Applied Science, http://www.yaleseas.com/fenn/ (accessed November 23, 2014).

(*10*) Dole, M.; Mack, L. L.; Hines, R. L.; Mobley, R. C.; Ferguson, L. P.; Alice, M. B. Molecular Beams for Macroions. *J. Chem. Phys.* **1968**, *49*, 2240–2249.

(*11*) Dole, M.; Cox, H. L.; Gieniec, J. Electrospray Mass Spectrometry. *Adv. Chem. Ser.* **1973**, *125*, 73–84.

(*12*) Fenn, J. B. *Engines, Energy, and Entropy: A Thermodynamics Primer*; W.H. Freeman: Gordonsville, VA, 1982. 293 pp.

(*13*) Fenn, J. B. Letter to the Editor Yale Alumni Magazine 2005 July/August, http://archives.yalealumnimagazine.com/issues/2005_07/letters.html (accessed 14.05.13).

(*14*) Fenn, J. B. Comments by John Fenn on Some of his Publications. In: *A Memorial Tribute to John Fenn (1917–2010)*. Yale School of Engineering and Applied Science. www.yaleseas.com/fenn/files/bibliographicalselfcommentsfenn.pdf (accessed October 21, 2014).

(*15*) The Nobel Prize in Chemistry 2002. Nobelprize.org. http://www.nobelprize.org/nobel_prizes/chemistry/laureates/2002/ (accessed May 13, 2013).

Michael Labowsky
Ansama Research, LLC, Wayne, NJ, USA

Michael A. Grayson
Washington University in St. Louis, St. Louis, MO, USA (ret.)

Catherine Clarke Fenselau

b.15 April, 1939

Catherine Clarke Fenselau is an American scientist who has been a pioneer in biomedical applications of mass spectrometry. Her contributions to fundamental mass spectrometry include the original measurement of the gas phase proton affinity of arginine, the first successful target capture experiments with peptides, extension of the theory of kinetic energy release to allow for entropic effects, and the first application of molecular dynamics calculations to gas phase peptide structures. Before many others she saw the possibility of extending mass spectrometry to biopolymers and promoted consideration of potential hurdles, e.g., the complexity of isotope envelopes in ions weighing more than 10,000 Da and fragmentation rates that change with increasing peptide mass.

Dr. Fenselau was born in York, Nebraska on April 15, 1939. She received the A.B. from Bryn Mawr College (1961), and her Ph.D. from Stanford University (1965) for fundamental research on fragmentation mechanisms using stable isotope labels (directed by Carl Djerassi). She was a post-doctoral fellow at the University of California at Berkeley (AAUW fellow) and the NASA Space Sciences Laboratory. She joined the Department of Pharmacology and Experimental Therapeutics at the Johns Hopkins University School of Medicine in 1967. One of the first formally trained mass spectroscopists to join a US medical faculty, she was given the career mandate to develop and exploit mass spectrometry in biomedical research. She moved to the University of Maryland system in 1987 and has served as Professor and Chair at both the University of Maryland Baltimore County and the University of Maryland, College Park. She raised two sons, Andrew and Thomas Fenselau. She was married to fellow mass spectrometrist Professor Robert Cotter until his death in 2012 (see this volume, *Robert J. Cotter*).

Early applications work at Hopkins included the first characterization of achromaphoric anticancer drugs by GCMS, synthesis and analyses of intact Phase II drug metabolites, and early work in structure elucidation of glycoconjugates. With a 1975 benchmark publication on the subject (*1*), she became a leader in the rapid detection of intact microorganisms by mass spectrometry. She has moved that method from phospholipid fingerprinting by fast atom bombardment to library-free proteomic analysis with matrix assisted laser desorption and bioinformatics. In 1982 Fenselau was funded by the National Science Foundation to establish a Regional Mass Spectrometry Center to stimulate applications of the technology to analyses of bioconjugates and biopolymers, primarily using high field and four-sector analyzers. For this work she coined the term 'Middle Molecule Mass Spectrometry.' The Center moved with her to the University of Maryland in 1987.

Fenselau's work at the University of Maryland has focused on protein biochemistry, seeking to define the mechanisms of drug resistance and immunosuppression in cancer. As innovators in protein analyses, her team developed kinetic and stoichiometric methods to study metal ion flux among metalloproteins, demonstrated mass spectrometry for identification of polypeptides fractionated by gel electrophoresis, and published the first application of mass spectrometry to analyze structures of protein complexes based on solvent (and reagent) accessibility. Her contributions to the post-genomic field of proteomics began in 1998. With funding from the National Institutes of Health, her team has developed ^{18}O-labeling, microwave-supported acid digestion and nanoparticle pellicles as reliable components of proteomic workflows.

Fenselau has been Visiting Professor at the Moscow Institute of Physics and Technology, Kansai University (Japan), and the University of Warwick (U.K.). She holds five patents on mass spectrometry technologies and has published more than 360 peer-reviewed scientific papers. She is a past member of the Board of Trustees of the Maryland Science Center, past president of the American Society for Mass Spectrometry, founding editor (1973) of the journal Biomedical Mass Spectrometry (now Journal of Mass Spectrometry) and Assoc. Editor for Analytical Chemistry. She was the founding president of the US Human Proteomics Organization, and a founding member of the Board of the international Human Proteomics Organization.

Her awards include the Field and Franklin award from the American Chemical Society for contributions to mass spectrometry (2008), the Thomson Medal from the International Mass Spectrometry Foundation (2009), the Ralph Adams Award for innovation in bioanalytical chemistry from the Pittsburgh Analytical Conference (2010), The ASMS Distinguished Contribution in Mass Spectrometry

Award (2012), and the American Chemical Society's Garvan Medal (1985). She is a fellow of the AAAS (2001) and the ACS (2009, the inaugural class).

The publications cited below are representative of the author's most significant contributions to the field (*1–10*).

References

(*1*) Anhalt, J. P.; Fenselau, C. Identification of Bacteria Using Mass Spectrometry. *Anal. Chem.* **1975**, *47*, 219–225.

(*2*) Fenselau, C.; Pallante, S.; Batzinger, R. P., *et al.* Mandelonitrile β-Glucuronide: Synthesis and Characterization. *Science* **1977**, *198*, 625–627.

(*3*) Fenselau, C.; Cotter, R. J.; Heller, D.; Yergey, J. Isotope Distributions in Mass Spectra of Large Molecules. *Anal. Chem.* **1983**, *55*, 353–356.

(*4*) Wu, Z.; Fenselau, C. Proton Affinity of Arginine Measured by the Kinetic Approach. *Rapid Commun. Mass Spectrom.* **1992**, *6*, 403–405.

(*5*) Henderson, L. E.; Bowers, M. A.; Sowder, R. C., *et al.* Gag Proteins of the Highly Replicative NM Strain of Human Immunodeficiency Virus Type 1: Posttranslational Modifications, Proteolytic Processings and Complete Amino Acid Sequences. *J. Virol.* **1992**, *66*, 1856–1865.

(*6*) Cheng, X.; Wu, Z.; Fenselau, C. Collision Energy Dependence of Proton-Bound Dimer Dissociation: Entropy effects, Proton Affinities and Intramolecular Hydrogen-Bonding in Protonated Peptides. *J. Am. Chem. Soc.* **1993**, *115*, 4844–4848.

(*7*) Kaltashov, I. A.; Fabris, D.; Fenselau, C. Assessment of Gas-Phase Basicities of Protonated Peptides by the Kinetic Method. *J. Phys. Chem.* **1995**, *99*, 10046–10051.

(*8*) Yao, X.; Freas, A.; Ramirez, J.; Demirev, P.; Fenselau, C. Proteolytic ^{18}O labeling for Comparative Proteomics: Model Studies with Two Serotypes of Adenovirus. *Anal. Chem.* **2001**, *73*, 2836–2842.

(*9*) Cannon, J.; Nakasone, M.; Fushman, D.; Fenselau, C. Proteomic Identification and Analysis of K63-Linked Ubiquitin Conjugates. *Anal. Chem.* **2012**, *84*, 10121–10128.

(*10*) Wynne, C.; Edwards, N.; Fenselau, C. Phyloproteomic Classification of Unsequenced Organisms by top-down Identification of Bacterial Proteins using capLC-MS/MS on an Orbitrap. *Proteomics* **2010**, *10*, 3631–3643.

Catherine Fenselau
Department of Chemistry & Biochemistry, University of Maryland, College Park, MD, USA

Frank Henry Field

b. February 27, 1922
d. April 4, 2013

Photo credit: Courtesy Rockefeller University, New York, NY.

Frank Field was an American chemist known for his pioneering work in ion chemistry, organic, and bioorganic mass spectrometry.

Field was born in Keansburg, New Jersey on February 27, 1922. His war-time education at Duke University (BS '43; MA '44; PhD '48 under the mentorship of W.C. Vosburgh, 'Magnetic Studies on Complex Ions of Nickel, Cobalt, and Iron') prepared him well for a long and distinguished career in both industry and academia. A Texas connection was established in 1947, when he became an instructor at the University of Texas in Austin. Despite promotion to Assistant Professor in 1949, he commenced the industrial phase of his career at Humble Oil and Refining Company (later Esso Research and Engineering) in Baytown, TX in 1952, rising to Section Head before a 1966 move to the Esso (later Exxon) Linden, NJ facility.

From the outset of his period in Baytown, Field collaborated with Joe Franklin (see this volume, *Joe Louis Franklin, Jr.*), with whom he was co-honored with the naming in 1983 of the American Chemical Society's Franklin and Field Award for Outstanding Achievement in Mass Spectrometry (an award that Field himself would win in 1988). Exemplary of the enlightened view of industrial research in the 1950s–1960s era, Field's work at Humble focused on fundamental studies of the thermodynamics of gas-phase ions, leading to numerous publications, including the 1957 monograph he co-authored with Franklin *Electron Impact Phenomena* (*1*). Field has described this book as "a work whose time had come because... it defined the discipline of gaseous ion chemistry," adding "It was more than a modest success." The work paved the way for new areas of research that included Field's own pioneering development with fellow Humble employee Burnaby Munson of chemical ionization mass spectrometry (see this volume, *Milan Stephen Burnaby Munson*). The latter was recognized in the 1996 American Society for Mass Spectrometry Award for Distinguished Contribution in Mass Spectrometry that Field and Munson shared. Frank's work in the biomedical realm, in part facilitated by the possibilities chemical ionization opened for bioorganic mass spectrometry, proceeded mostly at Rockefeller University, where in 1970 Field became Professor and Head of the Laboratory of Mass Spectrometry and Chemistry of Gaseous Ions, and later as Director of the Extended Mass Range Mass Spectrometry Research Resource. In 1988 he was appointed Rockefeller's first Dreyfus Professor.

Development and applications of biological mass spectrometry (a term that surely would have been an oxymoron at the outset of Field's career) at Rockefeller comprise just over half of Field's publications, which at the time of this writing total ca. 168 and average over 40 citations each. The first of these was a 1946 paper in the *Journal of the American Chemical Society* (*2*); the latest reviewed publication appeared in the inaugural issue of the *Journal of the American Society for Mass Spectrometry* in 1990 (*3*). A retrospective Letter to the Editor of *Mass Spectrometry Reviews* appeared in 1994 (*4*) and a classic *JACS* paper (*5*) was reprinted in the *Journal of Mass Spectrometry* in 2001 (*6*). Most cited was the 1966 *JACS* paper introducing chemical ionization MS (*7*), co-authored with Munson.

Field retired from Rockefeller in 1989 to pursue for a time his interests in music and travel from a base in Oak Ridge, Tennessee. He subsequently returned to his college haunts in Durham, North Carolina, where he and Carolyn, his wife of many years, resided until his death (*8*).

References

(*1*) Field, F. H.; Franklin, J. L. *Electron Impact Phenomena*; Academic Press: New York, 1957.

(*2*) Field, F. H.; Saylor, J. H. The Vapor Pressure of some Organic Fluorides. *J. Am. Chem. Soc.* **1946**, *68*, 2649–2650.

(*3*) Field, F. H. The Early Days of Chemical Ionization – A Reminiscence. *J. Am. Soc. Mass Spectrom.* **1990**, *1*, 277–283.

(*4*) Field, F. H. *Mass Spectrom. Rev.* **1994**, *13*, 99–101.

(*5*) Field, F. H.; Franklin, J. L.; Lampe, F. W. Reactions of Gaseous Ions. I. Methane and Ethylene. *J. Am. Chem. Soc.* **1957**, *79*, 2419–2429.

(6) Field, F. H.; Franklin, J. L.; Lampe, F. W. Reactions of Gaseous Ions. I. Methane and Ethylene. *J. Mass Spectrom.* **2001**, *36*, 3–14.
(7) Musnon, M. S. B.; Field, F. H. Chemical Ionization Mass Spectrometry Mass Spectrometry I: General Introduction. *J. Am. Chem. Soc.* **1966**, *88*, 2621–2630.
(8) Chait, B. Frank H. Field (1922–2013). *J. Am. Soc. Mass Spectrom.* **2014**, *25*, 702–704.

Kelsey D. Cook
Department of Chemistry, University of Tennessee, Knoxville, TN, USA

Robert E. Finnigan

b. May 27, 1927

Robert Finnigan is known for his leadership in the commercialization of the quadrupole mass spectrometer. He has been a leading advocate for and producer of computerized GC/MS systems, and a vital proponent for applications of computerized GC/MS instruments that not only promote the understanding of the chemistry of the environment but also its protection.

Bob, as he is universally known, Finnigan was born in Buffalo, New York, on May 27, 1927. He is a graduate of the US Naval Academy, Annapolis, MD. He earned a PhD in electrical engineering from the University of Illinois in 1957 while serving as a Captain in the US Air Force. That same year, he joined the Lawrence Livermore Laboratory where he led a group that developed computer controls for an experimental nuclear reactor. Leaving the Air Force in 1959, Finnigan stayed on at Livermore, continuing his efforts in the real-time computer control of nuclear reactors.

In 1962, he joined the Stanford Research Institute (SRI), Palo Alto, CA, a leading center for computer technology to initiate a process control group with his Livermore-colleague Mike Uthe. It was at SRI, Finnigan and Uthe first encountered quadrupole mass spectrometers: SRI researcher Ken Shoulders and his team were developing quadrupole mass spectrometers for use as leak detectors and process monitors in an advanced electron-beam lithography system for fabricating microelectronic devices. Finnigan was struck by what he perceived as the broad potential for quadrupole instruments in research as well as in process control applications. In 1963, Finnigan and Uthe left SRI to found the Scientific Instruments Division of Electronic Associates Incorporated's (EAI), perhaps the leading producer of electronic analog computers at this time. There in Silicon Valley Finnigan assembled a team that produced the first commercial quadrupole mass spectrometer in 1964 (see this volume, *The Development of the Quadrupole Mass Filter and the Quadrupole Ion Trap*).

In the mid-1960s, EAI's quadrupole mass spectrometers enjoyed quick and substantial market success, especially for residual gas analysis and detection of trace contaminants in ultra-high vacuum systems. During this time, Finnigan became increasingly intrigued and excited by the possibility of using minicomputers for instrument control and data handling and interpretation in a combined gas chromatography-mass spectrometry instrument. While GC/MS instrumentation could provide a wealth of useful measurements, interpretation of the data by manual means was quite time consuming. A single GC/MS run, performed in a day, required weeks and months for interpretation (see this volume, *Reflections on the History of Computers in Mass Spectrometry, The Development of Computer Data Systems in Quadrupole Mass Spectrometry, A History of the Development of Mass Spectral Databases*, and *Development of the Business of Mass Spectrometry from 1960 to 1975*). Failing to interest EAI in his vision, Finnigan decided to start a new company to produce GC/MS/DS instruments in 1966: the Finnigan Instrument Company. With a critical investment from Roger Sant, an executive at the pharmaceutical company Syntex who had previously been an executive at gas-chromatograph-maker Wilkins Instrument, Finnigan then recruited a team. T. Z. Chu, who came from the gas chromatography producer Varian Aerograph, joined as both an original investor and a top manager. William Fies came to Finnigan from SRI, where he had developed the quadrupole electronics. Michael Story joined Finnigan from EAI where he had been central in quadrupole design and vacuum chemistry. By early 1968, Finnigan's team delivered prototype quadrupole GC/MS instruments to Joshua Lederberg's Stanford Genetics lab and to the Purdue Chemistry Department. Within months, they then introduced a fully computerized GC/MS – the Model 1015/System 150 – at the American Chemical Society meeting in New York City. The instrument, with its advanced quadrupole mass spectrometer and customized minicomputer and software, cost on the order of $100,000. The first system was delivered to Evan Horning at Baylor Medical College in late 1968 (see this volume, *Evan Charles Horning and Marjorie Groothuis Horning*).

The company was soon able to sell a number of its GC/MS/DS units to the US Environmental Protection Agency for analysis of water pollutants (see this volume, *Development of Environmental Mass Spectrometry*). Finnigan Instrument Corporation also developed systems for detecting the illegal use of drugs by athletes and for pharmacokinetic studies of pharmaceutical products. The company went public in the early 1970s, and by 1988 it had grown to more than $130 million in revenues, mostly as a result of its dominance of the environmental market for computerized GC/MS. In 1990, Finnigan stockholders accepted an offer from Thermo Instrument Systems to buy the company. Finnigan consulted with Thermo Instrument Systems and its parent, Thermo Electron, for approximately a decade thereafter and served on the board of the Thermo Spectra Corporation for five years.

Finnigan married his wife Bette in 1950, and they raised seven children in their home in Los Altos, California. A handball and wine aficionado, Finnigan was an active member of the American Electronics Association where he spearheaded several environmental initiatives. He also served as a long-term advisor to a variety of leading investment banks and venture capital partnerships in the San Francisco Bay area.

Further information about Robert Finnigan's career can be found in the works cited here (*1*–*5*).

References

(*1*) Finnigan, R. E. Oral History. Interview by David C. Brock. Audio recording and transcript. Oral History Collection. Chemical Heritage Foundation. Philadelphia, PA.

(*2*) Finnigan, R. E.; Hoyt, D. W.; Smith, D. E. Priority Pollutants II – Cost Effective Analysis. *Environ. Sci. Technol.* **1979**, *13*, 534–541.

(*3*) Finnigan, R. E. 'Methods of Consensus: Computerized GC–MS and Environmental Chemistry.' Video interview with David C. Brock. 2008. http://vimeo.com/13075172 (accessed 30 November 2010).

(*4*) Finnigan, R. E. Quadrupole Mass Spectrometers: From Development to Commercialization. *Anal. Chem.* **1994**, *66*, 969A–975A.

(*5*) Brock, D. C. 'From Cold War Computing to Environmental Protection: Robert Finnigan and the Making of Computerized GC/MS.' Unpublished manuscript. 2010.

David C. Brock
Center for Contemporary History and Policy, Chemical Heritage Foundation, Philadelphia, PA, USA

Russell Elwell Fox

b. December 28, 1916
d. February 3, 2008

Russell E. Fox was an American atomic physicist who worked for the Westinghouse Corp. for his entire career. He carried out significant research on aspects of electron ionization and mass spectrometry.

Russell Fox was born in 1916 in Richmond, VA, but grew up and graduated from high school in Hampton, VA. Despite his strong interest in science, he had not planned on attending college until he won a scholarship sufficient to cover his initial tuition from a local church. He graduated from Hampden-Sydney College summa cum laude in 1938 and went on for graduate work in physics at the University of Virginia. As the US entered World War II, his work was absorbed into the Manhattan Project. His PhD dissertation, completed in 1942 under the direction of Jesse Beams, was on developing ultracentrifuge isotope separation and was classified secret from the start (see this volume, *Mass Spectrometry and the Separation and Enrichment of Isotopes*). He graduated without making a defense of his thesis; it was not declassified until 1969.

Upon finishing his doctorate, Fox married Thelma Todd; they would have three children. They moved immediately to Pittsburgh, PA, where Fox was appointed as a post-doctoral fellow in the Westinghouse Electric Corporation Research and Development Labs. There he entered the mass spectrometer group and a year later joined the permanent research staff. During the remainder of the war he worked on mass spectrometry in leak detection and in chemical analysis for the program developing neoprene as a replacement for natural rubber (*1*).

In the immediate post-war years, Fox worked with John Hipple and Ed Condon in the classic research that identified and characterized the appearance of metastable ions in mass spectrometry (*2*, see this volume, *John Alfred Hipple, Jr.*). This began a long series of investigations of electron ionization.

Fox and his colleagues devised a way to use retarding potentials on additional grids in an electron ionization source that became known as a 'Fox gun.' This made it possible to reduce the *apparent* energy spread in electron ionization, by use of differential potentials in the acceleration electron beam into the ion source creating the effect of a reduced energy spread thereby making it possible to determine appearance potentials with greatly improved precision (*3–5*).

Fox explored many aspects of electron ionization over several years; some examples of his research are listed below (*6–10*). He advanced to become the manager of his department at Westinghouse. In 1963 and 1964 he was the chairman of the E-14 Committee of the American Society for Testing and Materials, the organization that served as the scientific society for mass spectrometry at that time. He later became director of Atomic and Molecular Sciences Research and Development at Westinghouse and then research and development director for Industry Products until he retired in 1981.

Through his career, Russell Fox continued to serve the wider community, serving in various municipal, church, and nonprofit positions, and establishing and supporting several scholarships. After moving from the Pittsburgh area back to Hampton, Fox's first wife died in 1995. He remarried in 1996 and was survived by Margaret Louise Fox when he died in 2008 (*11–13*).

References

(*1*) Fox, R. E.; Hipple, J. A.; Williams, T. W. Mass Spectrometer with a Small Magnet. *Phys. Rev.* **1944**, *65*, 353.

(*2*) Hipple, J. A.; Fox, R. E.; Condon, E. U. Metastable Ions Formed by Electron Impact in Hydrocarbon Gas. *Phys. Rev.* **1946**, *69*, 347–356.

(*3*) Fox, R. E.; Hipple, J. The Use of Retarding Potentials with the Mass Spectrometer. *Rev. Sci. Instrum.* **1948**, *19*, 462–466.

(*4*) Fox, R. E.; Hickam, W. M.; Kjeldaas, T.; Grove, D. J. Ionization Potentials and Probabilities Using a Mass Spectrometer. *Phys. Rev.* **1951**, *84*, 859–860.

(*5*) Fox, R. E.; Hickam, W. M.; Grove, D. J.; Kjeldaas, T. Ionization in a Mass Spectrometer by Monoenergetic Electrons. *Rev. Sci. Instrum.* **1955**, *26*, 1101–1107.

(*6*) Schulz, G. J.; Fox, R. E. Excitation of Metastable Levels in Helium near Threshold. *Phys. Rev* **1957**, *106*, 1179–1181.

(*7*) Fox, R. E. Dissociative Attachment of Electrons in Iodine. II. Mass Spectrographic Determination of the Energy Dependence of the Cross Section. *Phys. Rev.* **1958**, *109*, 2008–2011.

(*8*) Fox, R. E. Study of Multiple Ionization in Helium and Xenon by Electron Impact. In: *Advances in Mass Spectrometry*; Waldron, J. D., Ed.; Pergamon Press: New York, etc., 1959; pp 397–412.

(*9*) Fox, R. E. Negative Ion Formation in NO_2 by Electron Attachment. *J. Chem. Phys.* **1960**, *32*, 285–287.

(*10*) Fox, R. E. Ionization Cross Sections near Threshold by Electron Impact. *J. Chem. Phys.* **1961**, *35*, 1379–1382.

(*11*) Anonymous. Russell Elwell Fox. *Pittsburgh Post–Gazette*, 10 February 2008.

(*12*) Anonymous. Russell Elwell Fox. *Daily Press* (Hampton Roads, VA), 6 February 2008.

(*13*) Dudley, J. Our Boys and the Bomb. *The Record* (Hampden-Sydney College) July 2010.

Keith A. Nier
Independent Historian, Madison, NJ, USA

Joe Louis Franklin, Jr.

b. August 11, 1906
d. August, 1982

Joe Franklin was born in Natchez, Mississippi in 1906, but became a naturalized Texan early in life. He received all his academic training at the University of Texas, earning a B.S. in 1926, an M.S. in 1930 (chemical engineering), and a Ph.D. in 1934 (physical chemistry). He spent two years as a fellow at MIT between his masters and doctoral work. Joe spent the early part of his career in research and development with the Humble Oil and Refining Company (later Esso then Exxon) in Baytown, Texas before accepting an appointment as the Robert A. Welch Professor of Chemistry at Rice University, where he remained for the rest of his career (*1*).

After 1949, Franklin's focus shifted from applied research toward a total dedication to fundamental science. While at Humble, he developed the Franklin group-equivalent method of estimating the standard enthalpies and free energies of formation of molecules, radicals and ions (*2*). He also began the fundamental studies of gaseous ion chemistry, work for which he was best known, that in 1957 resulted in the well-known book "Electrical Impact Phenomena and the Properties of Gaseous Ions," coauthored with F. H. Field (*3*, see this volume, *Frank Henry Field*). The substantial number of very prominent mass spectrometrists who spent early parts of their careers in Franklin's group at Humble includes Fred Lampe, Jean Futrell, and Burnaby Munson (see this volume, *Frederick W. Lampe, Jean H. Futrell*, and *Milam Stephen Burnaby Munson*).

At Rice, his reputation attracted scholars at all levels – graduate and postdoctoral students, and faculty members. He and his students and collaborators not only continued the research started at Humble but also branched out into other areas, such as ion-molecule reactions, ion- kinetic energy studies, and photoelectron spectroscopy (*4–6*).

Joe Franklin was named a Distinguished Alumnus of the University of Texas in 1973. He was instrumental in the transition of the ASTM Committee E-14 to the American Society for Mass Spectrometry, for which he served as the first President. It should also be noted that three of his former colleagues at Humble, Field, Munson, and Futrell, also served as presidents of ASMS. Franklin was honored with Field by The American Chemical Society with an award named in their honor (The Field and Franklin Award).

Joe exuded a genuine warmth and concern for people that, combined with his intellect, enthusiasm and broad interests, enabled him to provide a unique degree of scientific stimulation, guidance and inspiration to many. Among his close colleagues he was known as a connoisseur of fine food and was particularly fond of a number of restaurants in New Orleans. He and his wife Mildred with whom he raised 3 sons had been married for 50 years at the time of his death.

References

(*1*) Anon. Rice University,website http://chemistry.rice.edu/Content.aspx?id=495 (accessed June 6, 2014).
(*2*) Franklin, J. L. Prediction of Heat and Free Energies of Organic Compounds. *Ind. Eng. Chem.* **1949**, *41*, 1070–1076.
(*3*) Field, F. H. *Electron Impact Phenomena*; Academic Press: New York, 1957.
(*4*) Field, F. H.; Franklin, J. L. Reactions of Gaseous Ions. X. Ionic Reactions in Xenon–Methane Mixtures. *J. Am. Chem. Soc.* **1961**, *83*, 4509–4515.
(*5*) Field, F. H.; Franklin, J. L.; Munson, M. S. B. Reactions of Gaseous Ions. XII. High Pressure Mass Spectrometric Study of Methane. *J. Am. Chem. Soc.* **1963**, *85*, 3575–3583.
(*6*) Walder, Ray; Franklin, J. L. Proton Affinities of Neutral Molecules. *Int. J. Mass Spectrom. Ion Phys.* **1980**, *36*, 85–112.

Kelsey D. Cook
Department of Chemistry, University of Tennessee, Knoxville, TN, USA

Jean H. Futrell

b. October 20, 1933

Jean Futrell is a physical and analytical chemist from the United States noted for pioneering developments in mass spectrometry instrumentation and fundamental studies of ion reaction mechanisms.

Born in Grant Parish, Louisiana on October 20, 1933, Jean Futrell studied chemical engineering at Louisiana Tech University, graduating as valedictorian of his class with a BS in Chemical Engineering in 1955. He joined Nobel Laureate Glenn Seaborg's nuclear chemistry group at the Lawrence Berkeley Laboratory and received his Ph.D. in Physical Chemistry at the University of California in 1958. His thesis research in radiation chemistry was supervised by senior staff scientist Amos S. Newton.

While a graduate student Futrell pioneered gas chromatographic separation, a technique for which commercial instruments had not yet been marketed in the United States, of radiolysis products followed by mass spectrometric identification of separated fractions utilizing a then state-of-the-art Consolidated Electrodynamics 21-102C mass spectrometer. Upon graduation in 1958 he worked for half a year in Exxon's Baytown, Texas, Research Center with noted mass spectrometrists Joe Franklin, Frank Field and Fred Lampe (see this volume, *Joe Louis Franklin, Jr., Frank Henry Field, and Frederick W. Lampe*). As his first independent research endeavor he developed theoretical models for gas phase radiation chemistry of hydrocarbons in which ion-molecule reactions play a dominant role. In 1959 he reported for active duty as Air Force officer at the Aerospace Research Laboratory, Wright-Patterson Air Force Base, Dayton, Ohio, and an appointment he continued as civilian scientist from 1961–67.

His research in the decade 1959–1968 focused on fundamental studies of unimolecular and biomolecular reactions of gaseous ions. Determining the relevant experimental parameters that defined rates and mechanisms of these reactions required several innovations in mass spectrometry (*1*). His development of a tandem mass spectrometer that utilized Nier-Johnson and Mattauch-Herzog mass spectrometers for reagent ion selection and product ion mass analysis was a major instrumental advance (*2*, see this volume, *Magnetic and Electrostatic Analyzers before 1960*). The key enabling invention was strong-focusing rectangular tube lenses for decelerating ions to near-thermal energy; the integration of these lenses and their focal properties into magnetic and electrostatic sectors operated in unconventional ways enabled many unique experiments. His interest in both development of novel instrumentation and ion chemistry and physics has evolved more recently to ion-surface collisions, including both dissociation of recoiling ions and soft-landing of complex ions on self-assembled monolayer surfaces (*3–5*; see this volume, Chapter 9, *The Development of Tandem Mass Spectrometry*).

Jean launched his academic career at the University of Utah in 1967 as Associate Professor of Chemistry and was promoted to Full Professor in 1968. In 1987 the University of Delaware appointed him the Willis F. Harrington Professor and Chair of the Department of Chemistry and Biochemistry. In 1998 he was appointed Director of the Wiley Environmental Molecular Sciences Laboratory at the Pacific Northwest National Laboratory (PNNL), the recently dedicated first national user facility of the U. S. Department of Energy focusing on molecular scale phenomena. In 2002 he was named Battelle Fellow, the highest technical rank for scientists employed by Battelle Memorial Institute. His current position is Battelle Fellows and Senior Chief Scientist for Physical Sciences in the Fundamental and Computational Sciences Directorate at PNNL.

Jean has received numerous awards recognizing pioneering contributions to mass spectrometry and ion physics, including designation as Pioneer in Mass Spectrometry by the Chemical Heritage Foundation (2012), the American Chemical Society's Frank H. Field and Joseph L. Franklin Award for Outstanding Achievement in Mass Spectrometry (2007), the German Mass Spectrometry Wolfgang Paul Award (2007), the Pacific Northwest National Laboratory Director's Award for Lifetime Achievement in Science and Technology (2006), the Schroedinger Gold Medal of the Section on Atomic and Surface Physics of the Austrian Physics Society (2000), the University of Delaware Distinguished Scholar Award (1997), and the Delaware Section of the American Chemical Society Distinguished Research Award (1995),

Jean is a Fellow of the American Chemical Society (2009 inaugural class), the American Physical Society, and the American Association for the Advancement of Science. He served as President of the American Society for Mass Spectrometry and

Chair of the Council for Chemical Research professional societies. He is a member of Sigma Xi. He has many years of service on committees of the National Academy of Sciences, Department of Energy, National Science Foundation, many universities and foundations in the US and internationally.

His published works, comprising more than 600 research reports, includes over 300 refereed journal publications and three patents.

Jean is married to distinguished artist and designer Anne Krohn Graham; they have three children – Craig Futrell (deceased), Alison Futrell, and Tracy Wenzinger – and four grandchildren – Nathan and Virginia Futrell, Zara and Tiffany Wenzinger.

References

(*1*) Futrell, J. H. Crossed-Molecular Beam Studies of State-to-State Reaction Dynamics. *Adv. Chem. Phys.* **1992**, *82*, 501.

(*2*) Futrell, J. H. Development of Tandem Mass Spectrometry: One Perspective. *Int. J. Mass Spectrom.* **2000**, *200*(1–3), 495.

(*3*) Laskin, J.; Denisov, E.; Shukla, A. K.; Barlow, S. E.; Futrell, J. H. Surface-Induced Dissociation in a Fourier Transform Ion Cyclotron Resonance Mass Spectrometer: Instrument Design and Evaluation. *Anal. Chem.* **2002**, *74*, 3255.

(*4*) Laskin, J.; Bailey, T.; Futrell, J. H. Shattering of Peptide Ions on Self-Assembled Monolayer Surfaces. *J. Am. Chem. Soc.* **2003**, *125*(6), 1625.

(*5*) Hadjar, O.; Wang, P.; Futrell, J. H., *et al.* Design and Performance of an Instrument for Soft Landing of Biomolecular Ions on Surfaces. *Anal. Chem.* **2007**, *79*(17), 6566–6574.

Jean H. Futrell
Physical Sciences Division, Pacific Northwest National Laboratory, Richland, WA, USA

Simon James Gaskell

b. May 2, 1950

Simon James Gaskell, Ph.D., is a British chemist whose work has concerned the development of mass spectrometry for the characterization and quantitative analysis of compounds of biological importance including steroids, peptides, and proteins. He has developed new analytical approaches and has contributed to the understanding of the gas-phase ion chemistry of peptides, enabling contributions to the field of proteomics.

Gaskell was born on May 2, 1950, in Lytham St. Annes, Lancashire, England. He attended the University of Bristol from 1968–1974, where he received B.Sc. and Ph.D. degrees, the latter under the supervision of Professor Geoffrey Eglinton FRS (see this volume, *Geoffrey Eglinton*). On a post-doctoral research fellowship at the University of Glasgow, he worked with Professor Charles J. W. Brooks from 1974–1977 before becoming Head of the Mass Spectrometry Unit at the University of Wales College of Medicine in Cardiff. After a period at Baylor College of Medicine, Houston, Texas, Gaskell joined UMIST (Manchester, UK) as Professor of Mass Spectrometry in 1993, before becoming Head of Chemistry in 1999 and, following its merger with the University of Manchester, Associate Vice President for Research in the University of Manchester in 2004, and Vice President for Research in 2006. He joined Queen Mary University of London as Principal in October 2009, a position he continues to hold.

Under the guidance of two influential research mentors, Geoffrey Eglinton at the University of Bristol and Charles Brooks at the University of Glasgow, Gaskell's work was primarily on steroid analysis (*1*), a focus that he continued at the Tenovus Institute for Cancer Research in Cardiff. During the latter period he developed gas chromatography/mass spectrometry with metastable peak monitoring (*2*), a forerunner of selected reaction monitoring, implemented on a double-focusing electric sector/magnetic sector instrument. Borrowing ideas from the field of immunoassays, he used immunoaffinity isolation to improve the quantitative mass spectrometric analyses of steroids, facilitating the development of reference methods.

Gaskell's major methodological contributions to mass spectrometry were made during sixteen years as Director of the Michael Barber Centre for Mass Spectrometry in Manchester and while at Baylor College of Medicine. Although during this period (1990–2009) some of his work was on the analysis of eicosanoids and glutathione conjugates, his focus was increasingly on the analysis of peptides and proteins. In 2009, Gaskell and Professor Vicki Wysocki (see this volume, *Vicki Hopper Wysocki*) of the University of Arizona shared the American Society for Mass Spectrometry Award for Distinguished Contribution in Mass Spectrometry for their work (partly in parallel, partly in collaboration) on what has become known as the 'mobile proton' hypothesis (*3,4*). The ideas developed in the Wysocki and Gaskell laboratories aid understanding of the low-energy collision-induced fragmentations of protonated peptides. Recent work on coupled ion mobility spectrometry/tandem MS (in collaboration with the instrument company, Waters) has enabled further contributions to the understanding of peptide ion fragmentations.

In conjunction with Rob Beynon (University of Liverpool), Gaskell has developed new approaches in the use of mass spectrometry for the quantitative characterization of the proteome. This work has included a new approach to the determination of the turnover (the balance of synthesis and degradation) of large numbers of proteins in a single experimental series using stable isotope labeling. The development, also with Beynon, of the QconCAT methodology involved the design and production of artificial stable isotope-labeled proteins, yielding on proteolysis multiple peptide standards to enable the absolute quantification of multiple proteins (*5*).

Gaskell is a member of the American Chemical Society, the American Society for Mass Spectrometry, the British Mass Spectrometry Society, the British Proteomics Society, and is a Fellow of the Royal Society of Chemistry. He was an associate editor of the *Journal of the American Society for Mass Spectrometry* and served on the editorial boards of *Comparative and Functional Genomics*, the *International Journal of Mass Spectrometry*, and *Rapid Communications in Mass Spectrometry*. Gaskell received the Royal Society of Chemistry Award in Mass Spectrometry in 2004 and the previously mentioned ASMS award shared with Dr. Wysocki in 2009.

Gaskell is married to Deirdre Jane Ellison, an educational psychologist, and has two children: lawyer

Clare Fiona Gaskell Houston and architect Nicholas James Gaskell.

The publications cited below are representative of the author's most significant contributions to the field.

References

(*1*) Gaskell, S. J.; Eglinton, G. Rapid Hydrogenation of Sterols in a Contemporary Lacustrine Sediment. *Nature* **1975**, *254*, 209–211.

(*2*) Gaskell, S. J.; Millington, D. S. Selected Metastable Peak Monitoring: A New Specific Technique in Quantitative Gas Chromatography-Mass Spectrometry. *Biomed. Mass Spectrom.* **1978**, *5*, 557–558.

(*3*) Burlet, O.; Yang, C.-Y.; Gaskell, S. J. The Influence of Cysteine to Cysteic Acid Oxidation on the Collisionally Activated Decomposition of Protonated Peptides; Evidence for Intra-Ionic Interactions. *J. Am. Soc. Mass Spectrom.* **1992**, *3*, 337–344.

(*4*) Cox, K. A.; Morris, M.; Whiting, A.; Gaskell, S. J. Role of the Site of Protonation in the Low-Energy Decompositions of Gas-Phase Peptide Ions. *J. Am. Soc. Mass Spectrom.* **1996**, *7*, 522–531.

(*5*) Beynon, R. J.; Doherty, M. K.; Pratt, J. M.; Gaskell, S. J. Multiplexed Absolute Quantification in Proteomics Using Artificial QCAT Proteins of Concatenated Signature Peptides. *Nat. Methods* **2005**, *2*, 587–589.

Simon J. Gaskell
Queen Mary's College, London, UK

Johannes Geiss

b. September 4, 1926

Johannes Geiss is a German-Swiss astro- and geophysicist who has contributed greatly to the understanding of astronomical objects of all sorts through meticulous isotopic and elemental analysis and imaginative experimental design. He has been key in making Bern a center for international cooperation in the design of space missions and the understanding of their results (*1–15*).

Johannes Geiss was born in 1926 near Stolp (now Slupsk) in what then was part of Germany and now is in Poland and attended school there. He was inducted into the German army in 1944 but was still in training when captured by British forces. He managed to complete his secondary education and enroll in the University of Göttingen in 1946. He received his diploma in 1950 and joined the group of Wolfgang Paul (see this volume, *Wolfgang Paul*). In 1953 he completed his Ph.D. at Göttingen, which involved building two magnetic sector mass spectrometers and using one to measure lead isotopes to date some geological samples.

Geiss then had the job of installing the second mass spectrometer at the Physics Institute of the University of Bern, Switzerland. The instrument had been ordered by F. G. Houtermans who had been at Göttingen during Geiss' student days but had moved to Bern in 1952 as Professor of Experimental Physics and institute director. In 1946 Houtermans had been one of the first to take up Alfred Nier's measurements of uranium and lead isotopes in order to estimate the age of the earth and he wanted to be able to continue using a mass spectrometer in research in the earth sciences (see this volume, *The Development of Mass Spectrometry in the Earth and Planetary Sciences* and *Alfred Otto Carl Nier*). Geiss stayed at Bern for two years, helping to establish the field of nuclear geology. In 1955 he got married and with his wife Carmen left Bern for a post-doctoral position at the Institute for Nuclear Studies with Harold Urey at the University of Chicago (see this volume, *Harold Clayton Urey*). He returned to Bern for a year and produced his habilitation thesis. He then went back to the US as associate professor at the University of Miami, FL. There, he worked with Cesare Emiliani, whom he had met at Chicago, to create a paleoclimatology lab, measuring oxygen isotope ratios in Foraminifera in cores drilled from the sea floor to determine paleotemperatures. He again returned to Bern in 1960, and there he has stayed apart from brief periods at NASA's Goddard Space Flight Center and various other institutions in the US and elsewhere. He became full professor and director of the physics institute and served terms as Dean and Rector. He left the University in 1989, turning his attention to founding the International Space Science Institute in Bern. Once it was firmly established he retired in 2002 to the status of Honorary Director.

Geiss's research contributions primarily, though certainly not exclusively, have been focused on measuring characteristics of extraterrestrial objects large and small and working out what these reveal about the history as well as the current nature of those bodies. During his brief stay in Chicago he measured potassium-argon ages for more meteorites than had ever been done before and showed that they gave significantly different, but not random, results, and that one stood out from all others as very much younger. Many years later this was determined to be from Mars rather than the asteroids and to be of unusually recent origin. After creating one of the world's best labs for noble gas analysis in Bern, he and his colleagues investigated many more meteorites, working on their histories as well as their overall ages. He soon turned his attention to preparations for the lunar missions and designed the only experiment that went on every lunar landing mission – the solar wind collectors that were unfurled in the sunlight at each site, kept exposed to the solar wind as long as practicable, and then returned to earth for analysis of what had impacted on the foil. This allowed far better understanding of the solar wind and bears on our understanding of other phenomena, like the nature of planetary atmospheres.

Geiss has since been involved in the planning and implementation of measurements carried out on many space missions, around the sun, to comets, and other planets. His work has helped reveal the nature of the protosolar cloud, the interstellar medium, and even the baryonic density of the universe. His leadership was crucial to the development of George Gloeckler's new type of time-of-flight mass spectrometer in the 1970s that has radically improved the measurement of plasma composition in space.

Geiss has been instrumental in making Bern into an international center for space sciences and he has worked long and hard to shape science policy and promote public support for science generally and space exploration in particular. He has had long-term and significant influence both with NASA and the European Space Agency. His work has been recognized in many ways including the Emil award from the International Astronautical Federation, an honorary doctorate from the University of Chicago, and the 2005 Bowie Medal, the highest award of the American Geophysical Union.

References

(*1*) Begemann, F. Johannes Geiss: The Humble Beginnings of an Octogenarian. *Space Sci. Rev.* **2007**, *130*, 479–487.
(*2*) Zurbuchen, T. H. Johannes Geiss: Explorer of the Elements. *Space Sci. Rev.* **2007**, *130*, 515–526.
(*3*) Schlaepfer, H. Science First. *Spatium* **2001**, *27*, 3–23.
(*4*) Glockler, G. 2005 William Bowie Medal Winner Johannes Geiss: Citation. *Eos Trans. Am. Geophys. U.* **2006**, *87*, 28.
(*5*) Geiselmann, F. Entwicklung der Berner Weltraumforschung. *UNIPRESS* **1998**, 96.
(*6*) Begemann, F.; Geiss, J.; Hess, D. C. Radiation Age of a Meteorite from Cosmic-Ray-Produced He3 and H3. *Phys. Rev.* **1957**, *107*, 540–542.
(*7*) Geiss, J.; Hess, D. C. Argon-Potassium Ages and the Isotopic Composition of Argon from Meteorites. *Astrophys. J.* **1958**, *127*, 224–236.
(*8*) Emiliani, C.; Geiss, J. On Glaciations and their Causes. *Int. J. Earth Sci.* **1959**, *46*, 576–601.
(*9*) Geiss, J.; Hirt, P.; Leutwyler, H. On Acceleration and Motion of Ions in Corona and Solar Wind. *Solar Phys.* **1970**, *12*, 458–483.
(*10*) Geiss, J. Processes Affecting Abundances in the Solar Wind. *Space Sci. Rev.* **1982**, *33*, 201–217.
(*11*) Geiss, J.; Gloeckler, G.; von Steiger, R., *et al.* The Southern High Speed Stream: Results from SWICS/Ulysses. *Science* **1995**, *268*, 1033–1036.
(*12*) Altwegg, K., Ehrenfreund, P., Geiss, J., Huebner, W. F., Eds. *Composition and Origin of Cometary Materials*; Springer & ISSI: New York, 2000.
(*13*) Bleeker, J. A. M., Geiss, J., Huber, M. C. E., Eds. *The Century of Space Science*; Kluwer Academic: Dordrecht, Boston, 2001.
(*14*) Geiss, J.; Bühler, F.; Cerutti, H., *et al.* The Apollo SWC Experiment: Results, Conclusions, Consequences. *Space Sci. Rev.* **2004**, *110*, 307–335.
(*15*) von Steiger, R., Gloeckler, G., Mason, G. M., Eds. *The Composition of Matter: Symposium Honouring Johannes Geiss on the Occasion of his 80th Birthday*; Springer & ISSI: New York, 2007.

Keith A. Nier
Independent Historian, Madison, NJ, USA

Roland S. Gohlke

b. November 29, 1929
d. November 9, 2000

Courtesy of the Dow Chemical Company

Roland Gohlke was an early pioneer in mass spectrometry and gas chromatography instrumentation, applications, and reference mass spectra. He played a key role in developing the first linkage of a gas chromatograph with a mass spectrometer (GC-MS) in 1956 and in developing the first DTA-MS in 1962 (see this volume, *A History of Gas Chromatography Mass Spectrometry (GCMS)*).

He was born November 29, 1929 in Alpena, Michigan. He earned his M.Sc. at the University of Michigan in 1952. That year he married Jeane Anderson; they had four children, Mark, Linda, Jeffrey, and Anne.

In 1952, Gohlke joined Fred McLafferty's group in the Dow Chemical Co., Midland MI, where he applied mass spectrometry to the analysis of a wide variety of organic compounds at a time when there were only a few dozen people in this field (see this volume, *Fred W. McLafferty* and *The Beginnings of Organic Mass Spectrometry*). In 1954, he designed and developed a versatile gas chromatographic instrument (*1*). These were built and installed as process analyzers in over 200 Dow locations before suitable commercial versions were available.

In collaboration with McLafferty and Bendix Corp. scientists, in early 1956 he combined his GC with the new Bendix time-of-flight MS for the first GC-MS instrument (*2*). His 1959 publication described its unusual speed and versatility in a wide variety of analytical applications (*3*). At Dow's Eastern Research Laboratory, Framingham, MA, he collaborated with Horst Langer in inventing another 'hyphenated technique,' Differential Scanning Calorimetry-MS (*4*), demonstrating the analytical advantages of its greatly increased information content.

Gohlke and McLafferty measured approximately 2000 reference mass spectra, the world's largest collection when published by Gohlke in 1963 (*5*), and they developed an IBM Collator/Hollerith punched card system for matching an unknown spectrum against such an MS database (see this volume, *A History of the Development of Mass Spectral Databases*).

In 1966, he moved to Dow Corning in Midland, MI, to take charge of mass spectrometry there. He joined Finnigan Instruments, Palo Alto, CA, (1968–1970) where he was Project Leader on developing the Finnigan 3000, a low cost GC/MS instrument (see this volume, *The Development of the Business of Mass Spectrometry from 1960 to 1975*). He then headed Mass Spectrometry Section at Eastman Kodak, Rochester, NY, extending MS to new chemistry relevant to the photographic industry. After he retired in 1987, Roland and Jeane lived in Port Charlotte, FL, until his death.

Roland is especially remembered for his wonderful creativity – i.e., he would try anything. In his striking demonstration of the 'Gohlke Separator,' a GC/MS take-off, he would pull a vacuum on a right-hand finger from a valve near the TOF/MS ion source and dip a left-hand finger in liquids such as benzene and CCl_4. After about 25 seconds the corresponding mass spectra would appear sequentially on the oscilloscope screen, with intrabody passage of the samples achieving both pressure reduction and separation. Doubling the length of the 1 m flight tube of his Bendix TOF MS somewhat improved resolution, but made access to the rear of the instrument difficult. He invented 'exploding ionization' of nonvolatile compounds in which they were mixed with nitrocellulose (*6*). However, the method's adoption was discouraged when higher 'success' necessitated extensive ion source repair.

References

(*1*) Gohlke, R. S. Instrument Design for Gas-Liquid Partition Chromatography. *Anal. Chem.* **1957**, *29*, 1723–1724.

(*2*) Gohlke, R. S.; McLafferty, F. W. Early Gas Chromatography/Mass Spectrometry. *J. Amer. Soc. Mass Spectrom.* **1993**, *4*, 367–371.

(*3*) Gohlke, R. S. The Use of Time-of-Flight Mass Spectrometry and Vapor-Phase Chromatography in the Identification of Unknown Mixtures. Preprints, ACS Division. *Petroleum Chem.* **1957**, D77-D83; *Anal. Chem.* **1959**, *31*, 535–541. Reprint, *J. Am. Soc. Mass Spectrom.* **1993**, *4*, 367–371.

(*4*) Langer, H. G.; Gohlke, R. S. Mass-Spectrometric Differential Thermal Analysis. *Anal. Chem.* **1963**, *35*, 2, **1965**, *37*, 433–434.

(5) Gohlke, R. S. *Uncertified Mass Spectral Data*; Dow Chemical Co: Midland, MI, 1963.
(6) Gohlke, R. S.; Wakeman, L. S. Exploding-Film Sample Introduction for Mass Spectrometry of Involatile or Thermally Labile Substances. *Anal. Chem.* **1982**, *54*, 2114–2115.

Fred W. McLafferty
Chemistry Department, Cornell University, Ithaca, NY, USA

Michael L. Gross

b. November 6, 1940

Michael Gross is an American chemist known primarily for his work in ion chemistry and fagmentation mechanisms as well as for his conributions to instrumentation development, primarily in the area of Fourier Transform Ion Cyclotron Resonance (FTICR).

Michael Gross was born on November 6, 1940 in St. Cloud, MN, USA. He received his B.A. in 1962 from St. John's University in Minnesota, Collegeville, MN and his Ph.D. in 1966 in Organic Chemistry at the University of Minnesota, Minneapolis, working with Professor E.A. Hill. After a one-year postdoctoral fellowship at the University of Pennsylvania (E.R. Thornton) and another one-year fellowship at Purdue University (F.W. McLaffterty; see this volume, *Fred W. McLafferty*), he was appointed Assistant Professor of Chemistry at the University of Nebraska-Lincoln in 1968. He was appointed 3M Professor in 1983 and the C. Petrus Peterson Professor in 1988. In 1994, he assumed his present position at Washington University in St. Louis, where he is currently Professor of Chemistry, Internal Medicine, and Biochemistry and Principal Investigator of the National Institutes of Health Mass Spectrometry Research Resource at Washington University.

Michael Gross has worked in mass spectrometry research and development his entire career. His early interests were in ion chemistry and mechanism and in environmental chemistry, where he validated and utilized high performance mass spectrometry for ultratrace analysis of dioxins and related substances. His group was the first to find dioxin in human tissues of veterans of the Vietnam conflict. With C.L. Wilkins (see this volume, *Charles Lee Wilkins*), he built the second Fourier transform ion cyclotron resonance mass spectrometer in academia. His interest in the development of this instrumentation (notably electrically compensated traps to improve performance) and its use in biology and chemistry (notably top-down proteomics) continues to the present.

Following his move into the study of DNA adducts, lung, breast, and skin cancer, and oligonucleotide fragmentation mechanisms with mass spectrometry, he began in the early 2000s a program in structural proteomics. His group developed PLIMSTEX (protein ligand interactions by MS, titration and H/D exchange) as a means for determining protein/ligand affinities. They also developed FPOP (fast photochemical oxidation of proteins) as a means to follow protein folding/unfolding, to determine protein/protein and protein/ligand interfaces, and to assign structures to protein/ligand complexes. The FPOP approach is suitable as a means of following protein folding and unfolding on the sub-millisecond timeframe.

Dr. Gross has been a member of the American Society for Mass Spectrometry since 1968; he became the founding editor of the *Journal of the American Society for Mass Spectrometry* in 1990, a position he continues to hold. Prior to that service, he was editor of *Mass Spectrometry Reviews* from 1983 to 1990. He is also a member of the American Chemical Society and the Indian Society for Mass Spectrometry. He is Series Editor of *The Encyclopedia of Mass Spectrometry,* additionally editing several volumes in the encyclopedia. He has published over 500 scientific papers and book chapters and has served as editor of five books.

In 2006, Dr. Gross received the Thomson Medal Award from the International Mass Spectrometry Society for "for outstanding achievements in mass spectrometry and for distinguished service to international mass spectrometry." He received the Field and Franklin Award from the American Chemical Society in 1999, the Midwest Award from the American Chemical Society in 2002, and the Pioneer Award for his contributions to understanding the health effects of dioxin exposure from the Commonwealth of Massachusetts in 1987. He has been the named lecturer at a number of universities and industries, among them the Albright & Wilson Visiting Professor at the University of Warwick in 1988 and the Van't Hoff Visiting Professor of the International Graduate School of the University of Amsterdam in 1990. He has been identified by the Institute of Scientific Information as one of the most cited chemists worldwide.

The publications cited below are representative of the author's most significant contributions to the field (*1–8*).

References

(*1*) Chen, J.; Rempel, D. L.; Gau, B. C.; Gross, M. L. Fast Photochemical Oxidation of Proteins (FPOP) and Mass

Spectrometry Follow Sub-millisecond Protein Folding at the Amino-Acid Level. *J. Am. Chem. Soc.* **2012**, *134*, 18724–18731.
(*2*) Hambly, D. M.; Gross, M. L. Laser Flash Photolysis of Hydrogen Peroxide to Oxidize Protein Solvent-Accessible Residues on the Microsecond Timescale. *J. Am. Soc. Mass Spectrom.* **2005**, *16*, 2057–2063.
(*3*) Grese, R. P.; Cerny, R. L.; Gross, M. L. Metal Ion-Peptide Interactions in the Gas Phase: A Tandem Mass Spectrometry Study of Alkali Metal Cationized Peptides. *J. Am. Chem. Soc.* **1989**, *111*, 2835–2842.
(*4*) Gross, M. L. An Ion Cyclotron Resonance Study of the Structure of $C_3H_6^+$ and the Mechanism of its Reaction with Ammonia. *J. Am. Chem. Soc.* **1972**, *94*, 3744–3748.
(*5*) Jensen, N. J.; Tomer, K. B.; Gross, M. L. Gas-Phase Ion Decompositions Occurring Remote to a Charge Site. *J. Am. Chem. Soc.* **1985**, *107*, 1863–1868.
(*6*) Ledford, E. B. Jr.; Rempel, D. L.; Gross, M. L. Space Charge Effects in Fourier Transform Mass Spectrometry.II. Mass Calibration. *Anal. Chem.* **1984**, *56*, 2744–2748.
(*7*) Gross, M. L.; Sun, T.; Lyon, P. A., *et al.* Heath, Method Validation Study for Analysis of Tetrachlorodibenzodioxin (TCDD) at the Low Parts-Per-Trillion Level. *Anal. Chem.* **1981**, *53*, 1902–1906.
(*8*) Zhu, M.; Rempel, L.; Du, Zhaohui; Gross, L. Quantification of Protein-Ligand Interactions by Mass Spectrometry, Titration and H/D Exchange: PLIMSTEX. *J. Am. Chem. Soc.* **2003**, *125*, 5252–5253.

Alfred L. Yergey
National Institute of Child Health and Human Development, NIH, Bethesda, MD, USA

P. Jane Gale
Waters Corp., Milford, MA, USA

Michael Guilhaus

b. March 19, 1954
d. October 16, 2009

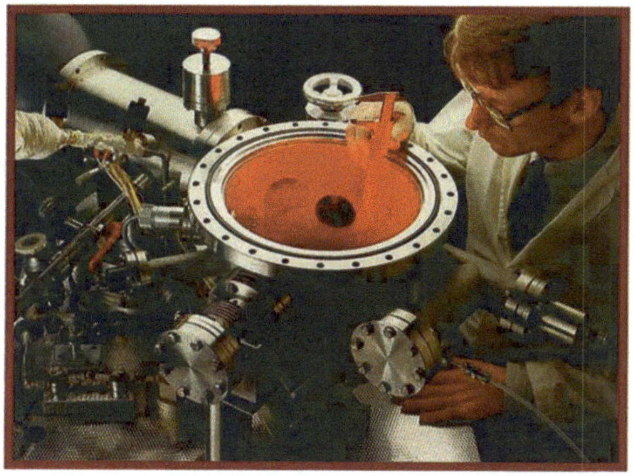

Photo from J. Am. Soc. Mass Spectrom. ***2012****. 23, 1302.*

Michael Guilhaus was an Australian chemist of German descent who is best known for rediscovery of the orthogonal acceleration (OA) technique, a means of coupling continuous ionization sources to time-of-flight (TOF) mass spectrometers (see this volume, *The Development of Time-of-Flight (TOF) Mass Spectrometry*). Unaware of earlier as well as parallel efforts when he first published his findings (*1*), he was quick to describe his own effort as a 'rediscovery' and to acknowledge both the 1960s work of O'Halloran (*2*) and contemporaneous efforts of Dodonov (*3*) that were unknown in the West at the time of Guilhaus' first publication (*4*). In developing and promoting the benefits of this technology – primarily sensitivity and speed of analysis, in the early 1990s – Guilhaus ignited a spark that led indirectly to the commercial (and academic) development of hybrid instruments involving TOF as the secondary mass analyzer (*5* and see this volume, *The Development of Tandem Mass Spectrometry*).

Michael was born in Wollongong, New South Wales, Australia, on March 19, 1954, the second of three children born to German parents who had recently immigrated to Australia (*6*). He was educated in Sydney at Randwick Boys High School and the University of New South Wales (UNSW), where he was awarded a BSc (Hons) in 1978 and in 1983 a PhD under the supervision of Ian Gregor. He spent a two-year post-doctoral appointment in John Beynon's laboratory at Swansea University (Wales) and then returned to Australia, taking up a post as lecturer at the University of Wollongong in 1985 (see this volume, *John Herbert Beynon*). He joined the faculty of UNSW as a lecturer in 1986, and there he remained for the rest of his career. He was later promoted as Professor, as well as Director of the Centre for Bioanalytical Mass Spectrometry. Michael and his first wife Lynn had two children, a son and a daughter. He and his second wife Gera were married in 2009.

Guilhaus' scientific career was built around mass spectrometry, often literally as he undertook the challenge of designing and building novel instrumentation at a time when the purchase of commercial instruments for academic institutions had become more common. His interest in designing apparatus customized to suit a purpose began as a graduate student at UNSW and was nurtured during his post-doc at Swansea. Capitalizing on the desire for new instrumentation in the School of Chemistry upon his return to the university as an academic, Guilhaus joined forces with UNSW colleague Jim Dawson, who had considerable skill in ion optics and electronic control, having worked with Keith Jennings at Warwick and Nico Nibbering (see this volume, *Keith R. Jennings* and *Nicolaas Martinus Maria Nibbering*). In 1986, the two embarked on the design and construction of TOF instrument that would offer substantial advantages in resolution and sensitivity over current instrumentation for continuous ionization sources, work they reported in 1989 (*1,7–9*). Information on O'Halloran's work on OA, ultimately determined to be sufficiently different from the Guilhaus-Dawson effort for US and UK patents to be issued to the latter team in 1992 (*7–9*), only came to light with the patenting due diligence. The first documentation of Dodonov's efforts occurred in as a USSR patent filed in 1987 (*3*), but descriptions of his invention did not appear in the open literature until several years later (*10,11*).

The explosion of interest in OA-TOF instrumentation, especially for use with electrospray sources that followed Guilhaus' and Dodnov's publications makes it clear that the full potential of OA had gone unrealized until the two teams had (separately) discovered how to obtain optimal mass resolution in linear and reflecting instruments (see this volume, *The Development of Electrospray Ionization (ESI) for Mass Spectrometry*). As Dodonov's biographers note, "The orthogonal injection principle revolutionized the field of TOF as it became possible to perform accurate and sensitive MS and MS/MS analysis of electrosprayed biomolecules" (*12*). For the remainder of his career, Guilhaus continued to develop OA instrumentation (*13,14*) but also took on the study of biological molecules, publishing a number of papers with colleagues and collaborators of the BMSF (*15*).

Guilhaus was a Fellow of the Royal Australian Chemical Institute. In recognition of the significance of his contributions to the development of OA and its relationship to time-of-flight mass spectrometry, he was awarded the 1990 Morrison Medal by the Australia and New Zealand Society for Mass Spectrometry and the 1997 Curt Brunnée Award given by the International Mass Spectrometry Society.

In his younger days, Michael was an avid wind surfer, boogie boarder, and cyclist. Toward the end of his life, he pursued the cycling with enhanced enthusiasm, saying that it allowed the drama of being a cancer patient to become secondary to the present reality of living (*6*). He died on October 16, 2009, in Sydney, Australia.

References

(*1*) Dawson, J. H. J.; Guilhaus, M. Orthogonal-Acceleration Time-of-Flight Mass Spectrometer. *Rapid Commun. Mass Spectrom.* **1989**, *3*, 155–159.

(*2*) O'Halloran, G. J., Aeronautical Systems Division (U.S.), Walker, L. W., Fluegge, R. A., Belts, J. F., Everett, W. L. Determination of Chemical Species Prevalent in a Plasma Jet. Technical Documentary Report, Aeronautical Systems Division (U.S.). Air Force Materials Laboratories, Research and Technology Division, Air Force Systems Command, United States Air Force, 1964.

(*3*) Dodonov, A. F., Chernushevich, I. V., Dodonova, T. F., Raznikov, V. V., Talrose, V. L. Mass Reflectron. Soviet Patent No. 1681340. 1987.

(*4*) Guilhaus, M.; Selby, D.; Mlynski, V. Orthogonal Time-of-Flight Mass Spectrometry. *Mass Spectrom. Rev.* **2000**, *19*, 65–107.

(*5*) Sheil, M. M. Critical Moments in Time: Reflections on the Development of Orthogonal Acceleration Time-of-Flight Mass Spectrometry. *J. Am. Soc. Mass Spectrom.* **2012**, *23*, 1301–1305.

(*6*) Guilhaus, P. Private communication. 30 August 2012.

(*7*) Dawson, J. H. J., Guilhaus, M.: US PCT No. PCT/AU88/00498 Filed 23 December 1988, Published 29 June 1989. Foreign Application Priority, Australian Provisional Patent PCT 16079, 24 December 1987.

(*8*) Guilhaus, M., Dawson, J. H. J. (inventors), Unisearch Ltd. (Assignee), Mass Spectrometer. US Patent No. 5, 117,107, 26 May 1992.

(*9*) Guilhaus, M., Dawson, J. H. J. (inventors), Unisearch Ltd. (Assignee), Mass Spectrometer. UK Patent No. 2233149, July 1992.

(*10*) Dodonov, A. F., Chernushevich, I. V., Laiko, V. V.: Extended Abstracts, 12th International Mass Spectrometry Conference, Amsterdam, 1991, pp 153.

(*11*) Dodonov, A. F.; Chernushevich, I. V.; Laiko, V. V. Time of Flight Mass spectrometry. In: *American Chemical Society Symposium Series*; Cotter, R., Ed.; American Chemical Society: Washington, DC, 1994; vol. 549, pp 108–123.

(*12*) Chernushevich, I. V.; Alexander, V.; Loboda, A. V.; Kozlovski, V. A.; Raznikov, V. V.; Vladislav, V.; Zelenov, V. V. Alexander F. Dodonov: 1939–2005. *J. Am. Soc. Mass Spectrom.* **2006**, *17*, 113–115.

(*13*) Coles, J.; Guilhaus, M. Orthogonal Acceleration – a New Direction for Time-of-Flight Mass Spectrometry. *Trends Anal. Chem.* **1993**, *12*, 203–213.

(*14*) Guilhaus, M. Spontaneous and Deflected Drift Trajectories in Orthogonal-Acceleration Time-of-Flight Mass Spectrometry. *J. Am. Soc. Mass Spectrom.* **1994**, *5*, 588–595.

(*15*) Williams, T. J.; Burg, D. W.; Raftery, M. J., *et al.* Global Proteomic Analysis of the Insoluble, Soluble, and Supernatant Fractions of the Psychrophilic Archaeon *Methanococcoides burtonii* Part I: The Effect of Growth Temperature. *J. Proteome Res.* **2010**, *9*, 640–652.

Margaret M. Sheil
University of Melbourne, VIC, Australia

P. Jane Gale
Waters Corp., Milford, MA, USA

Alexander Norman Halliday

b. August, 11, 1952

Alexander Norman Halliday is a British geochemist widely noted for work in the general areas of isotope geochemistry, mass spectrometry and planetary science. Currently he is the Head of the Mathematical, Physical and Life Sciences Division of Oxford University. An enthusiast for technological innovation Halliday uses mass spectrometry to elucidate the origin and early development of the solar system and recent earth processes such as continental erosion and climate. However, he is also engaged in studies as different as the mechanisms of volcanic eruptions, the formation of mineral and hydrocarbon deposits and the development of civilization.

Halliday was born in Penzance but spent his childhood in London before returning to Cornwall as an adolescent and eventually going to the same school as was formerly attended by Humphry Davy. Following a bachelors in geology he did his doctorate in the physics department (1977) at the University of Newcastle-upon-Tyne under the supervision of John Mitchell, one of the pioneers of $^{40}Ar/^{39}Ar$ dating. He spent 10 years at the Scottish Universities Research and Reactor Centre near Glasgow before moving to the University of Michigan in 1986, where he was a professor for 12 years. He then moved to the ETH in Zürich until 2004 when he took up the Chair in Geochemistry at the University of Oxford. He is married and has two children.

In 1992, Halliday proposed that one could make significant advances in geochemistry, cosmochemistry and paleoceanography with different kinds of isotopic measurements made possible by a new kind of mass spectrometry, Multi-Collector Inductively Coupled Plasma Mass Spectrometry (MC-ICPMS) (see this volume, *The Development of Inductively-Coupled Plasma Mass Spectrometry (ICPMS)* and *The Development of Mass Spectrometry in the Earth and Planetary Sciences*). He acquired the first such instrument in 1994 and in 1995 published the first of several papers that have paved the way for the development of this field. Initially, he made the first relatively precise isotopic measurements of small amounts of W and used the Hf-W system to provide powerful new constraints on the early evolution of the inner solar system. Second, he made in situ measurements of Sr and Pb isotopic compositions in natural materials at new levels of precision using laser ablation and produced detailed paleoceanographic records at high spatial resolution. Third, he showed that accurate concentration measurements of a range of poorly understood (geochemically speaking) elements were possible, allowing the best determinations yet of their abundances in the Earth. These include Cd, In, Sn and Te and the platinum group elements. He also developed Hf isotopic studies of the oceans and the measurement of U and Th isotopic compositions, and hence disequilibrium series dating, to new levels of precision.

At ETH Zurich he established a new research programme again using MC-ICPMS. He installed the first large geometry high resolution MC-ICPMS which he had built by Nu Instruments and with a sizeable team developed new programmes in Li, Mg, Si, Ti, Fe, Ni, Zr, Cd, Te, Tl, Hf, W, and U isotope geochemistry. He used these studies to tackle several kinds of problems. First, he placed new limits on the abundances of short-lived nuclides in the early solar system, in particular, ^{92}Nb, ^{126}Sn, ^{205}Pb and ^{247}Cm. Second, he developed stable isotopic measurements to show that the silicate Earth is slightly heavy in Fe and Si compared with other objects. The fact that Li and Mg show no such feature led to the view this was a product of high pressure core formation, providing evidence for Si in the core. Third he developed high precision U isotopic measurements and used these to date relatively old corals, and find the first reliable evidence of widespread variations in $^{235}U/^{238}U$ in low temperature environments. Fourth, he built a programme in oceans and climate focusing on the records in ferromanganese crusts, the stable isotopic compositions of soils and rivers and the ocean chemistry of key elements, most importantly Hf. Upon his move to Oxford in 2004, he began establishing new programmes in Si, Ti, Cr, V, Fe, Ni, Zr, Mo, Ru, W and Tl isotopes.

The publications cited below are representative of the author's most significant contributions to the field (*1–14*).

He is the recipient of the Bowen Award of the American Geophysical Union (1998) and the Murchison Medal of the Geological Society (2003). He was the Mineralogical Society's Hallimond Lecturer in 2003 and the American Geophysical Union's Daly Lecturer in 2008. He has been President of the Geochemical Society (1995–97), European Association of Geochemistry (2007–09) and the Volcanology,

Geochemistry and Petrology section of the American Geophysical Union (2008–10). He was elected a Fellow of the Royal Society in 2000, and has also been elected a Fellow of the American Geophysical Union (2000), Geochemical Society and European Association of Geochemistry (2001) and Meteoritical Society (2008).

References

(*1*) Halliday, A. N.; Lee, D.-C.; Christensen, J. N., *et al.* Recent Developments in Inductively Coupled Plasma Magnetic Sector Multiple Collector Mass Spectrometry. *Int. J. Mass Spec. Ion Process.* **1995**, *146/147*, 21–33.

(*2*) Lee, D.-C.; Halliday, A. N. Precise Determinations of the Isotopic Compositions and Atomic Weights of Molybdenum, Tellurium, Tin and Tungsten Using ICP Source Magnetic Sector Multiple Collector Mass Spectrometry. *Int. J. Mass Spec. Ion Process.* **1995**, *146/147*, 35–46.

(*3*) Lee, D.-C.; Halliday, A. N. Hafnium-Tungsten Isotopic Evidence for Rapid Accretion and Differentiation in the Early Solar System. *Science* **1996**, *274*, 1876–1879.

(*4*) Christensen, J. N.; Halliday, A. N.; Godfrey, L. V.; Hein, J. R.; Rea, D. K. Climate and Ocean Dynamics and the Lead Isotopic Records in Pacific Ferro-Manganese Crusts. *Science* **1997**, *277*, 913–918.

(*5*) Luo, X.; Lee, D.-C.; Rehkämper, M.; Halliday, A. N. High Precision $^{230}Th/^{232}Th$ and $^{234}U/^{238}U$ Measurements Using Energy-Filtered ICP Magnetic Sector Multiple Collector Mass Spectrometry. *Int. J. Mass Spec. Ion Process.* **1997**, *171*, 105–117.

(*6*) Halliday, A. N.; Lee, D.-C.; Christensen, J. N., *et al.* Applications of Multiple Collector ICPMS to Cosmochemistry, Geochemistry and Paleoceanography. (The 1997 Geochemical Society Presidential Address.). *Geochim. Cosmochim. Acta* **1998**, *62*, 919–940.

(*7*) Rehkämper, M.; Halliday, A. N. Accuracy and Long-Term Reproducibility of Lead Isotopic Measurements by MC-ICPMS Using an External Method for Correction of Mass Discrimination. *Int. J. Mass Spec. Ion Proc.* **1998**, *181*, 123–133.

(*8*) Halliday, A. N.; Christensen, J. N.; Lee, D.-C.; Rehkämper, M.; Hall, C. M.; Luo, X. Multiple Collector Inductively Coupled Plasma Mass Spectrometry. In: *Inorganic Mass Spectrometry, Fundamentals and Applications*; Barshick, C. B., Duckworth, D. C. Smith, D. H., Eds.; Marcel Dekker Inc.: New York, 2000; pp 291–328, chap. 8.

(*9*) Stirling, C. H.; Esat, T. M.; Lambeck, K., *et al.* Orbital Forcing of the Stage 9 Interglacial: Evidence from U-Series Dating of Henderson Island Corals. *Science* **2001**, *291*, 290–293.

(*10*) Andersen, M. B.; Stirling, C. H.; Potter, E.-K.; Halliday, A. N. Toward Epsilon Levels of Measurement Precision on $^{234}U/^{238}U$ by Using MC-ICPMS. *Int. J. Mass Spec.* **2004**, *237*, 107–118.

(*11*) Nielsen, S. G.; Rehkämper, M.; Norman, M. D.; Halliday, A. N. Thallium Isotopic Evidence for Ferromanganese Sediments in the Mantle Source of Hawaiian Basalts. *Nature* **2006**, *439*, 314–317.

(*12*) Newman, K.; Freedman, P. A.; Williams, J.; Belshaw, N. S.; Halliday, A. N. High Sensitivity Skimmer Cones and Non-Linear Mass Dependent Fractionation in ICP-MS. *JAAS.* **2009**, doi:10.1039/b819065h.

(*13*) Rickli, J.; Frank, M.; Halliday, A. N. The Hafnium-Neodymium Isotopic Composition of Atlantic Seawater. *Earth Planet. Sci. Lett.* **2009**, *280*, 118–127.

(*14*) Hendry, K. R.; Georg, R. B.; Rickaby, R. E. M.; Robinson, L. F.; Halliday, A. N. Deep Ocean Nutrients During the Last Glacial Maximum Deduced from Sponge Silicon Isotopic Compositions. *Earth Planet. Sci. Lett.* **2010**, *292*, 290–300.

Alex N. Halliday
University of Oxford, Oxford, UK

Alexander George Harrison

b. April 1, 1932

Photo ca. 1986 in front of ZAB-BEqQ instrument.

Alex (as he is called) Harrison is a Canadian mass spectrometrist known for his studies of ion thermodynamics and fragmentation mechanisms, as well as for his now classic book on chemical ionization first published in 1983. His career of more than 50 years spans studies of ionic and radical enthalpies of formation under conditions of electron ionization through his work on the uses of chemical ionization as a tool for fundamental studies (*1*) to his present work on mechanisms of peptide ion fragmentation, an area in which he remains active as of this writing (see this volume, *Mass Spectrometric Methods for the Determination of Thermodynamic Data* and *A History of Chemical Ionization (CI) in Mass Spectrometry*).

Alexander George Harrison was born on April 1, 1932 in Ontario, Canada. He obtained both his BS and MSc at the University of Western Ontario and his PhD at McMaster University in 1956 under the direction of Harry Thode with a dissertation entitled 'Isotope Effects in the Reduction of Sulphur Compounds' (see this volume, *Henry Thode*). Having completed his thesis work in only three years, he stayed on at McMaster for a fourth as a postdoctoral fellow. That year was productive: He published five papers on diverse subjects related to sulfur isotope distributions in the biosphere ranging from 'Sulphur Isotope Abundances in Hydrocarbons and Source Rocks of Uinta Basin, Utah' to 'Mechanism of the Bacterial Reduction of Sulphate from Isotope Fractionation Studies.' He spent the next two years in Fred Lossing's lab at the National Research Council in Ottawa, where he began his studies in thermodynamics, particularly ionic and radical enthalpies of formation using appearance potential measurements (see this volume, *Frederick Lossing*).

Harrison spent the whole of his academic career at the University of Toronto. In 1958, following his postdoctoral work, he was appointed a temporary lecturer in the Chemistry Department and was appointed the next year to a tenure-track position. He became a full professor in 1967 and Professor Emeritus in 1993.

Harrison's long career investigating ion energetics and fragmentation began with a series of papers coauthored with his postdoctoral mentor Fred Lossing, an example of which is given below (*2*). Upon assuming his appointment at the University of Toronto, he began several series of studies that included the development of a means for trapping ions formed by electron ionization using a sector field mass spectrometer (*3*) and following their fragmentation pathways (*4*). In the 1970s Harrison realized the utility of chemical ionization for fundamental studies of reaction pathways, a realization that led to publication of a long series of scientific papers, as well as the book for which he is perhaps best known (*1*,*5*).

Harrison's work has been recognized in a number of ways. He was chosen an Alfred P. Sloan Fellow from 1962–64 and a Killam Research Fellow from 1985–87; he has been a guest lecturer and distinguished visiting scientist at several institutions. He has the distinction of having been the first Canadian to be elected to the Board of Directors of the American Society for Mass Spectrometry. In 2005, his home Department at the University of Toronto established the Alex Harrison Graduate Fellowship in Analytical Environmental Mass Spectrometry.

Unlike many colleagues who slowly wind down existing research efforts upon achieving emeritus status, Alex Harrison celebrated this achievement with the launch of a series of studies in an entirely different class of ion chemistry. Beginning at that time and continuing to the present, his focus has shifted from the energetics and fragmentation of small molecules to those of peptides (*6*–*9*).

Harrison continues to live in Toronto with his wife Barbara and to enjoy the company of their children and grandchildren.

References

(*1*) Harrison, A. G. *Chemical Ionization Mass Spectrometry*; 2nd ed., CRC Press: Boca Raton, FL, 1992. (First edition published in 1983).

(*2*) Harrison, A. G.; Lossing, F. P. Free Radicals by Mass Spectrometry. 17. Ionization Potential and Heat of Formation of Vinyl Radical. *J. Am. Chem. Soc.* **1960**, *82*, 519–521.

(*3*) Harrison, A. G.; Lin, P. H.; Tsang, C. W. Proton-Transfer Reactions by Trapped-Ion Mass-Spectrometry. *Int. J. Mass. Spectrom. Ion Processes* **1976**, *19*, 23–51.

(*4*) Benoit, F. M.; Harrison, A. G. Hydrogen Migrations in Mass-Spectrometry. 2. Single and Double Hydrogen Migrations in Electron-Impact Fragmentation of Normal-Propyl Benzoate. *Org. Mass Spectrom.* **1976**, *11*, 1056–1062.

(*5*) Harrison, A. G.; Onuska, F. I. Fragmentation in Chemical Ionization Mass-Spectrometry and Proton Affinity of Departing Neutral. *Org. Mass Spectrom.* **1978**, *13*, 35–38.
(*6*) Yalcin, T.; Khouw, C.; Csizmadia, I. G.; Peterson, M. R.; Harrison, A. G. Why are B Ions Stable Species in Peptide Spectra? *J. Am. Soc. Mass Spectrom.* **1995**, *6*, 1165–1174.
(*7*) Harrison, A. G. The Gas-Phase Basicities and Proton Affinities of Amino Acids and Peptides. *Mass Spectrom. Rev.* **1997**, *16*, 201–217.
(*8*) Knapp-Mohammady, M; Young, A. B.; Paizs, B.; Harrison, A. G. Fragmentation of Doubly-Protonated Pro-His-Xaa Tripeptides: Formation of b_2^{2+} Ions. *J. Am. Soc. Mass Spectrom.* **2009**, *20*, 2135–2143.
(*9*) Harrison, A. G. Fragmentation Reactions of b(5) and a(5) Ions Containing Proline – The Structures of a(5) Ions. *J. Am. Soc. Mass Spectrom.* **2012**, *23*, 594–601.

Alfred L. Yergey
National Institute of Child Health and Human Development, NIH, Bethesda, MD, USA

P. Jane Gale
Waters Corp., Milford, MA, USA

John Michael Hayes

b. September 6, 1940

John M. Hayes is an American chemist who is known for his work in devising mass spectrometers expressly to resolve questions regarding the Earth's carbon cycle. Hayes's specialty has been the measurement and interpretation of variations in the abundances of the isotopes of carbon, hydrogen, nitrogen, and oxygen, using instrumentation allowing precise analyses of minimal quantities of samples.

John Michael Hayes was born in Seattle, WA, in 1940. He obtained a BS in chemistry at Iowa State University (1962) and a Ph.D. in analytical chemistry at the Massachusetts Institute of Technology with Klaus Biemann (see this volume, *Klaus Biemann*) in 1966. After graduation, Hayes spent a brief period as an Enrico Fermi Fellow at the University of Chicago before serving in the US Army from 1966–68, detailed to the Chemical Evolution Branch at NASA Ames Research Center. He was a NATO-NSF postdoctoral fellow at the University of Bristol, UK, during 1969 and then joined the faculty at Indiana University in 1970, where for the next twenty-six years he had an active career in biogeochemistry. In 1996, Hayes became a Senior Scientist at Woods Hole Oceanographic Institution (until 2006) and Professor of Biogeochemistry at Harvard (1997–2000).

An initial series of instruments designed with Dale Schoeller and Dwight Matthews described and employed pulse-counting detection for carbon isotope ratio measurements with a precision of 0.05 °/oo for sample sizes as small as 100 ng carbon (*1,2*). Further engineering studies in the Hayes lab concluded that Faraday cups with electrometer amplifiers could operate at the shot-noise limit for currents as low as10^{-12} amp (*3*). This resulted in further improvements in precision, and was coupled with a closed-loop control of a sample reservoir to provide complete consumption of small samples (*4*) and this instrumentation enabled seminal studies on the origins of organic materials in geological samples (*5,6*).

Prior to Hayes' work, isotope-ratio mass spectrometers conventionally used static sample reservoirs to maximize accuracy and precision of measurement. Matthews and Hayes shifted this paradigm in order to conduct isotopic analyses of individual organic compounds separated on a gas chromatographic time scale (*7*). They employed a post-column combustion system and used a permeable-membrane separator to transfer CO_2 and N_2 from the carrier gas stream to a computer-controlled, beam-switching isotope ratio mass spectrometer. The resulting $^{15}N/^{14}N$ or $^{13}C/^{12}C$ ratios were measured with a precision of 0.5% or better with 20 nmol of CO_2 or 100 nmol of N_2, detecting enrichments as low as 0.004 atom-% excess. Subsequently, following parallel developments in conventional GCMS instruments, the system was improved by switching to capillary chromatographic columns. The permeable-membrane interface was eliminated and, after removal of water, the combustion products and carrier gas were transmitted directly to the ion source of an isotope-ratio mass spectrometer. The first application of this system demonstrated its power as tool of molecular analysis and instruments of this type are now commonplace in biogeochemical laboratories. Technical details and relevant statistical have been fully documented (*8,9*).

In their research, Hayes and his coworkers unraveled intricate details of Earth's environmental machinery and the mechanisms of its control. For example, they provided the first evidence for the origin of oxygen-producing photosynthesis at least 2.8 billion years ago. They have also reconstructed pathways of carbon flow in ancient lakes, oceans, and sediments and shown that the concentration of CO_2 in oceanic surface water is a key factor controlling the abundance of ^{13}C in the organic matter produced by marine algae. Accordingly, isotopic analysis of ancient algal debris can aid estimation of former concentrations of CO_2. The ability to analyze very small samples made it possible to measure isotopic abundances at specific positions within molecules. These investigations showed that isotopic compositions of individual organic molecules were controlled by isotope effects and carbon budgets within metabolic systems (*10*). As a result, molecular-isotopic studies are now an important tool for microbial ecologists as well as biogeochemists (*11*).

At Woods Hole Oceanographic Institution, Hayes served as Director of the National Ocean Sciences Accelerator Mass Spectrometry facility. With the members of that laboratory, he took the first steps toward introducing continuous-flow techniques in accelerator mass spectrometric systems.

As a result of his achievements in applying mass spectrometry to biogeochemistry, John Hayes has been recognized by his peers with a number of honors: the Geochemistry Division Medal of The American Chemical Society, 2003; the Goldschmidt Medal of The Geochemical Society, 2002; selection as a Fellow of American Geophysical Union, 2001; elected member of the National Academy of Sciences, 1998; elected Fellow of American Academy of Arts and Sciences, 1998; recipient of the Treibs Medal from The Geochemical Society, 1998; and co-recipient (jointly with Geoffrey Eglinton) of the Urey Medal, European Association for Geochemistry, 1997.

References

(*1*) Schoeller, D. A.; Hayes, J. M. Computer Controlled Ion Counting Isotope Ratio Mass Spectrometer. *Anal. Chem.* **1975**, *47*, 408–415.

(*2*) Hayes, J. M.; Matthews, D. E.; Schoeller, D. A. Effective Deadtime of Pulse-Counting Detector Systems. *Anal. Chem.* **1978**, *50*, 25–32.

(*3*) Peterson, D. W.; Hayes, J. M. Signal to Noise Ratios in Mass Spectroscopic Ion-Current Measurement Systems. In: *Contemporary Topics in Analytical and Clinical Chemistry*; Hercules, D. M., Hieftje, G. M., Snyder, L. R., Evenson, M. E., Eds.; 1978; vol. 3, pp 217–252.

(*4*) Hayes, J. M.; DesMarais, D. J.; Peterson, D. W.; Schoeller, D. A.; Taylor, S. P. High Precision Stable Isotope Ratios from Microgram Samples. In: *Advances in Mass Spectrometry*; Daly, N. R., Ed.; Heyden and Son, 1977; vol. 7, pp 475–480.

(*5*) Hayes, J. M.; Takigiku, R.; Ocampo, R.; Callot, H. J.; Albrecht, P. Isotopic Compositions and Probable Origins of Organic Molecules in the Eocene Messel Shale. *Nature* **1987**, *329*, 48–51.

(*6*) Hayes, J. M.; Popp, B. N.; Takigiku, R.; Johnson, M. W. An Isotopic Study of Biogeochemical Relationships Between Carbonates and Organic Carbon in the Greenhorn Formation. *Geochimica et Cosmochimica Acta* **1989**, *53*, 2961–2972.

(*7*) Matthews, D. E.; Hayes, J. M. Isotope-Ratio-Monitoring Gas Chromatography-Mass Spectrometry. *Anal. Chem.* **1978**, *50*, 1465–1473.

(*8*) Merritt, D. A.; Hayes, J. M. Factors Controlling Precision and Accuracy in Isotope-Ratio-Monitoring Mass Spectrometry. *Anal. Chem.* **1994**, *66*, 2336–2347.

(*9*) Merritt, D. A.; Freeman, K. H.; Ricci, M. P.; Studley, S. A.; Hayes, J. M. Performance and Optimization of a Combustion Interface for Isotope-Ratio-Monitoring Gas Chromatography/Mass Spectrometry. *Anal. Chem.* **1995**, *67*, 2461–2473.

(*10*) Monson, D. K.; Hayes, J. M. Biosynthetic Control of the Natural Abundance of Carbon 13 at Specific Position Within Fatty Acids in *Saccharomyces cerevisiae*. Isotope Fractionations in Lipid Synthesis as Evidence for Peroxisomal Regulation. *J. Biol. Chem.* **1982**, *257*, 5568–5575.

(*11*) Hayes, J. M. Fractionation of the Isotopes of Carbon and Hydrogen in Biosynthetic Processes. In: *Stable Isotope Geochemistry, Reviews in Mineralogy and Geochemistry*; Valley, J. W., Cole, D. R., Eds.; Mineralogical Society of America: Washington, DC, 2001; vol. 43, pp 225–278.

Sanford P. Markey
Laboratory of Neurotoxicology, National Institute of Mental Health, Bethesda, MD, USA

William J. Henzel

b. April 5, 1953

William Henzel is an American chemist, now retired from a position as Senior Scientist with Genentech. He directed the operation of a protein structural research facility and was responsible for protein characterization for all of the research departments of Genentech and for developing new state-of-the-art methodologies for protein analysis. His research focused on primary structure determination and posttranslational modifications using mass spectrometry and Edman degradation.

Henzel was born in Schenectady, NY. He received a Bachelor of Science in Chemistry (with ACS accreditation) in 1976 from the State University of New York at Geneseo. He worked in research labs at Children's Hospital, Boston, and the University of Massachusetts biology department before joining Genentech in 1982. His interest in the use of mass spectrometry as an analytical tool for the structural elucidation of protein primary structure began in 1980 through a collaboration with Klaus Biemann's laboratory at MIT on a project involving insect cuticle cross-linking that confirmed the presence of dityrosine cross linked peptides using fast atom bombardment mass spectrometry (FABMS).

His next opportunity to utilize FABMS for peptide analysis occurred in the late 80's when Genentech purchased a JEOL tandem four sector mass spectrometer. Initial results were disappointing due to the high sample quantity required for analysis (>1 nmol). Edman degradation allowed routine determination of the amino acid sequence with peptide quantities at or below 1 pmol. The coupling of continuous flow FABMS with capillary HPLC significantly improved sensitivity and contributed materially to the structural elucidation of a number of proteins. FABMS allowed Henzel and colleagues to develop the concept of peptide mass fingerprinting in 1989; however, the large amounts of peptide required for FABMS analysis precluded making this application a routine method.

The arrival of a matrix-assisted laser desorption/ionization (MALDI) time-of-flight mass spectrometer in 1992 resulted in the ability to obtain mass measurements of peptides at the low picomole to femtomole levels (see this volume, *Matrix-Assisted Laser Desorption Ionization Mass Spectrometry: Observations on its Advantages, Shortcomings, and Intricacies*). This permitted the use of MALDI-MS to obtain peptide masses from tryptic digests of proteins that had been separated by two-dimensional electrophoresis and enabled routine protein identification by peptide mass fingerprinting. The commercialization of ion traps in the 1990s enabled Henzel and colleagues to analyze highly complex peptide mixtures in identifying coeluting 2-D gel spots (*1*). For this work, he and Genentech colleagues John Stults and Colin Watanabe shared the ASMS Award for Distinguished Contribution in Mass Spectrometry (2002) (see this volume, *John T. Stults*). In announcing the award, ASMS noted, "Their landmark study reported in 1993 in the Proceedings of the National Academy of Sciences USA (*2*) convincingly showed that MS approaches alone could be used to analyze proteins from two-dimensional polyacrylamide gels. Further developments by a number of researchers, including the completion of a number of genome sequences, the incorporation of MS/MS peptide sequencing, and improvements in sensitivity, mass accuracy, and sample throughput have made this the method of choice for the identification of proteins from gels. This methodology is a key component that links mass spectrometry to post-genome biology (*3*).

Dr Henzel is a member of the American Association for the Advancement of Science, the Association of Biomolecular Resource Facilities, and the American Society for Mass Spectrometry. In addition to sharing the ASMS Award in 2002, he was recognized in 2012 by his alma mater with the Geneseo Medal of Distinction (*4*). Some representative papers are listed below (*5–8*). He retired from Genentech in 2003 in order to teach high school and middle school science.

References

(*1*) Henzel, W. T.; Watanabe, C.; Stults, J. T. Protein Identification: The Origins of Peptide Mass Fingerprinting. *J. Am. Soc. Mass Spectrom.* **2003**, *14*, 929–930.

(*2*) Henzel, W. J.; Billeci, T. M.; Stults, J. T.; Wong, S. C.; Grimley, C.; Watanabe, C. Identifying Proteins from Two-Dimensional Gels by Molecular Mass Searching of Peptide Fragments in Protein Sequence Databases. *Proc. Natl. Acad. Sci. USA* **1993**, *90*, 5011–5015.

(*3*) Gross, M. L.; Robinson, C. Focus on Proteomics in Honor of the 2002 Distinguished Contribution in Mass Spectrometry Award to W. J. Henzel, J. T. Stults, and C. Watanabe. *J. Am. Soc. Mass Spectrom.* **2003**, *14*, 931–942.

(*4*) Anon. Retired Research Scientist William Henzel '76 Receives Geneseo Medal of Distinction. Available at: http://www.geneseo.edu/news_events/mretired-research-scientist-william-henzel-%E2%80%9976-receives-geneseo-medal-distinction (accessed 2012).

(*5*) Henzel, W. J.; Stults, S. T.; Hsu, C.; Aswad, D. W. The Primary Structure of a Carboxyl Methyltransferase from Bovine Brain that Selectively Methylates L-Isoaspartyl Sites. *J. Biol. Chem.* **1989**, *264*, 15905–15911.
(*6*) Henzel, W. J.; Bourell, J. H.; Stults, J. T. Analysis of Protein Digests by Capillary High-Performance Liquid Chromatography and On-Line Fast Atom Bombardment Mass Spectrometry. *Anal. Biochem.* **1990**, *187*, 228–233.
(*7*) Arnott, D.; Henzel, W. J.; Stults, J. T. Rapid Identification of Comigrating Gel-Isolated Proteins by Ion Trap Mass Spectrometry. *Electrophoresis* **1998**, *19*, 968–980.
(*8*) Henzel, W. J.; Tropea, J.; Dupont, D. Protein Identification Using 20-Minute Edman Cycles and Sequence Mixture Analysis. *Anal. Biochem.* **1999**, *267*, 148–160.

William J. Henzel
Genentech, Inc., South San Francisco, CA, USA (ret.)

David Michael Hercules

b. August 10, 1932

David Michael Hercules, Ph.D., is an American analytical chemist whose research career has been a random walk through analytical spectroscopy. At various times he has been involved in research on optical spectroscopy, molecular luminescence, x-ray photoelectron spectroscopy (ESCA, XPS), Auger electron spectroscopy, secondary-ion mass spectrometry (SIMS), low energy ion scattering spectroscopy, x-ray absorption spectroscopy (EXAFS), surface analysis, heterogeneous catalysis, laser-desorption mass spectrometry, MALDI, electrospray mass spectrometry and ion-mobility spectrometry (see this volume, *The Early History of Secondary Ion Mass Spectrometry (SIMS), The Development of Matrix Assisted Laser Desporption (MALDI) Mass Spectrometry, and The Development of Electrospray Ionization (ESI) for Mass Spectrometry*). He is perhaps best known for demonstrating the analytical utility of the laser micro analysis (LAMMA) technique in the early 1980s.

Hercules was born in Somerset, Pennsylvania, USA on August 10, 1932. He received a B.S. degree was from Juniata College in 1954 and a Ph.D. from M.I.T. in 1957. His doctoral research was overseen by the late L. B. (Buck) Rogers. Others who contributed significantly (but informally) to his scientific education were R. C. Lord, D. N. Hume, K. Siegbahn, R. L. Burwell, W. K. Hall, K. Biemann, F. W. McLafferty, and A. Benninghoven (see this volume, *Klaus Biemann, Fred W. McLafferty, and Alfred Benninghoven*). Doing research in mass spectrometry came late in his scientific career: His first scientific publication was in 1957 (as a graduate student) but his first paper in mass spectrometry wasn't published until 1979, 22 years later. Hercules has held faculty positions at Lehigh University, Juniata College, M.I.T., University of Georgia, University of Pittsburgh and Vanderbilt University. His current position is Centennial Professor Emeritus, Department of Chemistry, and Special Assistant to the Associate Provost for Research, Vanderbilt University, Nashville, Tennessee.

As an undergraduate Hercules developed an interest in the correlation between physical measurements and molecular structure, and in graduate school he studied correlations between molecular fluorescence and structure under Professors Rogers and Lord. For the first decade of his academic career he remained involved with various aspects of electronic spectroscopic phenomena. He became interested in chemiluminescence, particularly that derived from aromatic hydrocarbon anions (*1*), and in 1964 while perusing *Zeitschrift für Physik* he noticed a paper from Kai Siegbahn's group describing chemical shifts in the X-Ray Photoelectron spectra (XPS, ESCA) of S and N compounds and their correlation with nominal oxidation state. He became instantly interested in the possibility of correlating such shifts with molecular structure. (note that this work precedes the availability of heteronuclear NMR.) With the essential assistance of Siegbahn, Carl Nordling (both from Uppsala University, Sweden) and their associates, an ESCA instrument was produced with a particular emphasis on its use in XPS for quantitative analysis.

As it became apparent even to these enthusiasts that XPS used alone would not solve all of the world's surface analysis problems, they began to investigate the use of other surface techniques to supplement XPS. Their studies ultimately led them to low-energy ion scattering and secondary-ion mass spectrometry (SIMS) – work that would lead Hercules into a major research program in mass spectrometry. Hercules notes ruefully, "I should have recognized the next instrument addition as a harbinger of things to come - we added a laser-desorption mass spectrometer, the LAMMA-1000."

In his investigation of the use of SIMS for studying polymers, Hercules soon found that it was possible to obtain structural information (*2*); he was prevented from investigating oligomer distributions, however, by the limited mass range of the quadrupole instruments in use at the time. Learning that Prof. A. Benninghoven had just developed a time-of-flight SIMS (TOF-SIMS) instrument, Hercules began what became a long and productive collaboration (see this volume, *The Development of Time-of-Flight (TOF) Mass Spectometry*). With TOF-SIMS he was able to obtain oligomer distributions up to number average molecular weight (Mn) ~ 7500, establish repeat units, and, by studying fragmentation caused by SIMS, investigate polymer architecture (*3*).

When MALDI came on the scene, Hercules and co-workers initially studied the capability of MALDI for quantification to demonstrate that quantitative analysis with MALDI was possible (*4*). They also developed a practical way to couple MALDI with planar chromatography that used conventional TLC supports. They then shifted their polymer program from SIMS to MALDI with particular emphasis on the use of controlled chemical reactions to assist polymer characterization (*5*). They were able to use selective reactions to measure soft blocks and hard blocks in polyurethanes. They also began an extensive program of using MS/MS of polymers as a means for end-group analysis and learning other aspects of polymer architecture. Most recently they have used the evaporation-grinding method, developed by Dr. A. Gies, to obtain MALDI spectra of insoluble polymers. They have begun studies on the combined use of MALDI and ion mobility spectrometry as a means of studying polymer branching and to resolve isobars.

Hercules has served the American Chemical Society, the National Science Foundation, the Gordon Research Conferences, and various other organizations in many capacities. He has served on the editorial boards of many scientific periodicals and has been editor or co-editor for three series from Plenum Press. He was a John Simon Guggenheim Memorial Fellow (1973–74), a Fellow of the Society for Applied Spectroscopy (2009). He has received the Society for Applied Spectroscopy's Lester W. Strock Medal (1981), the Alexander von Humboldt Foundation Presiträger (1983), the ACS Fisher Award in Analytical Chemistry (1986), and the Spectroscopy Society of Pittsburgh Award (1996).

Hercules is married to Shirley (Hoover) Hercules, having just celebrated their 40th wedding anniversary. They have two children, Sherri (Sokolovich) and Kevin, two grandsons, Zack and Nicola Sokolovich and a granddaughter, Katie Correia.

References

(*1*) Hercules, D. M. Chemiluminescence Resulting from Electrochemically Generated Species. *Science* **1964**, *145*, 808–809.

(*2*) Gardella, J. A. Jr.; Hercules, D. M. Static Secondary Ion Mass Spectrometry of Polymer Systems. *Anal. Chem.* **1980**, *52*, 226–232.

(*3*) Bletsos, I. V.; Hercules, D. M.; Greifendorf, D.; Benninghoven, A. Time- of-Flight Secondary Ion Mass Spectrometry of Nylons: Detection of High Mass Fragments. *Anal. Chem.* **1985**, *57*, 2384–2388.

(*4*) Muddiman, D. C.; Gusev, A. I.; Proctor, A.; Hercules, D. M.; Venkataraman, R.; Diven, W. Direct Quantitation of Cyclosporin A in Blood by Time-of-Flight Mass Spectrometry. *Anal. Chem.* **1994**, *66*, 2362–2368.

(*5*) Murgasova, R.; Brantley, E. L.; Hercules, D. M.; Nefzger, H. Characterization of Polyester-Polyurethane Soft and Hard Blocks by a Combination of MALDI, SEC and Chemical Degradation. *Macromolecules* **2002**, *35*, 8338–8345.

David M. Hercules

Department of Chemistry, Vanderbilt University, Nashville, TN, USA

Zdeněk Herman

b. March 24, 1934

Zdeněk Herman is a Czech chemist whose work in gas phase chemistry has contributed greatly to our understanding of the dynamics and kinetics of chemical reactions, ion-molecule reactions, reactive scattering of ions, elementary collision processes of multiply-charged ions, and ion-surface interactions (Figs 1 and 2).

Zdeněk Herman was born in Libušín, Bohemia, in Czechoslovakia (now the Czech Republic). He attended Gymnasium in Kladno, graduating in 1952, and then studied chemistry at the School of Mathematics and Physics, Charles University in Prague, receiving his diploma in physical and radiochemistry in 1957. Herman joined the Institute of Physical Chemistry, Czechoslovak Academy of Sciences (now J. Heyrovsky Institute of Physical Chemistry, Academy of Sciences of the Czech Republic), where he has remained throughout his career. His thesis research, supervised by Vladimir Čermák, involved the study of chemical kinetics and chemical dynamics in the ion source of a mass spectrometer (*1,2*). He was awarded his PhD for this work in 1962. Following a post-doctoral appointment at Yale University, where he studied with Richard Wolfgang, he returned to Prague to serve as Scientist, Senior Research Fellow and Research Team Head in the Čermák Laboratory at the J. Heyrovský Institute. He later became Professor of Physical Chemistry at the Institute of Chemical Technology, Prague and Honorary Professor of Ion Physics at the University of Innsbruck, Innsbruck, Austria. He was a Visiting Fellow at the Joint Institute for Laboratory Astrophysics (JILA) in Boulder, CO, in 1980 and again in 2001 (*3*).

Figure 1
Zdeněk Herman with Vladimír Čermák in 1962 (Taken from Ref. *1*).

Figure 2
Zdeněk Herman with Richard Wolfgang and the Yale EVA machine, 1968 (Taken from Ref. *1*).

Zdeněk's abilities as a chemist have been well served by another of his talents: he is a first-rate artist (*4*), and it is his ability as a sculptor to fashion objects of the finest detail with exquisite care that has enabled him to build the apparatus necessary for studying collision dynamics. His instruments are legendary: to study the trajectory of ions at very low – thermal – energies, where even the smallest of stray fields can perturb the trajectory resulting from a collision, he built impossibly small instruments in which such forces could be controlled (*2*).

'Fiddling around,' as he described it, with the voltage between the ion source and the trap, Herman observed 'crazy results' that many would have ignored. Instead, he and Čermák investigated the phenomena and eventually applied them to what was the first experiment in chemical dynamics designed to answer the question, "Was the reactive collision a fly-by event or did it proceed through long-lived complex?" Conducted behind the Iron Curtain at a time that scientific findings there were not readily accessible to or known in the West, this work actually preceded similar work done by Herschbach, for which Herschbach, Lee, and Polanyi won the Nobel Prize in 1986 (*5*).

Following his graduate research, Zdeněk was able to spend a couple of years in the US. Amid many invitations, he chose to work in the laboratories of Richard Wolfgang at Yale, and there he built EVA, a crossed-beam apparatus to study the chemical dynamics of ion molecule reactions at near thermal energies (*6*). Together Herman and Wolfgang applied their research efforts to the question, "What happens in a chemical reaction when the Coulombic energy is more powerful than the kinetic energy?" – work they documented in a series of papers that provided a foundation for chemical dynamics work that followed in the 1970 s (see, for example *7–9*).

Zdeněk's work, a selection of which is listed in references (*6–16*), has been widely recognized, and he has been honored with special issues of two journals: in 1995, on the occasion of his 60th birthday, the *Journal of Physical Chemistry* (*2*) and in 2009, on the occasion of his 75th birthday, the *International Journal of Mass Spectrometry* (*1*).

Zdeněk's awards are many and varied. They include the Czechoslovak Academy of Sciences Research Award in 1972, 1975, 1977; the Special Award of the Czechoslovak Academy of Sciences in 1980 and 1987; the Joint Prize of the Czechoslovak Academy of Sciences and the Academy of Sciences of GDR for research in reaction dynamics in 1982; The Marcus Marci Medal of the Czechoslovak Spectroscopic Society in 1989, The Alexander von Humboldt Research Award in 1992, the E. Schrodinger Award of the Symposium on Atomic and Surface Physics (SASP) in 1994, the National Prize "Czech Intellect 2003), MOLEC Award from the European Conference on the Dynamics of Molecular Collisions in 2006. In 2009 he was awarded a Dr. h.c. (Honorary Degree in Natural Sciences) by Leopold Franzens University, Innsbruck.

Zdeněk's legacy, both personal and scientific, is perhaps best summarized by the introduction to his 1995 Festschrift, entitled "Bohemian at Large," (*2*) that says in part,

"Few countries have an ambassador quite as beloved. For a host of colleagues throughout the world Zdeněk Herman has been the introduction not only to the mechanism of ionic reactions but to the culture of Bohemia and Prague – its Golden Capital. On the ceiling of the old laboratory in Máchova Street one could read the signatures of over 250 of the world's best known atomic, molecular, and chemical physicists. They came – many on repeated visits – to pay court to Vladimir Čermák and Zdeněk Herman, inspired by the quality of their science and charmed by the warmth of their friendship."

References

(*1*) Bierbaum, V. Zdeněk Herman: An Ambassador of Science. Forward to Volume Honoring Zdeněk Herman on the Occasion of his 75th Birthday. *Int. J. Mass Spectrom.* **2009**, *280*, 1–3.

(*2*) Friedrich, B.; Henchman, M.; Herschbach, D. Introduction to Zdeněk Herman Festschrift. *J. Phys. Chem.* **1995**, *99*, 15317–15326.

(*3*) Anon. Curriculum Vitae Zdeněk Herman in Volume Honoring Him on the Occasion of his 75th Birthday International. *J. Mass Spectrom.* **2009**, *280*, 4–5.

(*4*) A collection of his drawings from 1969-2005 can be found at http://www.jh-inst.cas.cz/~herman/files/. See in particular drawings 4 and 5 in "Occasional Drawings." The originals of these depictions of molecular dynamics hung on the walls of Dudley Herschbach's office at Harvard for many years (accessed September 2012).

(*5*) http://www.nobelprize.org/nobel_prizes/chemistry/laureates/1986/ (accessed August 2014).

(*6*) Herman, Z.; Kerstetter, J. D.; Rose, T. L.; Wolfgang, R. A Crossed Beam Apparatus for Investigation of Ion-Molecule Reactions. *Rev. Sci. Instrum.* **1969**, *40*, 538–545.

(*7*) Herman, Z.; Kerstetter, J.; Rose, T.; Wolfgang, R. Crossed Beam Studies of Ion-Molecule Reaction Mechanisms. *Disc. Faraday Soc.* **1967**, *44*, 123–136.

(*8*) Herman, Z.; Lee, A.; Wolfgang, R. Crossed-Beam Studies of Energy Dependence of Intermediate Complex Formation in an Ion–Molecule Reaction. *J. Chem. Phys.* **1969**, *51*, 452–455.

(*9*) Hierl, P.; Herman, Z.; Wolfgang, R. Chemical Accelerator Studies of Isotope Effects on Collision Dynamics of Ion–Molecule Reactions: Elaboration of a Model for Direct Reactions. *J. Chem. Phys.* **1970**, *53*, 660–674.

(*10*) Hierl, P. M.; Paček, V.; Herman, Z. Kinematics of Charge Transfer: $Ar^+ + H_2$. *J. Chem. Phys.* **1977**, *67*, 2678–2687.

(*11*) Herman, Z. Ion-Molecule Reactions: Collision Dynamics and Energy Partitioning. *Int. J. Mass Spectrom. Ion Phys.* **1982**, *45*, 293–304.

(*12*) Friedrich, B.; Vančura, J.; Herman, Z. Crossed-Beam Investigation of the Single-Electron Charge Transfer Process $Kr^{2+} + He \rightarrow Kr^+ + He^+$ at Sub-eV Collision Energies. *Int. J. Mass Spectrom. Ion Proc.* **1989**, *80*, 177–185.

(*13*) Hrušák, J.; Žabka, J.; Dolejšek, Z.; Herman, Z. A DFT/HF Study of the Potential Energy Surface of Protonated Ethane $C_2H_7^+$. *Int. J. Mass Spectrom. Ion Proc.* **1997**, *167–168*, 675–687.

(*14*) Schröder, D.; Loos, J.; Schwarz, H.; Thissen, R.; Roithova, J.; Herman, Z. Does Ionized Diacetylene have a Positive Proton Affinity? *Int. J. Mass Spectrom.* **2003**, *230*, 113–121.

(*15*) Pysanenko, A; Žabka, J.; Zappa, F.; Märk, T. D.; Herman, Z. Scattering of Very Slow (3–10 eV) Hydrocarbon Ions CD_3^+, $CD_4^{\bullet +}$, and CD_5^+ from Room-Temperature Carbon (HOPG) Surfaces. *Int. J. Mass Spectrom.* **2008**, *273*, 35–47.

(*16*) Keim, A.; Rasul, B.; Endstrasser, N.; Scheier, P.; Märk, T. D.; Herman, Z. Interaction of Small Hydrocarbon Ions and Ar^+ with Carbon-Fibre-Composite Surfaces at Room Temperature. *Int. J. Mass Spectrom.* **2011**, *306*, 204–209.

P. Jane Gale

Waters Corp., Milford, MA, USA

Richard Franz Karl Herzog

b. March 13, 1911
d. September 26, 1999

Richard FK Herzog was an Austrian and American physicist whose work was of fundamental importance for two different aspects of mass spectrometry. He was born and educated in Vienna, Austria, and earned his PhD at the University of Vienna in 1933. He emigrated to the USA in 1953 and became a US citizen in 1958 (*1*).

In the 1934, when mass spectrometers (including spectrographs) were new and rare, Herzog published an improved and more general treatment of the ion optics involved in mass spectrometry (*2*). That same year brought the first report of a double-focusing mass spectrograph employing what soon became known as Mattauch–Herzog geometry (*3*, and see this volume, *Magnetic and Electrostatic Analyzers before 1960* and *Josef Heinrich Elisabeth Mattauch*).

Herzog was at the University of Vienna in various capacities during and after World War II. He patented in 1942 a new source design for a mass spectrometer which in 1948 was embodied in an instrument constructed by his doctoral student FP Vieböck and briefly reported in joint letter in 1949 (*4*,*5*). The phenomenon of secondary ions had been noted before and studied to a minimal extent, but this was the earliest mass spectrometer specifically designed as (though not yet called) a SIMS instrument (*6*–*8* and see this volume, *The Early History of Secondary Ion Mass Spectrometry*).

In the USA, Herzog first worked for some years at the Air Force Cambridge Research Center. He shifted to the Geophysics Corporation of America, first as manager of their ion physics department and later as chief scientist for space science operations until 1973. He then was a professor of physics and astronomy at the University of Southern Mississippi until his retirement in 1978 (*1*).

Herzog further contributed to the development of SIMS after his move to the US, notably through improvements in ion source and optics improvements, published in 1963, that enabled a significant increase in the rate at which these techniques progressed (*9*). That work was done together with Helmut Liebl, who in turn significantly further advanced one line of SIMS instrumentation and application (*10*–*11*).

References

(*1*) *Who's Who in America, 1978/1979*, s.v. Herzog, Richard Franz Karl. Marquis Who's Who: Chicago, 40th ed., **1978**.

(*2*) Herzog, R. Ionen- und elektronenoptische Zylinderlinsen und Prismen. I. *Z. Physik* **1934**, *89*, 447–473.

(*3*) Mattauch, J.; Herzog, R. Uber einen neuen Massenspektrographen. *Z. Physik* **1934**, *89*, 786–795.

(*4*) Vieböck, F. P. Doctoral Dissertation, University of Vienna: Vienna, Austria, 1948.

(*5*) Herzog, R. F. K.; Vieböck, F. P. Ion Source for Mass Spectrography. *Phys. Review* **1949**, *76*, 855L–856L.

(*6*) Vickerman, J. C. Introducing Secondary Ion Mass Spectrometry. In: *Secondary Ion Mass Spectrometry*; Vickerman, J. C., Brown, A., Reed, N. M., Eds.; Oxford University Press: Oxford, 1989; pp 1–8.

(*7*) Honig, R. E. The Development of Secondary Ion Mass Spectrometry (SIMS): A Retrospective. *Int. J. Mass. Spec. Ion Proc.* **1985**, *66*, 31–54.

(*8*) Honig, R. E. The Growth of Secondary Ion Mass Spectrometry (SIMS): A Personal View of Its Development. In: *Secondary Ion Mass. Spectrometry SIMS V: Proceedings of the Fifth International Pro October 4, 1985*; Benninghoven, A., Colton, R. J., Simons, D. S., Werner, H. W., Eds.; Springer: Berlin, Heidelberg, New York, Tokyo, 1985; pp 2–15.

(*9*) Liebl, H. J.; Herzog, R. F. K. Sputtering Ion Source for Solids. *J. Appl. Phys.* **1963**, *34*, 2893–2896.

(*10*) Liebl, H. J. Ion Microprobe Mass Analyzer. *J. Appl. Phys.* **1967**, *38*, 5277–5283.

(*11*) Liebl, H. J. J. *Applied Charged Particle Optics*; Springer: Berlin, Heidelberg, New York, 2007.

Keith A. Nier
Independent Historian, Madison, NJ, USA

Franz Hillenkamp

b. March 18, 1936
d. August 22, 2014

Franz Hillenkamp was a German scientist who, together with Professor Michael Karas, was best known for the invention and development of Matrix Assisted Laser Desorption/Ionization Mass Spectrometry (MALDI-MS) (see this volume, *Micahel Karas* and *The Development of Matrix Assisted Laser Desorption Ionization (MALDI) Mass Spectrometry*). In recent years his interests have focused mainly on investigations of MALDI mechanisms and on MALDI applications for the analysis of nucleic acids. He is also known for his work in the field of clinical laser applications (Fig. 1).

Hillenkamp was born in Essen, Germany on March 18, 1936. He received his higher education at the Technical University of Munich where he obtained the degrees of Dipl.-Ing. in Electrical Engineering (1962) and a PhD in Physics in 1966. He also received a degree of Master of Science in Communications from Purdue University, West Lafayette, Ind. USA (1961). From 1963 until 1976 he was scientist and later senior scientist at the Gesellschaft für Strahlen- und Umweltforschung in Munich, a German National Laboratory in the Life Sciences. From 1976 to 1986 he was Associate Professor of Medical Physics at the University of Frankfurt, Germany, before moving on to his final position as chair of Medical Physics and Biophysics at the University of Muenster, Germany, from which he retired in 2002. He was Professor emeritus and past director of the Institute of Medical Physics and Biophysics at the Medical School of the University of Muenster, Germany.

Figure 1
Franz Hillenkamp at the LAMMA 1000 instrument that was used for the early development of MALDI-MS around 1992.

Hillencamp's first involvement with mass spectrometry was around 1970, when he met the late physiologist Prof. Raimund Kaufmann. Together they developed a method to measure the distribution of inorganic ions of metals such as Na, K, Fe and other trace elements in biological cells and tissue. An optical microscope was used to focus a UV-laser beam to a spot of minimally 0.5 μm diameter onto a cell or thin tissue section. The material in the irradiated volume was evaporated, and the ions generated in the process analyzed in a time-of-flight mass spectrometer. The laser spot was then scanned across the whole tissue section to generate an image of the ion distributions. The development was later taken up by the company Leybold-Heraeus in Cologne, Germany, further developed into a commercial instrument, and marketed as LAMMA 500 (LAser Micro Mass Analyzer). In a later version, the LAMMA 1000, the trans-illumination, typically used in histology was replaced by a top-illumination to allow the analysis of solid surfaces such as semiconductor wafers and chips. It was the background chemical noise signals in the LAMMA spectra that triggered Hillenkamp's interest in the laser desorption mass spectrometry of bio-organic molecules, after the 'noise' signals had been proven to originate from the organic matrix of the tissues and the epoxy embedding material. The systematic study of all frequently encountered amino acids and some small peptides led to the discovery of the MALDI principle: a soft desorption and efficient ionization of molecules of biological interest ensues if the analyte molecules are embedded in a large excess of small organic molecules (called the matrix) that absorb the laser energy. MALDI, it was later shown, generates abundant ions of biological (macro)molecules such as peptides and proteins up to 1 Million Da, nucleic acids, carbohydrates as well as synthetic polymers with very limited fragmentation (*1*). The advent of MALDI-MS and Electrospray Ionization mass spectrometry (ESI-MS) have revolutionized the application of mass spectrometry in the life sciences (see this volume, *The Background to Electrospray* and *The Development of Electrospray Ionization (ESI) for Mass Spectrometry*).

Hillenkamp received a number of prestigious awards, among them the Award for Distinguished

Contributions in Mass Spectrometry of the American Society for Mass Spectrometry (1998), the Thomson Medal of the International Mass Spectrometry Society (2003), the Fresenius Award of the German Chemical Society (GDCh) (2003) and the Torbern Bergman Medal of the Swedish Chemical Society (2006). He has held Visiting Professorships at a number of universities around the world, including Harvard University Medical School (Boston, USA), Secondá Facoltá di Medicina, Universitá di Napoli, (Italy), Texas A&M University (College Station, Texas, USA), Universität Innsbruck (Austria), and the University of Illinois (Champagne-Urbana, IL, USA).

References cited below are representative of the author's most significant contributions to the field (*1–8*).

Franz Hillenkamp married to Annemarie Hillenkamp née Nobbe since 1963. He was the father of three sons and the grandfather of four grandchildren (as of 2009).

References

(*1*) Hillenkamp, F.; Karas, M. MALDI – An Experience. *Int. J. Mass Spectrom.* **2000**, *200*, 71–77.

(*2*) Karas, M.; Bachmann, D.; Hillenkamp, F. The Influence of the Wavelength in High Irradiance Ultraviolet Laser Desorption Mass Spectrometry of Organic Molecules. *Anal. Chem.* **1985**, *57*, 2935–2939.

(*3*) Karas, M.; Hillenkamp, F. Laser Desorption Ionization of Proteins with Molecular Mass Exceeding 10.000 Da. *Anal. Chem.* **1988**, *60*, 2299–2301.

(*4*) Hillenkamp, F.; Chait, B. T.; Beavis, R. C.; Karas, M. Matrix-Assisted Laser Desorption/Ionization Mass Spectrometry of Biopolymers. *Anal. Chem.* **1991**, *63*, 1193A–1203A.

(*5*) Berkenkamp, S.; Kirpekar, F.; Hillenkamp, F. IR-MALDI of Large Nucleic Acids. *Science* **1998**, *281*, 260–262.

(*6*) Gross, J.; Hahner, S.; Karas, M., *et al.* Investigations of the Metastable Decay of DNA Under UV-MALDI Conditions with Post-Source-Decay Analysis and H/D Exchange. *J. Am. Soc. Mass Spectrom.* **1998**, *9*, 866–878.

(*7*) Horneffer, V.; Forssmann, A.; Strupat, K.; Hillenkamp, F.; Kubitschek, U. Localization of Analyte Molecules in MALDI Preparations by Confocal Laser Scanning Microscopy. *Anal. Chem.* **2001**, *73*, 1016–1022.

(*8*) Horneffer, V.; Strupat, K.; Hillenkamp, F. Localization of Non-covalent Complexes in MALDI Preparations by CLSM. *J. Am. Soc. Mass Spectrom.* **2006**, *117*, 1599–1604.

Alfred L. Yergey
National Institute of Child Health and Human Development, NIH, Bethesda, MD, USA

Heinrich Hintenberger

b. February 7, 1910
d. December 5, 1990

Source: http://www.mpic.de/en/history/chronicle-of-the-directors/chronic-2.html

Heinrich Hintenberger was an Austrian/German physicist who was among the founders of modern mass spectrometry in Germany. His work included studies in ion optics, instrument design, determination of nuclear masses, spark source mass spectrometry, radioactive dating, and analysis of lunar samples and meteorites.

Heinrich Hintenberger was born in Obergrafendorf in Lower Austria. He completed his higher education in physics in 1937 at the University of Vienna, where he worked with Prof. Josef Mattauch (see this volume, *Josef Heinrich Elisabeth Mattauch*). He then worked in the Siemens company's research laboratory in Berlin, which was headed by Gustav Hertz, the cowinner of the 1925 Nobel Prize for physics. In 1943, Hintenberger moved to the Kaiser-Wilhelm-Institut für Chemie in Berlin–Dahlem and joined the physics section, which had been headed by Mattauch since 1939. In the following years, the Institute was moved twice and had its name changed, ending up as the Max-Planck-Institut für Chemie in Mainz associated the new Johannes-Gutenberg-University there, with Mattauch as its overall director. Hintenberger continued with the Institute through the changes and, apart from a two-year sabbatical in Bern, Switzerland, for the entire remainder of his career. In 1959 he became director of the department of mass spectrometry at the MPI für Chemie. He retired in 1978. He and his wife had one son. He died in Mainz in 1990 (*1–4*).

During 1951, Hintenberger and his MPI colleague Heinz Ewald wrote the first German language general text on mass spectrometry, *Methoden und Anwendungen der Massenspektroskopie*, which finally appeared in 1953 (*5*). This was the first of a cluster of such general texts that appeared in the middle of the 1950s (see this volume, *Heinz Ewald*).

Much of his work up to the later 1950s concentrated on question of ion optics, which he addressed both through detailed calculation and precise measurement (*6*). From the end of the 1950s, his research was primarily on cosmochemical topics with a focus on meteorites and lunar samples (see this volume, *The Development of Mass Spectrometry in the Earth and Planetary Sciences*).

In describing Hintenberger's research, his colleague Ludolf Schultz noted in 1991: "The basis of his publications was always measurement carried out with mass spectrometers designed by him and constructed under his continual guidance. The flight tube of his first all-metal machine for the measurement of noble gases, built in 1956, is still in use" (*7*).

Hintenberger and colleagues developed capabilities for measuring minute amounts of substances very exactly. These capabilities facilitated determination of isotopic abundances and precise nuclear masses (*8*). They enabled dating using ^{187}Re decay (*9*). A method to determine exposure ages of iron meteorites was developed (*10*). Primordial noble gases were identified in some meteorites and various trace elements in many others (*11,12*). The groundwork was laid for a generation of improved spark source mass spectrometry (*13,14*). It was a matter of course that Hintenberger was among the original principal investigators – and one of the very few outside the US – who initially received lunar material that had been brought to earth from the first Apollo landing (*15*, see this volume, *The Development of Mass Spectrometry in the Earth and Planetary Sciences*).

References

(*1*) Schultz, L. Heinrich Hintenberger: 7.2.1910–5.12.1990. *MPG Berichte und Mitteilungen* **1991**, *4*, 96–98.

(*2*) Hintenberger, H.; Mattauch, J. Über die im Massenspektrometer ohne Magnetfeld beobachteten "Geister". *Z. Physik* **1937**, *106*, 279–290.

(*3*) Elbert, W. History of the Institute. 2007. In: Max-Planck-Institut für Chemie website, Home/Institute/History: http://www.mpic.de/en/history/overview.html (accessed June 22, 2014).

(*4*) Anon. Directors and research fellows from 1959 until 1967. In: Max-Planck-Institut für Chemie website, Home/Institute/History/Chronicle of the Directors/Chronic 2: http://www.mpic.de/en/history/chronicle-of-the-directors/chronic-2.html (accessed June 22, 2014), page 4.

(*5*) Ewald, H.; Hintenberger, H. *Methoden und Anwendungen der Massenspektroskopie*; Ver. Chemie: Weinheim, 1953.

(*6*) Hintenberger, H.; König, L. A. Mass Spectrometers and Mass Spectrographs Corrected for Image Defects. In: *Advances in Mass Spectrometry*; Waldron, J. D., Ed.; Pergamon Press: London, New York, Paris, Los Angeles, 1959; pp 16–35.

(*7*) Schultz, L. Memorial for Heinrich Hintenerger. *Meteoritics* **1991**, *26*, 172.
(*8*) Hintenberger, H., Ed. *Nuclear Masses and Their Determination, Proceedings of the Mainz Conference*; Pergamon Press: London, New York, Paris, Los Angeles, 1957.
(*9*) Hintenberger, H.; Herr, W.; Voshage, H. Radiogenic Osmium from Rhenium Containing Molybdenite. *Phys. Rev.* **1954**, *95*, 1690–1691.
(*10*) Voshage, H.; Hintenberger, H. Massenspektrometrische Isotopenhäufigkeitsmessungen an Kalium aus Eisenmeteriten und das Problem der Bestimmungen der ^{41}K–^{40}K Strahlungsalter. *Z. Naturforsch.* **1961**, *16a*, 1042–1053.
(*11*) Hintenberger, H.; König, H.; Wänke, H. Uredelgase im Meteoriten Breitscheid. *Z. Naturforsch.* **1962**, *17a*, 306–309.
(*12*) Hintenberger, H.; Jochum, K. P.; Seufert, M. The Concentration of Heavy Metals in Four New Antarctic Meteorites. *Earth Planet Sci. Lett.* **1973**, *20*, 391–394.
(*13*) Franzen, J.; Schuy, K. D.; Hintenberger, H. Sources of Fluctuations in the Mass Spectroscopic Analysis of Solids by Vacuum Techniques. In *Twelfth Annual Conference on Mass Spectrometry and Allied Topics, June 7–12, 1964, Montreal, Canada*; King, A. B., Ed., American Society for Testing and Materials Committee E-14: n. p., 1964, pp 251–270.
(*14*) Cornides, I. Mass Spectrometric Analysis of Inorganic Solids – The Historical Background. In: *Inorganic Mass Spectrometry*; Adams, F., Gijbels, R., Van Grieken, R., Eds.; John Wiley & Sons: New York, 1988; pp 1–15.
(*15*) Hintenberger, H.; Weber, H. W.; Voshage, H., *et al.* Rare Gases, Hydrogen and Nitrogen: Concentrations and Isotopic Composition in Lunar Material. *Science* **1970**, *167*, 543–545.

Keith A. Nier
Independent Historian, Madison, NJ, USA

John Alfred Hipple, Jr.

b. June 22, 1911
d. May 29, 1985

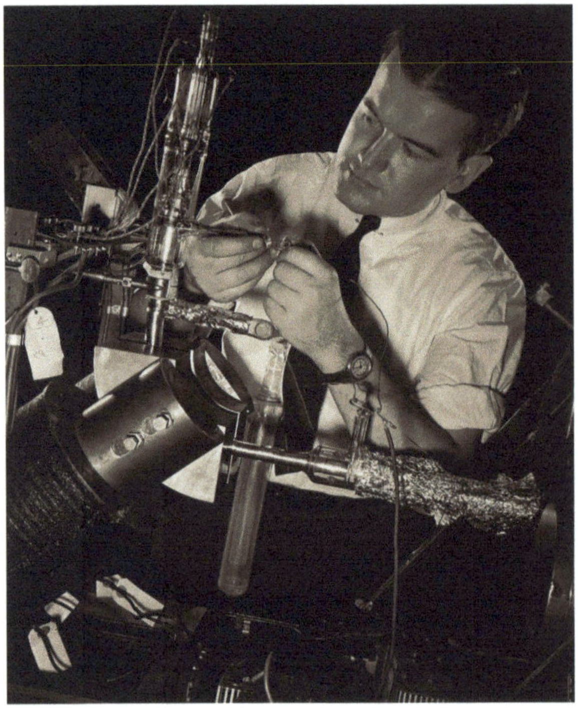

John Hipple in August 1943 with the prototype of his 90° magnetic sector mass spectrometer.

John Hipple was an American physicist who made significant contributions to both the design of mass spectrometers and the understanding of molecular spectra produced by electron ionization. He led the way in the creation of ion cyclotron resonance mass spectrometers.

John Alfred Hipple was born Lancaster, Pennsylvania, US in 1911. His father was a lawyer and he intended to go to law school but became interested in science during his undergraduate education, starting at Penn State and finishing at Franklin and Marshall College in Lancaster, where he received his BS in 1933, with a major in physics (*1,2*). He did his graduate work in physics at Princeton where Walker Bleakney was his advisor (see this volume, *Walker Bleakney*). He published several papers with Bleakney and others and earned his PhD in 1937. Together with Bleakney, he showed that a combination of crossed uniform magnetic and electric fields in principle produces perfect double-focusing in the plane normal to the magnetic field with the ions moving on a trochoidal path. They built two small instruments that were the first to demonstrate this principle, each embodying a different version (*3–5*).

In 1938 Hipple became the first research fellow at the Westinghouse Electric Co., under the supervision of Edward Condon. After the fellowship ended he became a part of the regular scientific staff until 1947, when he left for the National Bureau of Standards, then headed by Condon. While at Westinghouse, he designed a 90° magnetic sector mass spectrometer that was used in analytical work in the wartime program to produce synthetic rubber and that became the basis for Westinghouse's efforts to sell mass spectrometers commercially (*6–8*, and see this volume, *Magnetic and Electrostatic Analyzers before 1960, The Beginnings of Organic Mass Spectrometry*, and *Mass Spectrometry Manufacturers and Markets from the Beginning to 1960*). With D. P. Stevenson he made the first successful direct measurement of ionization potentials of free radicals and with R. E. Fox and Condon he first identified and began research on the phenomenon of metastable ions (*9–11* and see this volume, *David Paul Stevenson* and *Russell Elwell Fox*).

At the Bureau of Standards (now the National Institute of Standards and Technology) in Washington, DC, he became the head of the atomic physics section. He continued his research on electron ionization and metastable ions while directing work on the measurement of isotopic masses. He also continued his interest in innovative instrumental techniques and designs. He explored use of retarding potentials as a means of simulating monochromatic electron energies though the use of differential potentials, thus eliminating the tailing effect on appearance potential measurements, and he was the first to combine spark source analysis of solids with a double-focusing instrument (see this volume, *The Development of Thermal, Spark Source, and Glow Discharge Ionization*). He built a large-scale trochoidal-path mass spectrometer and developed the omegatron, an ion cyclotron resonance mass spectrometer, initially for precise measurements on protons and for determining the value of the Faraday (*12–17*).

Hipple supported and promoted the advance of mass spectrometry in other ways as well. He wrote surveys and introductions highlighting the potentials of such methods for chemists and industrial engineers. He chaired the organizing and editorial committee that held the Bureau's international conference on mass spectrometry in 1951 and published the proceedings. And he began the mass spectrometry series in the decades-long run of regular reviews of fundamental areas of analytical chemistry that started in 1949 (*18–20*).

Hipple left the Bureau in 1953 to become the head of the Mineral Industries Experiment Station at Penn State. In 1957 he became director of research at the Philips Labs Division of the No. American Philips Corp. He retired in 1975.

John Hipple was married; he and his wife had four children. He died in Atlanta, GA, in 1985.

References

(*1*) Biographical Records for John A. Hipple. Archives, National Institute of Standards and Technology, Gaithersburg, MD, USA, 1960 and earlier.

(*2*) *American Men and Women of Science*; R. R. Bowker:New Providence, NJ, USA, 1976; 13th ed., s.v. "Hipple, John Alfred."

(*3*) Bleakney, W.; Hipple, J. A. A New Mass Spectrometer with Improved Focusing Properties. *Phys. Rev.* **1938**, *53*, 521–529.

(*4*) Hipple, J. A. The Dissociation of Ethane by Electron Impact. *Phys. Rev.* **1938**, *53*, 530–533.

(*5*) Kiser, R. *Introduction to Mass Spectrometry and its Applications*; Prentice-Hall: Englewood Cliffs, NJ, USA, 1965. pp 72–81.

(*6*) Hipple, J. A. Portable Mass Spectrometer. *Nature* **1942**, *150*, 111–112.

(*7*) Hipple, J. A. Gas Analysis with the Mass Spectrometer. *J. Appl. Phys.* **1942**, *13*, 551–560.

(*8*) Hipple, J. A.; Grove, D. J.; Hickam, W. M. Electronic Problems Involved in the Practical Application of the Mass Spectrometer. *Rev. Sci. Instr.* **1945**, *16*, 69–75.

(*9*) Stevenson, D. P.; Hipple, J. A. Ionisation and Dissociation by Electron Impact: Normal Butane, Isobutane, and Ethane. *J. Am. Chem. Soc.* **1942**, *64*, 1588–1594.

(*10*) Hipple, J. A.; Condon, E. U. Detection of Metastable Ions with the Mass Spectrometer. *Phys. Rev.* **1945**, *68*, 54–55.

(*11*) Hipple, J. A.; Fox, R. E.; Condon, E. U. Metastable Ions Formed by Electron Impact in Hydrocarbon Gas. *Phys. Rev.* **1946**, *69*, 347–356.

(*12*) Fox, R. E.; Hipple, J. A. Effect of Temperature on the Mass Spectra of Hydrocarbons. *J. Chem. Phys.* **1947**, *15*, 208.

(*13*) Langer, A.; Hipple, J. A.; Stevenson, D. P. Ionization and Dissociation by Electron Impact: Methylene, Methyl, and Methane. *J. Chem. Phys.* **1954**, *22*, 1836–1844.

(*14*) Fox, R. E.; Hipple, J. The Use of Retarding Potentials with the Mass Spectrometer. *Rev. Sci. Instr.* **1948**, *19*, 462–466.

(*15*) Gorman, J. G.; Jones, E. J.; Hipple, J. A. Analysis of Solids with the Mass Spectrometer. *Anal. Chem.* **1951**, *23*, 438–440.

(*16*) Hipple, J. A.; Sommer, H.; Thomas, H. A. A Precise Method of Determining the Faraday by Magnetic Resonance. *Phys. Rev.* **1949**, *76*, 1877–1878.

(*17*) Hipple, J. A.; Sommer, H.; Thomas, H. A. Omegatron. *Phys. Rev.* **1950**, *78*, 332.

(*18*) Hipple, J. A. Applications of the Mass Spectrometer. Frontiers in Chemistry. In: *Recent Advances in Analytical Chemistry*; Burk, R. E., Grummitt, O., Eds.; Interscience: New York, London, 1949; vol. 7, pp 183–203.

(*19*) Hipple, J. A., Ed. Mass Spectrometry in Physics Research, National Bureau of Standards Circular 522; United States Government Printing Office: Washington, DC, 1953.

(*20*) Hipple, J. A.; Shepherd, M. Mass Spectrometry. *Anal. Chem.* **1949**, *21*, 32–36.

Keith A. Nier
Independent Historian, Madison, NJ, USA

John Leonard Holmes

b. November 29, 1931

Photograph taken from University of Ottawa Chemistry Department website: http://www.chemistry.uottawa.ca/details.php?lang=eng&id=209

John Leonard Holmes is an English-born Canadian chemist best known for his work in gas phase ion and neutral thermochemistry. In particular, his focus on elucidation of ion structures through experimentation and theory has led to a better understanding of ion energetics and reactivity. Holmes is also well known for the active role he took in shaping the journal *Organic Mass Spectrometry* as its North American editor over the period 1976–1992 (*1–3*).

John Holmes was born in London in 1931. He obtained his PhD in 1957 from University College London, in the chemistry department that boasted such luminaries as Sir Christopher Ingold, Edward Hughes, Sir Ronald Nyholm, and David Craig. He had received his London BSc in 1954. His PhD studies were on the gas phase pyrolyis of alkyl iodides, directed by Allan Maccoll. The university awarded him a DSc in 1983. He spent a two-year post-doctoral appointment at the National Research Council (NRC) of Canada in Ottawa (1958–1960), followed by two years at the University of Edinburgh. Responding to what he called, 'the delights of Canada and its great outdoors' he took up an academic appointment at the University of Ottawa in 1962, where he remains today, now holding the rank of emeritus professor of chemistry, and still (albeit slowly, he asserts) publishing research (*1*).

Holmes' interest in mass spectrometry began in November, 1964, when he accepted responsibility for the University of Ottawa chemistry department's first mass spectrometer, a Hitachi RMU-6D. The decision to volunteer for this position would have far-reaching consequences: Holmes soon observed that a deeper understanding of the chemistry that lay behind the mass spectra was achievable by experiment. His own work at that time was still focused on the kinetics of free radical reactions and the thermal decomposition of organic compounds, and his connection to mass spectrometry remained only 'service-oriented' until he was asked by Neil Isaacs at Reading University to obtain a spectrum of fumaric acid. Out of curiosity, Holmes decided to compare its spectrum to that of its *cis*-isomer maleic acid. The differences in their fragmentation patterns were marked and clearly demonstrated not only that the molecular ions of the two species did not interconvert but that the losses observed in the spectra reflected the *neutral* [emphasis Holmes'] geometries of the species and so were indeed characteristic of structure. The RMU-6D was particularly well suited for the examination of metastable peaks; their intensities in a normal mass spectrum could be significantly increased by raising the repeller electrode voltages in order to reduce the ions' source residence time. Exploiting this capability, Holmes and co-workers were, for example, able to show that the addition of DCl to norbornene produced 56% of the 3-*exo*-d_1 isomer and 44% of the 7-*syn*-d_1 isomer (*4*), a result then only achievable by high resolution nmr spectroscopy. Holmes was, as he noted some 30 years later, 'hooked' by the fumaric/maleic results (*5*), and these, coupled with the metastable ion findings, set the course for his laboratory for the next 30 or so years.

Success bred success, and in 1971 the NRC soon supported purchase for the department of a double-focusing high resolution mass spectrometer, the AEI (later Kratos) MS902S. A VG-Micromass 7070 (later Waters) and ZAB-2F were granted later that same decade, together with generous operating funds. The MS902S provided Holmes with the ability to examine selected metastable ion peaks under conditions of high energy resolution, leading to the deduction that *inter alia* metastable peak shapes could likely be related to precursor ion structure (*6,7*). This MS9 period also marked the beginning of the long-continuing collaboration and friendship with Hans (J. K.) Terlouw, then at the University of Utrecht, and who spent 1974 in the Holmes laboratory.

Their ZAB in its variously in-house modified forms, allowed mass-selected ions to be subjected to a wide variety of collision-based experiments, including the emission of radiation from collisionally excited ions (*8*). The instrument was now, in Holmes' words, 'an almost complete gas phase ion chemistry laboratory housed in one apparatus.' These instruments were crucial to furthering the central focus of Holmes' research, the structure of ions, because, as he observed, 'Unless one 'knows' the structure of an ion of interest, all conclusions concerning its energy, reactivity or other behaviour must be speculative and are likely to be wrong, especially in view of the perverse habit of ions to rearrange extensively prior to their experimental examination' (*5*).

During the early 1970s' another important and enduring collaboration began with Fred Lossing (see this volume, *Frederick Pettit Lossing*), and between 1976 and Lossing's death in 1998 they coauthored, often with Terlouw, some 35 papers describing their joint work on ion structures and energetics (e.g., *9,10*).

In a 1986 paper, Holmes reviewed and summarized the experimentalist's approach to the problem of assigning structures to ions (*11*). The methods described fall into two main categories – the measurement of ion enthalpies and transition state energies for their fragmentations and the detailed examination of the unimolecular and collision-induced fragmentation behavior of cations, anions and neutral species – and Holmes argued that in general, none 'of those methods *alone can* [emphasis Holmes] suffice for an ion structure determination, but that in combination these techniques provide a powerful tool by means of which ion structures may confidently be assigned.' Some years later, however, he more correctly observed that 'In truth we do not determine ion *structures* [emphasis Holmes'] by experiment, rather we infer a *connectivity* or arrangement of atoms which is consistent with the experimental results, modifying these connectivities as new results are achieved' (*5*). Thus, experimental progress, with its heavy reliance on isotopic labeling and mass spectral interpretation, coupled with qualitative and quantitative (e.g., ionic heats of formation via appearance energy measurements) information, cannot alone provide definitive answers to questions of ionic structure; rather, experimentation must be accompanied by information provided by *ab initio* molecular orbital theory calculations that greatly expand the range of interconverting isomeric forms made plausible by their location on potential energy surfaces. References (*12–17*) exemplify the productive interplay between experiment and theory to which Holmes was pointing.

Retirement has not resulted in a cessation of enjoyable research in gas phase ion chemistry. In particular, the unexpected observation of the *anions* of the alkali metal atoms and a complete revision of the thermochemical additivity tables (pioneered by S. W. Benson) have provided both stimulus and entertainment (*18,19*).

Holmes' work is well documented in over 300 publications. He is an editor of the *Encyclopaedia of Spectroscopy and Spectrometry* and was for over 17 years North American editor of *Organic Mass Spectrometry* in its various incarnations. About the latter he is said to have mused that it was 'a chore that brought new insights into human nature' (*1*). This task generated only one threatening letter, but also resulted in several entertaining series of editor-author communications, entirely in rhyming couplets.

Numerous honors and awards have been accorded Holmes over the course of his career. He has held visiting fellowships and professorships in Australia, Ghana, the Netherlands, and Switzerland. He was made a Fellow of the Royal Society of Canada, and he was the recipient of the Chemical Institute of Canada Medal. In a particularly fitting tribute, the mass spectrometry laboratory at the University of Ottawa now bears his name.

Holmes is an avid sailor, participating actively in the sport (including an Atlantic crossing in 2005), as well as acting as an international judge and assisting in the training of judges. He is a judge emeritus of the Canadian Yachting Association.

References

(*1*) Holmes, J. L.; Aubry, C.; Mayer, P. I. *Author Notes in Assigning Structures to Ions in Mass Spectrometry*; CRC Press, Taylor and Francis Group: Boca Raton, FL, 2007. pp 446.

(*2*) Maccoll, A. John L. Holmes – Editor 1976–1992. *Org. Mass Spectrom.* **1993**, *28*, 995.

(*3*) Mommers, S. Tribute to John Holmes. *Org. Mass Spectrom.* **1993**, *28*, 997–999.

(*4*) Holmes, J. L.; McGillivray, D. L.; Isaacs, N. S. The Mass Spectrum of Exo-2-Norbornyl Chloride; the Elucidation of the Fragmentation Mechanism and Ion Structures by Metastable Ion Analysis. *Can J. Chem.* **1970**, *48*, 2791–2797.

(*5*) Holmes, J. L. Mass Spectrometry and the Pleasures of Science. *Org. Mass Spectrom.* **1993**, *28*, 1388–1394.

(*6*) Holmes, J. L.; Terlouw, J. K. Metastable Ion Studies. V. The Identification of $C_2H_4O^{+\bullet}$· Ion Structures from their Characteristic Kinetic Energy Releases. *Can J. Chem.* **1975**, *53*, 2076–2083.

(*7*) Holmes, J. L.; Terlouw, J. K. Metastable Ion Studies. VI. The Identification of $C_2H_5N^{+\bullet}$ Ion Structures from their Characteristic Kinetic Energy Releases. *Can. J. Chem.* **1976**, *54*, 1007–1014.

(*8*) Holmes, J. L.; Meyer, P. M.; Mommers, A. A. The Emission of Radiation form Collisionally Activated Ions. *J. Am. Chem. Soc.* **1991**, *113*, 9405–9406.

(*9*) Holmes, J. L.; Lossing, F. P.; Terlouw, J. K.; Burgers, P. C. The Radical Cation $[CH_2OH_2]^{+\bullet}$ and Related Stable Gas-Phase Ion-Dipole Complexes. *J. Am. Chem. Soc.* **1982**, *104*, 2931–2932.

(*10*) Holmes, J. L.; Lossing, F. P.; Terlouw, J. K.; Burgers, P. C. Novel Gas-Phase Ions. The Radical Cations $[CH_2XH]^{+\bullet}$ F, Cl, Br, I, OH, NH = (X_2, SH) and $[CH_2CH_2NH_3]^{+\bullet}$. *Can. J. Chem.* **1983**, *61*, 2305–2309.

(*11*) Holmes, J. L. Assigning Structures to Ions in the Gas Phase. *Org. Mass Spectrom.* **1985**, *20*, 169–183.

(*12*) Bouma, W. J.; McLeod, J. K.; Radom, L. An Ab Initio Molecular Orbital Study of the Structures and Stabilities of the $C_2H_4O^{+\bullet}$. Isomers. *J. Am. Chem. Soc.* **1979**, *101*, 5540–5545.

(*13*) Buschek, J. M.; Holmes, J. L.; Terlouw, J. K. $CH_3OCH^{+\bullet}$, A New Stable $C_2H_4O^{+\bullet}$ Isomer, and a Reassessment of the [oxirane]$^{+\bullet}$ Potential Surface. *J. Am. Chem. Soc.* **1987**, *109*, 7321–7325.

(*14*) Yates, B. F.; Bouma, W. J.; Radom, L. Detection of the Prototype Phosphonium (CH_2PH_3), Sulfonium

(CH_2SH_2) and Chloronium (CH_2ClH) Ylides by Neutralization-Reionization Mass Spectrometry: A Theoretical Prediction. *J. Am. Chem. Soc.* **1984**, *106*, 5805–5808.
(*15*) Burgers, P. C.; Holmes, J. L.; Hop, C. E. C. A.; Terlouw, J. K. Gas Phase Ion Chemistry of Methyl Acetate, Methyl Propanoate and their Enolic Tautomers. An Experimental Approach. *Org. Mass Spectrom.* **1986**, *21*, 549–555.
(*16*) Holmes, J. L.; Hop, C. E. C. A.; Terlouw, J. K. Does the Unimolecular Dissociation of Ionized Methyl Acetate Produce $CH_3O^•$ OR $^•CH_2OH$ Radicals? *Org. Mass Spectrom.* **1986**, *21*, 776–778.
(*17*) Heinrich, N.; Schmidt, J.; Schwarz, H.; Apeloig, Y. On the Mechanism of $(C,H_3,O)^•$ Loss from Ionized Methyl Acetate. An Ab Initio Molecular Orbital Study. *J. Am. Chem. Soc.* **1987**, *109*, 1317–1322.
(*18*) Curtis, S.; Renaud, J.; Holmes, J. L.; Mayer, P. M. Old Acid, New Chemistry. Negative Metal Ions Generated from Metal Oxalates and Others. *J. Am. Soc. Mass. Spectrom.* **2010**, *21*, 1944–1946.
(*19*) Holmes, J. L.; Aubry, C. Group Additivity Values for Estimating the Enthalpy of Formation of Organic Compounds: An Update and Reappraisal. 2. C, H, N, O S, and Halogen. *J. Phys. Chem. A.* **2012**, *116*, 7196–7209.

P. Jane Gale
Waters Corp., Milford, MA, USA

Richard Edward Honig

b June 4, 1917
d July 31, 2001

Richard Honig was a physicist who spent the greatest part of his career in the pursuit of materials analysis for the semiconductor industry. In the course of that work he made major contributions to understanding gas phase cluster formation, elemental vapor pressures, and the production and analysis of secondary ions from the bombardment of solids by ions and light (see this volume, *The early history of secondary ion mass spectrometry (SIMS)*). The world class Materials Characterization Group he created and nurtured at RCA Laboratories in the 1970s and '80s set the standard for surface and sub-surface analysis techniques in the industry for many years.

Richard E. Honig was born in Göttingen, Germany, on June 4, 1917, the eldest of three sons of Kaete Heilfron and Richard Martin Honig. Honig senior, a professor of law at the University of Göttingen, was among the first group of professors dismissed from the university by the Nazi regime in 1933. Shortly thereafter, upon receiving an invitation from the Turkish government to help westernize the Turkish educational system, he moved the family to Istanbul. There the young Richard attended a German-language high school, augmenting the classical education he had received in Germany with a math and science curriculum. He went on to attend Robert College, an American college in Istanbul, from which he was graduated with a Bachelor of Science degree in Electrical Engineering in 1938.

Moving to the US with his father later that year, he began a doctoral program in Physics at the Massachusetts Institute of Technology. He studied and taught at MIT from 1938 to 1946 receiving an MS in 1939 and a PhD in 1944. The family later reunited in the US, and his brothers went on to have distinguished academic careers, Jurgen M. Honig as a Professor of Chemistry at Purdue University and Werner K. Honig as a Professor of Psychology at Dalhousie University.

Soon after finishing graduate school, Honig married Doris Jareckie. They raised two children, Daniel and Kaete, in Princeton, NJ, where the family moved in the early 1950s at the beginning of Honig's career at RCA Laboratories. In 1975, he married Joan Huizengen Javay, who remained his partner for the rest of his life. After Rick's retirement from RCA, the couple moved to Haverford, PA. Honig died there on July 31, 2001.

Honig's interest in mass spectrometry began at MIT. To aid in his study of the effects of alpha particles on petroleum-related compounds, he built a 60° magnetic sector mass spectrometer (resolving power of 100). A classic paper that came out of this work entitled 'Gas Flow in the Mass Spectrometer' describes the gas introduction system he designed for that system (*1*). Continuing to use mass spectrometry coupled with different modes of ionization, Honig began to study other materials under conditions of bombardment, an area of research on which he would focus throughout his long and productive career. He published the first paper on cluster formation in desorption from solids (*2*), the first on the effects of contamination on solid surfaces under low energy positive ion bombardment (*3*), and the first on laser desorption from solids (*4*).

In a much-cited paper on sputtering of 1958, he detailed the design of an instrument for the field that would become secondary ion mass spectrometry (SIMS) (*5*). Describing the parameters involved in performing the 'ideal' sputtering experiment, he laid out in this early paper a research agenda that would be pursued over the next 30 years world-wide by laboratories interested in SIMS. As an outgrowth of the *in vacuo* sputtering work, he became interested in measuring elemental vapor pressures as a function of temperature. The data he published first in 1957 (*6*) and then updated in 1963 (*7*) and 1967 (*8*) reprinted as wall charts that soon could be found in laboratories on every continent.

Early on in his career at RCA, Honig realized the importance of producing solid-state electronic devices from well characterized materials, and in the 1960s and '70's, he organized a laboratory to serve this purpose. With his reputation for conducting the highest quality science, he was able to attract first-class experimentalists as staff members. Insisting that the group be at the forefront of developing new techniques, he obtained funds for the group to purchase one of the first commercially available spark source mass spectrometers, to add Auger and photoelectron spectroscopy as tools in their surface analysis arsenal, to design and build a SIMS instrument in 1976 that set the standard for inorganic depth profiling for the next 10 years, and to design and build an instrument in the early '80's that would bring to organic SIMS the same rigor he and his colleagues had imposed on inorganic SIMS a decade earlier. Honig's assessment of the RCA group's contributions to SIMS can be found in three retrospective accounts he published between 1985 and 1995 (*9–11*). Throughout his 25 year tenure as Head of the Materials Characterization Group at RCA, Honig continued to be intimately involved with the science undertaken by the group, actively collaborating with its members in all areas of research.

Honig took professional responsibilities outside his employment as seriously as he did his commitment to the pursuit of scientific research at RCA. He served on the editorial review boards of the *Review of Scientific Instruments, Chemical Instrumentation, Spectrochimica Acta,* and *The International Journal of Mass Spectrometry and Ion Physics.* An active member of the ASTM Committee E-14 on Mass Spectrometry, he chaired the Subcommittee VII on Solids Studies from 1964–1968. He was a founding father of the American Society for Mass Spectrometry (ASMS) and served as its first Vice-President for Programs from 1968–1970, its President from 1970–1972 and its past President from 1972–1974. A naturalized US citizen, Honig remained fluent in German and French and served as Director of the Language Program at the Princeton (NJ) Adult School for nearly 30 years.

Honig was the recipient of the Science Medal from the Vrije Universiteit Brussel, he was elected to membership in the Böhmische Physical Society, he was made a Fellow of the American Physical Society, and he served as Adjunct Research Professor at Rensselaer Polytechnic Institute. In 1990 a special issue with Honor Biography of *The International Journal of Mass Spectrometry and Ion Processes* was dedicated to him. An oral history interview with Honig, supported by ASMS and the Chemical Heritage Foundation, was published by CHF (*12*).

References

(*1*) Honig, R. E. Gas Flow in the Mass Spectrometer. *J. Appl. Phys.* **1945**, *16*, 646–654.

(*2*) Honig, R. E. Mass Spectrometric Study of the Molecular Sublimation of Graphite. *J. Chem. Phys.* **1954**, *22*, 126–131.

(*3*) Honig, R. E. Sputtering of Surfaces by Positive Ion Beams of Low Energy. *Bull. Am. Phys. Soc. Ser. II* **1957**, *2*, 34–35.

(*4*) Honig, R. E. Laser-Induced Emission of Electrons and Positive Ions from Metals and Semiconductors. *Appl. Phys. Lett.* **1963**, *3*, 8–11.

(*5*) Honig, R. E. Sputtering of Surfaces by Positive Ion Beams of Low Energy. *J. Appl. Phys.* **1958**, *29*, 549–555.

(*6*) Honig, R. E. Vapor Pressure Data for the More Common Elements. *RCA Rev.* **1957**, *18*, 195–204.

(*7*) Honig, R. E. Vapor Pressure Data for the Solid and Liquid Elements. *RCA Rev.* **1962**, *23*(4), 567–586.

(*8*) Honig, R. E.; Kramer, D. A. Vapor Pressure Data for the Solid and Liquid Elements. *RCA Rev.* **1969**, *30*(2), 285–305.

(*9*) Honig R. E., The Development of Secondary Ion Mass Spectrometry (SIMS): A Retrospective, 1985, *66*, pp 31–54.

(*10*) Honig, R. E. The Growth of Secondary Ion Mass Spectrometry (SIMS): A Personal View of Its Development. In: *Springer Series in Chemical Physics (Secondary Ion Mass Spectrometry SIMS 5)*; Benninghoven, A., Colton, R. J., Simons, D. S., Werner, H. W., Eds.; Springer: New York, 1986; pp 2–15.

(*11*) Honig, R. E. Stone-Age Mass Spectrometry: The Beginnings of SIMS at RCA Laboratories, Princeton. *Int. J. Mass Spectrom. Ion Processes* **1995**, *143*, 1–10.

(*12*) Honig R. E., Interview by Grayson M. A. at The Quadrangle, Haverford, Pennsylvania, 27 April 1996 (Philadelphia: Chemical Heritage Foundation, Oral History Transcript # 0678).

P. Jane Gale
Waters Corp., Milford, MA, USA

Bryan L. Bentz
Waters Corp., Milford, MA, USA

Evan Charles Horning and Marjorie Groothuis Horning

Evan C. Horning
b. 6 June, 1916
d. 14 May, 1993

Marjorie G. Horning
b. 23 August, 1917

Evan Charles and Marjorie Groothuis Horning were chemists/biochemists whose technique development in chromatography and mass spectrometry was pivotal to the analysis of biological compounds and metabolic systems.

Evan Charles Horning was born in Philadelphia, PA. He received his BS from the University of Pennsylvania in 1937. His graduate work at the University of Illinois under the supervision of R. C. Fuson earned him a PhD in 1940. He was an instructor at Bryn Mawr College (1940–1941) before becoming an instructor (1941) and then a research associate in the Department of Chemistry at the University of Michigan. In 1945 he joined the faculty of the University of Pennsylvania and in 1950 became Chief of the Laboratory of the Chemistry of Natural Products at the National Heart Institute, National Institute of Health. In 1961 he accepted the position of Professor of Chemistry at the Baylor College of Medicine, and in 1966 the Director of the Institute for Lipid Research there. In 1971 he also accepted the post of Adjunct Professor of Biochemistry at Rice University.

Marjorie J. Groothuis, later Horning, was born in Detroit, Michigan. She completed her undergraduate studies at Goucher College, Baltimore, MD earning a BS in 1938. Her graduate work on phospholipid metabolism was mentored by Henry C. Ekstein in the Department of Biological Chemistry in the University of Michigan Medical School, where she earned an MS in 1940 and a PhD in 1943. She worked as a research chemist at the University of Pennsylvania between 1945 and 1951 and at the National Heart Institute at the National Institute of Health between 1951 and 1961. She joined the Department of Biochemistry at the Baylor College of Medicine as an associate professor in 1961 and was promoted to full professor in the Institute of Lipid Research in 1969. In 1981, she became Adjunct Professor of Biochemistry at the University of Houston.

When Evan and Marjorie married in the 1940s, they established a partnership that would serve them well both in life and in their scientific pursuits. The two were life-long collaborators, publishing more than 200 papers together over a period of 50 years. Evan's initial interest in organic chemistry and Marjorie's in biochemistry laid the foundation for many of the insights they would bring to lipid and steroid research as well as to metabolic studies. Their interest and support facilitated many important developments in analytical tools, and they were early adopters of the new technologies of gas chromatography (GC) and mass spectrometry (MS) (see this volume, *A History of Gas Chromatography Mass Spectrometry (GC/MS)*). Evan ensured production of the first commercial instrument for gas chromatography in 1957 by producing orders for 3 in ten minutes' time (one for himself and one each for David Turner at Mt. Sinai Hospital and S. R. Lipsky at Yale School of Medicine) (*1*). In 1969, the first digital computer-controlled GC/MS commercially produced in the U.S. and the world, a Finnigan Model 1015 GC/MS etc. and Systems Industries System 150, was introduced at the fall ACS exhibit in New York City and then delivered to the Hornings' laboratory at Baylor College of Medicine Institute of Lipid Research following the meeting (*2*). In the early 1970s, the Horning group pioneered the development of the atmospheric pressure ionization source (API), used together with the quadrupole mass spectrometer to carry out GC/MS and LC/MS research analyses. This technique is used routinely today in many biomedical and biopharmaceutical applications throughout the world. During much of their careers, the Hornings were very helpful to start-up analytical instrument companies, both in helping to define required instrument capabilities and by purchasing these advanced instruments for their research programs.

With these tools, the Hornings and their colleagues developed procedures for separation of sugars as acetyl derivatives and steroids as trifluoroacetoxy compounds (*3*). They also were successful in using gas chromatography for studying cholesterol esters and separation of sapogenins (*4,5*). These investigations produced new liquid phases and derivatives for gas chromatography separations of steroids (*6*). GC/MS analysis of steroids in human fluids – urine, bile and spinal liquids – and amino acid derivatives followed (*7*). They also isolated acids, alcohols, and waxes from human skin, and their techniques were later extended to investigations of the metabolism (in the cited article, they coin the term 'metabolic profile') of many different compounds in humans (*8,9*).

Evan received numerous honors and awards including a Guggenheim Memorial Fellowship at the University of London Post Graduate Medical School, the Torbern Bergman Award of the Swedish Society of Chemists, Dr. h.c., at the Karolinska Institute, Stockholm, Sweden, M. S. Tswett Medical in Chromatography, Warner-Lambert (General Diagnostics) Award of the American Association of Clinical Chemists, C. W. Scheele Award, Pharmaceutical Society of Sweden; Dr. h.c. at the University of Ghent, Ghent, Belgium; ACS Award in Chromatography; Founders Award, Chemical Industry Institute of Toxicology; Tswett Commemorative Medal, Soviet Academy of Sciences; and the S. S. Dal Nogare Award in Chromatography.

Marjorie was also much honored during her long career. She received an honorary doctorate from Goucher College in 1977, the Warner Lambert (General Diagnostics) Award of the American Association of Clinical Chemists in 1976, The Garvan Medal of the American Chemical Society in 1977, the Founders' Award of the Chemical Industry Institute of Toxicology in 1979, the Alumnae Athena Award of the University of Michigan in 1980, and the Tswett Chromatography Medal in 1987.

In 1990 Marjorie and Evan shared the Frank H. Field and Joe Franklin Award in Mass Spectrometry of the American Chemical Society.

The two were members of many scientific and honorary societies and served on several editorial boards. Among these are Organic Syntheses, Journal of Medicinal Chemistry, Chemical Reviews, Analytical Biochemistry, Amlytical Letters, Journal of Chromatography, Advances in Lipid Research, Life Sciences, and Journal of Atherosclerosis Research.

References

(*1*) Ettre, E. S.; Zlatkis, A. 75 Years of Chromatography. In: *Journal of Chromatography Library*; Elsevier: Amsterdam, The Netherlands, 1979; vol. 17, pp. 141–150.

(*2*) See this volume: Finnigan, R. E. Development of the Business of Mass Spectrometry from 1960 to 1975.

(*3*) Vandenheuvel, W. J.; Horning, E. C. Gas Chromatographic Separation of Sugars and Related Compounds as Acetyl Derivatives. *Biochem. Biophys. Res. Commun.* **1961**, *4*, 399–403.

(*4*) Horning, M. G.; Williams, E.; Horning, E. C. Separation of Tissue Cholesterol Esters and Triglycerides by Silicic Acid Chromatography. *J. Lipid Res.* **1961**, *1*, 482–485.

(*5*) Maume, B. F.; Wilson, W. E.; Horning, E. C. Gas Chromatographic and Mass Spectrometric Study of Trimethylsilyl Ethers of Cardiac Aglycones. *Anal. Lett.* **1968**, *1*, 401–415.

(*6*) Chambaz, E. M.; Horning, E. C. Conversion of Steroids to Trimethylsilyl derivatives for Gas Phase Analytical Studies: Reactions of Silylating Reagents. *Anal. Biochem.* **1969**, *30*, 7–24.

(*7*) Horning, M. G.; Horning, E. C.; Carroll, D. I.; Dzidic, I.; Stillwell, R. N. New Picogram Detection System Based on a Mass Spectrometer with an External Ionization Source at Atmospheric Pressure. *Anal. Chem.* **1973**, *45*(6), 936–943.

(*8*) Haathi, E.; Horning, E. C. Isolation and Characterization of Saturated and Unsaturated Fatty Acids and Alcohols of Human Skin Surface Lipids. *Scand. J. Clin. Lab. Invest.* **1963**, *15*(1), 73–78.

(*9*) Horning, M. G.; Horning, E. C. Human Metabolic Profiles Obtained by GC and GC/MS. *J. Chromatogr. Sci.* **1971**, *9*(3), 129–140.

Robert Finnigan
Thermo Electron Corporation (ret.)

P. Jane Gale
Waters Corp., Milford, MA, USA

Robert Samuel Houk

b. November 23, 1952

Robert Samuel 'Sam' Houk is an American chemist specializing in elemental and isotopic analysis by mass spectrometry. In his work with inductively coupled plasma-mass spectrometry (ICP-MS) he has made pivotal contributions to understanding the fundamentals of the technique, the ion extraction process, the use of ICPMS with chemical separations and to the development of instrumental improvements. Other areas of work include electrospray ionization MS; laser ablation ICP-MS and laser desorption MS; and ion-ion reactions and ion mobility MS.

Houk was born in 1952 in Energy, Pennsylvania, USA. He graduated from Slippery Rock State College in 1974 with a B.S. in Secondary Education. His student teaching, done at Union Area High School under Socrates Ruossos, kindled an interest in teaching that has remained important to him throughout his career. He then did graduate work at Iowa State University under the joint supervision of Velmer A. Fassel and Harry J. Svec, culminating in the PhD in analytical chemistry in 1980. He continued work on ICP-MS as a postdoctoral associate at Ames Laboratory until he joined the faculty at Iowa State in 1982.

Houk built the first ICP-MS device for elemental analysis (see this volume, *The Development of Inductively-Coupled Plasma Mass Spectrometry (ICPMS)*). He did the initial work on many of the important subsequent developments in this area, particularly GC-ICP-MS, LC-ICP-MS, collision cells, and low-flow nebulizers. This body of work is widely used in the various scientific applications of ICP-MS. He also did important work in ICP emission spectroscopy and the initial work on ion association (since renamed covariance mapping). Recent areas of interest are laser ablation ICP-MS, atmospheric pressure laser desorption MS, inorganic electrospray ionization MS, and the combination of ion–ion reactions with ion mobility MS.

Houk was an associate editor in the early days of the *Journal of the American Society for Mass Spectrometry*, served on the editorial boards of the *Journal of Analytical Atomic Spectrometry* and *Spectrochimica Acta Part B*, and was a guest editor for the V. A. Fassel Memorial Issue of the latter publication. He was a Fellow of the Society for Applied Spectroscopy (2007), and has received the Society for Applied Spectroscopy's Lester W. Strock Award (1986), the Spectroscopy Society of Pittsburgh's Maurice F. Hasler Award (1993), the ACS Award for Chemical Instrumentation (1993), the Association for Analytical Chemistry's Anachem Award (2000), and the ACS Award for Spectrochemical Analysis (2012). He has held guest professorships and lectureships at Wuhan University, Masaryk University, and Concordia College.

Houk is married to Linda Lembke, whom he met in the lab when he was a graduate student at Iowa State. They have one son and one daughter. He is interested in sports, fishing, and history. His current position is Professor of Chemistry and Senior Scientist at Ames Laboratory, Iowa State University, in Ames, Iowa.

The publications cited below are representative of the author's most significant contributions to the field (*1*–*5*).

References

(*1*) Houk, R. S.; Fassel, V. A.; Flesch, G. D.; Svec, H. J.; Gray, A. L.; Taylor, C. E. Inductively Coupled Argon Plasma as an Ion Source for Mass Spectrometric Determination of Trace Elements. *Anal. Chem.* **1980**, *52*, 2283–2289.

(*2*) Thompson, J. J.; Houk, R. S. Inductively Coupled Plasma Mass Spectrometric Detection for Multielement Flow Injection Analysis and Elemental Speciation by Reversed Phase Liquid Chromatography. *Anal. Chem.* **1986**, *58*, 2541–2548.

(*3*) Rowan, J. T.; Houk, R. S. Attenuation of Polyatomic Ion Interferences by Gas Phase Collision Processes. *Appl. Spectrosc.* **1989**, *43*, 976–980.

(*4*) Wiederin, D. R.; Smith, F. G.; Houk, R. S. Direct Injection Nebulization for ICP-MS. *Anal. Chem.* **1991**, *63*, 219–225.

(*5*) Zhao, Q.; Schieffer, G. M.; Soyk, M. W.; Anderson, T. J.; Houk, R. S.; Badman, E. R. Effects of Ion/Ion Proton Transfer Reactions on Conformation of Gas-Phase Cytochrome c Ions. *J. Am. Soc. Mass Spectrom.* **2010**, *21*, 1208–1217.

R. Sam Houk
Chemistry Department, Iowa State University, Ames, IA, USA

Donald Frederick Hunt

b. April 25, 1941

Donald (Don) F. Hunt was born on April 25, 1941 in Hyannis, USA. He attended the University of Massachusetts where he received BS (1962) and Ph.D. (1967) degrees for studies on organotransition metal chemistry with Marvin Rausch and Peter Lillya. Hunt's introduction to mass spectrometry came during an NIH postdoctoral fellowship (1967–68) with Klaus Biemann at MIT Hunt (see this volume, *Klaus Biemann*).

In 1968, Hunt accepted an assistant professorship in the Chemistry Department at the University of Virginia (UVa) in Charlottesville. He was promoted to associate professor in 1973, full professor in 1978, and University Professor with appointments in Chemistry and Pathology in 1993. Hunt has maintained his UVa affiliation throughout the course of his innovative career. In a bucolic setting just west of Charlottesville where they continue to reside, he and his wife psychologist Dr. Linda C. Hunt, raised two daughters, Amanda and Caroline.

Upon arrival in Charlottesville as an assistant professor in 1968, Hunt also took on responsibility for and supervision of the UVa Chemistry Department mass spectrometry facility. As part of that arrangement, the university agreed to acquire a chemical ionization source to go along with their new high resolution mass spectrometer. In the early years of his long association with UVa, Hunt productively utilized this instrumentation to describe new reagents for the characterization of diverse functional groups in model organic molecules (*1*).

When Hunt's graduate student George Stafford implemented negative ion detection on a quadrupole mass spectrometer in Hunt's laboratory at the University of Virginia, Hunt's group continued along this productive line of investigation and instrument development using a technique to measure simultaneously positive and negative ions (*2*, see this volume, *George C. Stafford, Jr.*). With their results, Hunt soon convinced the Finnigan Corporation of the potential commercial value of negative chemical ionization coupled with gas chromatography on their quadrupole mass analyzer. Hunt's relationship with the instrument manufacturer would continue to ripen, with the Hunt lab devising new instruments and techniques over next thirty years, many of which were rapidly implemented commercially by the Finnigan Corporation (now ThermoFisher).

During a 1975 sabbatical in Professor Dudley William's laboratory at the University of Cambridge, Hunt began to develop peptide sequence strategies using negative CI, then a graduate student of Chris Enke at the University of Michigan reported on Jim Morrison's new multi-pentaquadrupole instrumentation in Australia, Hunt determined to implement this strategy at the University of Virginia (see this volume, *Richard A. Yost, Christie George Enke, James Douglas Morrison, and The Development of Tandem Mass Spectrometry*). Jeff Shabanowitz, Hunt's long time collaborator and instrumentation specialist, constructed a functional triple stage quadrupole (TSQ) in less than 2 months by disassembling (much to the dismay of other graduate students) three working single stage instruments. At the 1979 ASMS meeting, Hunt demonstrated the value of analyzing derivatized peptides with isobutane CI on a TSQ (*3*). The Finnigan Corporation introduced its commercially successful TSQ series in 1980.

Hunt quickly realized that fast atom bombardment (FAB), introduced by Michael Barber in 1981, should be ideal for determination of underivatized peptide structures when combined with tandem mass spectrometry and implemented this strategy to sequence peptides (*4*, see this volume, *Michael Barber*). Hunt, in collaboration with R. T. McIver's laboratory, implemented external ionization ion optics by coupling a TSQ onto an FTICR to capitalize on the potential of using underivatized peptides (*5*). After a sabbatical year in the Howard Morris lab at Imperial College, Hunt concentrated fully on protein biochemistry with a new graduate student, John Yates (*6*, see this volume, *Howard Redfern Morris*). Hunt quickly added electrospray ionization (ESI) to the triple quadrupole mass spectrometers and through collaboration in 1991 with Arthur Moseley in Jim Jorgenson's laboratory, adapted microcolumn liquid chromatography to ESI. This combination of technologies led to significant increases in sensitivity and capability for the analysis of complex peptide mixtures and was widely adopted throughout the world of protein biochemistry (*7*). Hunt's lab had a significant role in teaching and popularizing peptide sequence analysis to protein biochemists worldwide through semi-annual 3½ day courses they offered at UVa fifteen times between 1989 and 1987, eventually reaching more than 400 participants.

Beginning in 1990, Hunt sought direct clinical application of the power of mass spectrometric protein analyses in the identification of cell surface antigens, the nine residue peptides that trigger a patient's cytotoxic T-cells. His laboratory has studied various aspects of the cellular immunologic response and discovered many antigens associated with a variety of diseases (*8*,*9*).

The introduction of electron capture dissociation as a tool for protein analysis demonstrated the significant advantage of odd electron dissociation chemistry. Hunt sought to capitalize on his experience with negative ion chemical ionization, and initiated a successful project to couple a CI source with a tandem linear ion trap (*10*). The resulting instrumentation provides facile complementary structure information to collision-induced dissociation and is especially powerful for characterizing phospho- and glycopeptides and proteins (*11*).

The diverse achievements throughout Don Hunt's career have been recognized with many fellowships and awards along the way. He was chosen as a recipient of both an NIH Fogarty Senior International Fellowship and a John Simon Guggenheim Fellowship in 1981–82. In 1990, he received the Charles H. Stone Award sponsored by the American Chemical Society. In 1992 he was named Virginia's Outstanding Scientist and also received the Pehr Edman Award for outstanding achievements in the application of mass spectrometry to the contemporary microsequence analysis of proteins. The Distinguished Contribution Award from the American Society for Mass Spectrometry was presented to Hunt in 1994 for his development of electron-capture, negative-ion mass spectrometry. In 1996 he was the first recipient of the Christian B. Anfinsen Award from the Protein Society for development of new technology in the field of protein chemistry. He received the Chemical Instrumentation Award sponsored by the American Chemical Society in 1997, recognizing his development of instrumentation capable of sequencing peptides and proteins at the attomole level. In 2000, Hunt was the recipient of both the Frank F. Field and Joe L. Franklin award presented by the American Chemical Society for outstanding achievement in the field of mass spectrometry and the Thomson Medal from the International Mass Spectrometry Society. Hunt received Distinguished Accomplishment Awards from the Human Proteome Organization (HUPO) in 2006 and the Association of Biomolecular Resource Facilities in 2007. He received the Distinguished Scientist Award at the University of Virginia in 2010 and the UVA School of Medicine Award for Excellence in Team Science in 2012.

Hunt's contributions to many aspects of mass spectrometry over the course of his long and extremely productive career were celebrated on the occasion of his 65th birthday with an honor volume of the International Journal of Mass Spectrometry (*12*).

References

(*1*) Hunt, D. F.; Ryan, J. F. Chemical Ionization Mass Spectrometry Studies I. Identification of Alcohols. *Tetrahedron Lett.* **1971**, 4535–4538.

(*2*) Hunt, D. F.; Stafford, G. C. Jr; Crow, F. W.; Russel, J. W. Pulsed Positive Negative Ion Chemical Ionization Mass Spectrometry. *Anal. Chem.* **1976**, *48*, 2098–2105.

(*3*) Hunt, D. F.; Shabanowitz, J.; Giordani, A. B. Collision Activated Decompositions of Negative Ions in Mixture Analysis with a Triple Quadrupole Mass Spectrometer. *Anal. Chem.* **1980**, *52*, 386–390.

(*4*) Hunt, D. F.; Bone, W. M.; Shabanowitz, J.; Rhodes, G.; Ballard, J. M. Sequence Analysis of Oligopeptides by Secondary Ion Collision Activated Dissociation Mass Spectrometry. *Anal. Chem.* **1981**, *53*, 1704–1706.

(*5*) Hunt, D. F.; Shabanowitz, J.; McIver, R. T.; Hunter, R. L.; Syka, J. E. P. Ionization and Mass Analysis of Nonvolatile Compounds by Particle Bombardment/Tandem Quadrupole Fourier Transform Mass Spectrometry. *Anal. Chem.* **1985**, *57*, 768–771.

(*6*) Hunt, D. F.; Yates, J. R. III; Shabanowitz, J.; Winston, S.; Hauer, C. R. Protein Sequencing by Tandem Mass Spectrometry. *Proc. Natl. Acad. Sci. USA* **1986**, *83*, 6233–6237.

(*7*) Hunt, D. F.; Shabanowitz, J.; Yates, J. R. III; Griffin, P. R.; Zhu, N. Z. Protein Sequence Analysis by Tandem Mass Spectrometry: New Methods and Instrumentation. In: *Analysis of Peptides and Proteins*; McNeal, C., Ed.; John Wiley and Sons: New York, NY, 1988; pp 151–165.

(*8*) Hunt, D. F.; Henderson, R. A.; Shabanowitz, J., *et al.* Characterization of Peptides Bound to the Class I MHC Molecule, HLA-A2.1. *Science* **1992**, *255*, 1261–1263.

(*9*) Cox, A. L.; Skipper, J.; Chen, Y., *et al.* Identification of a Peptide Recognized by Five Melanoma Specific Human Cytotoxic T-Cell Lines. *Science* **1994**, *264*, 716–719.

(*10*) Syka, J. E. P.; Coon, J. J.; Schroeder, M. J.; Shabanowitz, J.; Hunt, D. F. Peptide and Protein Sequence Analysis by Electron Transfer Dissociation Mass Spectrometry. *Proc. Natl. Acad. Sci. USA* **2004**, *101*, 9528–9533.

(*11*) Coon, J. J.; Ueberheide, B.; Syka, J. E. P., *et al.* Protein Identification Using Sequential Ion/Ion Reactions and Tandem Mass Spectrometry. *Proc. Natl. Acad. Sci. USA* **2005**, *102*, 9463–9468.

(*12*) Donald, F. Hunt Honour Issue. *Int. J. Mass Spectrom.* **2007**, *259*, 1–208.

P. Jane Gale
Waters Corp., Milford, MA, USA

Sanford P. Markey
Laboratory of Neurotoxicology, National Institute of Mental Health, Bethesda, MD, USA

Mark G. Inghram

b. November 13, 1919
d. September 29, 2004

Mark G. Inghram was an American physicist who greatly influenced the development and application of mass spectrometry in several fields. He was born in Livingston, Montana, the son of Mark Gordon Inghram and Luella McNay Gallagher. He obtained a B.A. degree from Olivet College in Michigan in 1939 and a PhD from the University of Chicago in 1947. In 1946, he was married to Evelyn Mae Dyckman; the couple had two children, Cheryl Ann, Mark Gordon III.

During his PhD studies, he was engaged in the Manhattan Project at Columbia from 1942 to 1945, using his specialty of mass spectrometry to assist the project in studies of transuranic elements. His early work, done at the University of Chicago and Argonne National Laboratory, focused on applying mass spectrometric techniques to problems in nuclear physics and geochemistry. This included an early study with John Reynolds of double beta-decay in Te^{130} (*1*, and see this volume, *John H. Reynolds*).

Returning to the University of Chicago as a Physics instructor in 1947, Inghram and his group went on to develop techniques to ascertain the decay products in meteorite and other rock samples from which they were able to place an early estimate of the age of the Earth, work that in 1957 earned him the J. Lawrence Smith Medal of the National Academy of Sciences. Along with Clair Patterson and George Tilton, Inghram ascertained lead isotope ratios in uranium samples from various locations from which they estimated the Earth's age to be approximately 4.5 billion years (see this volume, *Clair Cameron Patterson, George Robert Tilton*, and *The Development of Mass Spectrometry in the Earth and Planetary Sciences*). They reported their result in the January 1955 issue of *Science* magazine (*2*). This result was commonly used for decades as the accepted age of the Earth (*3*) and retains validity to this day.

Building on the techniques developed for the meteorite work and age of the Earth effort, Inghram's group helped develop high temperature sources for mass spectrometers that allowed the study of refractory materials such as aluminum oxide, gallium oxide and carbon (*4*). "This technique was first applied to carbon vapor, resulting in the unambiguous determination of the heat of vaporization of the carbon atom from graphite. At the time, this quantity was a matter of much contention and importance since it was required for the determination of bond energies of all organic molecules," remembered collaborator Bill Chupka (*5*). This work provided tools to determine the heats of formation of the high temperature species that result from the vaporization of the metallic oxides in the vacuum, an important contribution to the knowledge and understanding of high temperature chemistry.

During the same time, the late 1950s and early 1960s, Inghram's group at Chicago in collaboration with scientists at Argonne National Laboratory, developed other types of ionizations sources to create ions for mass spectrometric analysis. The photoionization techniques they pioneered allowed them to study a wide range of atomic and molecular systems. Gas discharge tubes coupled to a monochromator with appropriate windows or differential pumping to allow far ultraviolet light (down to around 60 nm) served as the ionization source, providing monochromatic light that ionized over a narrow range of energies and thus gave much better energy resolution in the energetics of the ionization process (*6*). This in turn provided better and more accurate information on gas phase ion chemistry and the energetics of fragmentation and molecular dissociations. Inghram and his students studied a number of different molecular and atomic systems over the years using the photoionization technique and also developed specialized spectroscopies for studying the threshold electron emitted at ionization onset and the molecular fragments in coincidence. These techniques, called threshold photoelectron spectroscopy and threshold photoelectron photo ion coincidence mass spectroscopy, were useful and successful tools in exploring the energetics of molecular ionization and of excited state dynamics (*7*).

During this same period Inghram and his collaborators, including Robert Gomer of the University of Chicago's Chemistry Department, developed field ionization sources for mass spectrometric application. The field ionization technique provided the ability to study the energetics of surface phenomena and to explore the nature of the molecule surface interaction in a high electric field (*8*). It also afforded the opportunity to examine the ionization phenomena from localized regions of a surface on the scale of atomic dimension, as well as to examine exotic species formed in high electric fields (see this volume, *A Historical Perspective on Field Ionization (FI) and Field Desorption (FD) Mass Spectrometry*).

Inghram was well respected as an experimental scientist and was credited by his collaborators with many innovations in mass spectrometry, including the use of particle detection techniques that greatly increased the sensitivity of mass spectrometer studies (*9*). Gerald Wasserburg said, "Leaders of the new fields of cosmochemistry and geochemistry were trained in his laboratories under his rigorous guidance and mentorship. His methods and approaches and instrument designs governed the research approach in laboratories throughout the world" (*10*).

In the later part of his career Inghram was the chair of the Physics Department at Chicago (1959–1970) and held other positions in the University, including the Associate Dean of the Physical Sciences Division (1964–1971). He was appointed the Samuel Allison Distinguished Service Professor in Physics in 1969 and retired in 1985. He was a member of the National Academy of Sciences (an honor also accorded to five of his former students) and served on two committees of its committees: Nuclear Geophysics (1953–1960), and Exploration of the Moon and Planets (1958–1961). He was also a fellow of the American Academy of Arts and Letters, the American Physical Society and the American Association for the Advancement of Science. He received the University's Llewellyn John and Harriet Mancheer Quantrell Award for Excellence in Undergraduate Teaching in 1981 (*5*).

References

(*1*) Inghram, M. G.; Reynolds, R. J. On the Double Beta Decay Process. *Phys. Rev.* **1949**, *78*, 1265–1266.
(*2*) Patterson, C.; Tilton, G.; Inghram, M. Age of the Earth. *Science* **1955**, *121*, 69–75.
(*3*) Patterson, C. Age of Meteorites and the Earth. *Geochim. Cosmochim. Acta* **1956**, *10*, 230–237.
(*4*) Chupka, W. A.; Inghram, M. G. Direct Determination of the Heat of Sublimation of Carbon. *J. Chem. Phys.* **1954**, *22*, 1472.
(*5*) Anonymous. In Memoriam: Mark G. Inghram (1920–2003) [sic]. *The Geochemical News*, **2004**, *118*, 7.
(*6*) Hurzeler, H.; Inghram, M. G.; Morrison, J. D. Photon Impact Studies of Molecules Using a Mass Spectrometer. *J. Chem. Phys.* **1958**, *28*, 76–82.
(*7*) Villarejo, D.; Herm, R.; Inghram, M. G. Measurement of Threshold Electrons in the Photoionization of Ar, Kr, and Xe. *J. Chem. Phys.* **1967**, *46*, 4995–4997.
(*8*) Inghram, M. G.; Gomer, R. Mass Spectrometric Investigation of the Field Emission of Positive Ions. *Z. Naturforsch.* **1955**, *10a*, 863–872.
(*9*) Inghram, M. G.; Hayden, R. J. *A Handbook on Mass Spectroscopy*; Nuclear Science Series Report No. 14; National Research Council Publication No. 311, National Academy of Sciences: Washington, DC, 1954.
(*10*) University of Chicago News Office. Mark G. Inghram, Physicist, 1919–2003. http://www-news.uchicago.edu/releases/03/031002.inghram.shtml

Albert Parr
National Institute of Science and Technology, Washington, DC, USA

Roger Stockbauer
Louisiana State University, Baton Rouge, LA, USA

Keith R. Jennings

b. December 5, 1932

Keith Jennings is a British chemist who has used multiple types of mass spectrometry in many areas of chemistry. He has continually sought to devise, adopt, and promote new methods and instrumentation. He was a pioneer in the deliberate use of collision induced dissociation and in the study and use of negative ions. He was one of the first to spread ion cyclotron resonance (ICR) mass spectrometry outside the US. And he has been very active in spreading an understanding of the capabilities and techniques of mass spectrometry beyond the community of specialists.

Keith Jennings was born in Sheffield, UK, in 1932. He was a student at the King Edward VII Grammar School there and won a scholarship to Queen's College in Oxford University. He earned his BA with First Class Honors in chemistry in 1956 and his D.Phil. in 1958 under the supervision of Wilfrid (Jack) Linnett. After some further work with Linnett, Jennings spent two years as a post-doctoral researcher in Ottawa, Canada. He returned to England in 1960 to take a position as lecturer in chemistry at the University of Sheffield, where he stayed for a dozen years. In 1972 he moved the University of Warwick and stayed there through his official 1997 retirement and beyond, though he changed his position from Professor of Physical Chemistry to Emeritus Research Professor in the Department of Biological Sciences. Despite his extensive administrative and other activities, he remained active in research throughout the decades (*1–3*).

Jennings first began to work with a mass spectrometer at Sheffield when he obtained an MS10 in 1963. Spurred in part by the characteristics of the instrument, he gave much attention to metastable ions. This led, once an MS9 double-focusing instrument was at hand, to explorations of the production of metastables through collisions with gas in the field-free region between the electrostatic and magnetic analyzers, and to a multitude of further studies by him and by others.

Over the decades that followed Jennings has studied a host of substances and explored many modes of mass spectrometry; a modest sample from his hundreds of publications is listed below to suggest the scope of his investigations (*4–22*). His early involvement with ion cyclotron resonance mass spectrometry is noteworthy as are his emphasis on chemical ionization and negative ions, his steady concern with reaction kinetics, and his more recent additions of four-sector and electrospray technologies to his repertoire (see this volume, *A History of Ion Cyclotron Resonance (ICR) and Fourier Transform (FTICR) Mass Spectrometry*). An overall shift has been from physical chemistry to structural studies of proteins of biological import. The citations listed here do not come close to exhausting the list of his topics of interest (*23*).

Throughout his career, Jennings has also devoted much time in service to the mass spectrometry community through educational efforts both in and outside regular university coursework, through administrative responsibilities, and work on the editorial boards of many journals. He has been widely honored for his research and service, most obviously through one of the inaugural, Thomson Medals at the International Mass Spectrometry Conference in 1985, the Distinguished Contribution in Mass Spectrometry Award from the American Society for Mass Spectrometry, the Field and Franklin Award from the American Chemical Society, and the Aston Medal from the British Mass Spectrometry Society.

References

(*1*) Maccoll, A. Professor K.R. Jennings – American Society for Mass Spectrometry Award for Distinguished Contributions to Mass Spectrometry, 1995. *Euro. Mass Spectrom.* **1995**, *1*, 221.

(*2*) Maccoll, A. K.R. Jennings – the Man and his Work. *Int. J. Mass Spectrom. Ion Proc.* **1997**, *165/166*, xix–xxi.

(*3*) Jennings, K. R. Interview by Michael A. Grayson at Leamington Spa, Warwickshire, United Kingdom, 24–25 April 2008 (Philadelphia: Chemical Heritage Foundation, Oral History Transcript # 0419); available at: http://www.asms.org/docs/oral-histories/jennings-oral-history.pdf?sfvrsn=2 (accessed June 12, 2014).

(*4*) Jennings, K. R. Metastable Transitions in the Mass Spectrum of Benzene. *J. Chem. Phys.* **1965**, *43*, 4176–4177.

(*5*) Barber, M.; Jennings, K. R.; Wolstenholme, W. A. Metastable Ions in a Double-Focussing Mass Spectrometer. *Nature* **1967**, *214*, 664–666.

(*6*) Jennings, K. R. Collision-Induced Decompositions of Aromatic Molecular Ions. *Int. J. Mass Spectrom. Ion Phys.* **1968**, *1*, 227–235.

(7) O'Malley, R. M.; Jennings, K. R. Ion Cyclotron Resonance Mass Spectrometry of Acetylene. *Int. J. Mass Spectrom. Ion Phys.* **1969**, *2*, 257–263.
(8) Goode, G. C.; O'Malley, R. M.; Ferrer-Correia, A. J.; Jennings, K. R. Ion Cyclotron Resonance Mass Spectrometry. In: *Advances in Mass Spectrometry, Volume 5*; Quayle, A., Ed.; Institute of Petroleum: London, 1971; pp 195–201.
(9) Goode, G. C.; Jennings, K. R. Reactions of O^- Ions with Some Unsaturated Hydrocarbons. In: *Advances in Mass Spectrometry, Volume 6*; West, A. R., Ed.; Applied Science Pubs.: Barking, UK, 1974; pp 797–803.
(10) Meisels, G. G.; Illies, A. J.; Stradling, R. S.; Jennings, K. R. Time Resolved Measurements in Mass Spectrometry. *J. Chem. Phys.* **1978**, *68*, 866–871.
(11) Jennings, K. R. Negative Ions. *Phil. Trans. R. Soc. Lond. A.* **1979**, *293*, 125–133.
(12) Burke, P. D.; Jennings, K. R.; Morgan, R. P.; Gilchrist, C. A. Identification of Crude Oils by Selective Chemical Ionisation Mass Spectrometry. *Anal. Chem.* **1982**, *54*, 1304–1308.
(13) Mason, R. S.; Farncombe, M. J.; Jennings, K. R.; Cooks, R. G. Angle-Resolved Mass Spectrometry by Z-Deflected Scanning. *Int. J. Mass Spectrom. Ion Phys.* **1982**, *43*, 327–330.
(14) Jennings, K. R.; Mason, R. S. Tandem Mass Spectrometry Utilising Linked Scanning of Double Focusing Instruments. In: *Tandem Mass Spectrometry*; McLafferty, F. E., Ed.; John Wiley & Sons: New York, 1983; pp 197–222.
(15) Farncombe, M. J.; Jennings, K. R.; Mason, R. S.; Schlunegger, P. Peptide Sequencing by Metastable Mapping. *Org. Mass Spectrom.* **1983**, *18*, 612–616.
(16) Hazell, S. J.; Bowen, R. D.; Jennings, K. R. The Origin of M^+ in the Chemical Ionisation Mass Spectra of Polymethylbenzenes. *Org. Mass Spectrom.* **1988**, *23*, 597–600.
(17) Bordas-Nagy, J.; Despeyroux, D.; Jennings, K. R.; Gaskell, S. J. Experimental Aspects of the Collision-Induced Decomposition of Ions in a Four-Sector Tandem Mass Spectrometer. *Org. Mass Spectrom.* **1992**, *27*, 406–415.
(18) Summerfield, S. G.; Dale, V. C. M.; Despeyroux, D.; Jennings, K. R. A Study of the Interactions of Group 1 Metals and the C Terminus of Peptides by Means of High Energy Tandem Mass Spectrometry. *J. Am. Chem. Soc.* **1995**, *117*, 10093–10097.
(19) Jasieczeck, C. B.; Buzy, A.; Haddleton, D. M.; Jennings, K. R. Electrospray Ionisation Mass Spectrometry of Polystyrenes. *Rapid Comm. Mass Spectrom.* **1996**, *10*, 509–514.
(20) Gill, A. C.; Jennings, K. R.; Wyttenbach, T.; Bowers, M. T. Conformations of Biopolymers in the Gas Phase: a New Mass Spectrometric Method. *Int. J. Mass Spectrom.* **2000**, *195/196*, 685–697.
(21) Jennings, K. R. The Changing Impact of the Collision Induced Decomposition of Ions on Mass Spectrometry. *Int. J. Mass Spectrom.* **2000**, *200*, 479–493.
(22) Fernandez, M. T.; Mira, M. L.; Florencio, M. H.; Jennings, K. R. Iron and Copper Chelation by Flavonoids – an Electrospray Mass Spectrometry Study. *J. Inorg. Biochem.* **2002**, *92*, 105–111.
(23) Jennings, K. R., Ed. *A History of European Mass Spectrometry*; IM Publications: Chichester, UK, 2012.

Keith A. Nier
Independent Historian, Madison, NJ, USA

Michael Karas

b. 1952

Michael Karas is a German chemist, born in 1952, best known for his invention and development of Matrix Assisted Laser Desorption/Ionization Mass Spectrometry (MALDI-MS) together with Prof. Franz Hillenkamp (see this volume, *Franz Hillenkamp* and *The Development of matrix Assisted Laser Desorption Ionization (MALDI) Mass Spectrometry*). In recent years he has devoted his interest to investigations of the MALDI mechanisms, matrices and applications. He is also known for his work in the field proteomics.

Karas received his higher education at the University of Bonn, where he studied chemistry and received his Ph.D. in Physical Chemistry (1982). From 1983 to 1986 he was a postdoc Franz Hillenkamp's group at the Institute of Biophysics of the University of Frankfurt. In 1987, he joined Hillenkamp at the Institute of Medical Physics and Biophysics, Medical Faculty, University of Münster and finalized Habilitation in Physical Chemistry in 1992. He returned to Frankfurt in 1995 as a full Professor for Instrumental Analytical Chemistry. Karas has been Professor at the Institute of Pharmaceutical Chemistry, School of Pharmacy, of the Johann Wolfgang Goethe-University Frankfurt, Germany, since 2001. He is also a member of the Center for Drug Research, Development and Safety (ZAFES), a member of the 'Scientific Society of the Johann Wolfgang Goethe-University Frankfurt' and a member of the supervisory board of the Society for Innovation Services Innovectis.

Michael Karas first became involved with mass spectrometry during his graduate work with Hillenkamp. Their development of the MALDI started in 1984, initially in Frankfurt and later in Münster (*1*). The starting point of their investigations was the interactions of organic molecules with ultraviolet laser irradiation (*2*). Normally molecules would decompose due to the local overheating that accompanies a laser shot; the addition of UV-active low-molecular weight substances, however, appeared to protect the analyte and thus allow its transfer into the gas phase for its subsequent mass analysis as an ion. Karas and Hillenkamp were eventually able to detect singly charged ions of proteins such as lysozyme (14 kDa), trypsin (23 kDa) and albumin (67 kDa) (*3*). This development triggered numerous research projects worldwide, with the technique quickly being applied with great success to the analysis of biopolymers (*4*,*5*).

Michael Karas has received a number of prestigious awards, many of them related to MALDI and shared with Franz Hillenkamp. Among them is the Mattauch-Herzog award of the Arbeitsgemeinschaft Massenspektrometrie (today German Society of Mass Spectrometry) (1990), the Award for Distinguished Contributions in Mass Spectrometry of the American Society for Mass Spectrometry (1997), the Molecular Bioanalytics award from the German Society for Biochemistry and Molecular Biology (2000), the ABRF (Association of Biochemical Research Facilities) award for 'Recognition of the outstanding contribution to biomolecular technologies' (2003), and the Karl-Heinz Beckurts Award (2003).

References

(*1*) Hillenkamp, F.; Karas, M. MALDI – an Experience Int. *J. Mass Spectrom.* **2000**, *200*, 71–77.

(*2*) Karas, M.; Bachmann, D.; Hillenkamp, F. The Influence of the Wavelength in High Irradiance Ultraviolet Laser Desorption Mass Spectrometry of Organic Molecules. *Anal. Chem.* **1985**, *57*, 2935–2939.

(*3*) Karas, M.; Hillenkamp, F. Laser Desorption Ionization of Proteins with Molecular Mass Exceeding 10 000 Da. *Anal. Chem.* **1988**, *60*, 2299–2301.

(*4*) Hillenkamp, F.; Chait, B. T.; Beavis, R. C.; Karas, M. Matrix-Assisted Laser Desorption/Ionization Mass Spectrometry of Biopolymers. *Anal. Chem.* **1991**, *63*, 1193A–1203A.

(*5*) Gross, J.; Hahner, S.; Karas, M., *et al.* Investigations of the Metastable decay of DNA under UV-MALDI Conditions with Post-Source-Decay Analysis and H/D Exchange. *J. Am. Soc. Mass Spectrom.* **1998**, *9*, 866–878.

Simone Koenig
University of Muenster, Muenster, NRW, Germany

Paul Kebarle

b. September 21, 1926

Paul Kebarle is a Canadian (né Bulgarian) chemist widely noted for his development of an ion source and reaction chamber in which ion-molecule reactions occur and reach equilibrium. The equilibrium constant K can be determined from data obtained with a mass spectrometer, thermochemical information that is of great use in understanding the fundamental chemistry of gas phase reactions for comparisons with the analogous reactions in solution, and for deduction of the role of the solvent (see this volume, *Mass Spectrometric Methods for the Determination of Thermodynamic Data*).

Paul's primary education coincided with some perilous times in the Second World War including occupation of and the installation of a communist regime in his native Bulgaria by the USSR. Paul managed to leave Bulgaria to study chemistry at the Eidgenossische Technische Hochschule in Switzerland (ETH, the Swiss Federal Institute of Technology) and ended up in Vancouver, Canada. There, working on his PhD at The University of British Columbia, he had his first encounter with a mass spectrometer. It was 'home made' and thus excellent for learning MS 'from the ground up'. After postdoctoral work with Fred Lossing at the Canadian Research Council, he accepted an academic position at the Chemistry Department, University of Alberta, Edmonton where he has remained. Paul is now retired but still engages in scientific endeavors on a small scale (see this volume, *Frederick Pettit Lossing*).

The nuclear end of the Second World War led also to a new science, 'Radiation Chemistry' that investigated the reactions induced by high energy particles. Paul decided to study the reactions induced by high-energy electrons that encounter gas at near atmospheric pressures. Kebarle and his graduate students designed an ion source that could contain gas at such pressures. The gas was irradiated with a high-energy electron beam that could be pulsed so that the time dependence of the reactions could be observed. The resulting major processes were ion molecule reactions that reached equilibrium and allowed determination of the equilibrium constants (*1*). These determinations led to the ability to calculate the standard Gibbs Free Energy, ΔG^0, as well as the standard enthalpy, ΔH^0, and entropy, ΔS^0 changes from the reactions. Abundant data for the proton affinity, gas phase acidity, electron affinity and ion-solvent molecule interactions were also obtained.

Comparisons of the gas phase data with the corresponding data in solution provided special insights on the role of the solvent on the reactions. Comparisons could be made not only for water but also a variety of other solvents such as ethanol and dipolar aprotic solvents such as acetone, acetonitrile (*2*). This led to a deeper understanding of the role of the solvent in reactions involving ions and had a significant impact not only to gas phase ion chemistry but also to physical organic chemistry.

Paul has received a number of awards for his work including the ACS Franklin and Field Award for Outstanding Achievement in Mass Spectrometry in1994 and the Canadian Institute for Chemistry (CIC) Medal for Outstanding Contributions to the Science of Chemistry in 1987. He was the first recipient of the Fred Lossing Award of the Canadian Mass Spectrometry Society in 1994.

In 1991, the International Journal of Mass Spectrometry and Ion Processes published a volume in Paul's honor on the occasion of his 65th birthday and in recognition of the outstanding contributions he has made to mass spectrometry, gas phase ion chemistry and chemistry in general throughout his scientific career.

The publications cited below are representative of additional significant contributions to the field made by the author (*3–6*).

Paul met Beverly Harris as a graduate student and she became his wife. They have three daughters who live in Vancouver.

References

(*1*) Kebarle, P.; Hogg, A. M. Heats of Hydration and Solvation by Mass Spectrometry. *J. Chem. Phys.* **1965**, *42*, 798–799.

(*2*) Durden, D. A.; Kebarle, P.; Good., A. Thermal Ion–Molecule Reaction Rate Constants at Pressures up to 10 Torr with a Pulsed Mass Spectrometer. Reactions in Methane, Krypton, and Oxygen. *J. Chem. Phys.* **1969**, *50*, 805–814.

(*3*) Chowdhurry, S.; Grimsrud, E. P.; Kebarle, P. Bonding of Charge Delocalized Anions to Protic and Dipolar Aprotic Solvent Molecules. *J.Phys. Chem.* **1987**, *91*, 2551–2556.

(*4*) Blades, A. T.; Klassen, J. S.; Kebarle, P. Free Energies of Hydration in the Gas Phase of the Anions of Some Oxoacids of C, N, S, P, Cl, and I. *J. Am. Chem. Soc.* **1995**, *117*, 10563–10571.

(*5*) Kebarle, P. Gas Phase Ion Thermochemistry Based on Ion-Equilibria. From the Ionosphere to the Reactive Centers of Enzymes. *Int. J. Mass Spectrom.* **2000**, *200*, 313–330.

(*6*) McMahon, T. B., Ed. Honor Biography Paul Kebarle. *Int. J. Mass Spectrom. and Ion Proc.* **1991**, *109*, 358 pp.

Paul Kebarle
University of Alberta, Edmonton, AB, Canada

Charles Kemball

b. March 27, 1923
d. September 4, 1998

Charles Kemball was a British chemist most noted for his studies of catalysis and chemistry on surfaces. He pioneered the use of mass spectrometry and other physical instrumentation to monitor reactions in this area (*1–3*).

Charles Kemball was born in Edinburgh. He was an only child; his father was a dental surgeon and Fellow of the Royal Society of Edinburgh, as his son would also become. He was educated at Edinburgh Academy before going to Trinity College in Cambridge University in 1940. He earned an Honors degree in Chemistry in 1943 and then did his graduate research in the colloid science department, working on adsorption topics with support from the Ministry of Aircraft Production. He became a Research Fellow at Trinity in 1946 and spent a year with H. S. Taylor at Princeton, where he worked on hydrocarbon reactions over nickel catalysts. Back at Cambridge, Kemball launched a career-long course of research in catalysis. He moved to the physical chemistry department in 1949. He worked on hydrocarbon exchange reactions with deuterium over evaporated metal films, using mass spectrometry to determine the products. This proved to be a powerful approach for studying catalytic reaction mechanisms and investigating intermediates adsorbed on catalyst surfaces. Over the course of his career he was ever alert to bringing in additional techniques along with mass spectrometry to improve the results he could obtain. Kemball's research was honored by the Meldola Medal of the Royal Institute of Chemistry and a College Lectureship in 1951. Meanwhile he also began to play a role in academic administration; this kind of work would also characterize the rest of his career.

In 1954, Kemball was appointed to the Chair of Physical and Inorganic Chemistry at the Queen's University in Belfast, N. Ireland. He built up an internationally renowned research school on catalysis, established strong links with industry, and simultaneously took on major posts in university administration. He also met and in 1956 married Kathleen Lynd.

Kemball was elected a Fellow of the Royal Society of London in 1965 and the next year moved to the Chair of Chemistry at Edinburgh University. He again developed a major school of researchers on heterogeneous catalysis while taking on administrative duties in difficult times and serving in significant ways in several scientific organizations as well. He also became a Director of his old school, the Edinburgh Academy. He was elected a Fellow and, after retirement, President of the Royal Society of Edinburgh. He was honored with the rank of Commander in the Order of the British Empire.

Kemball pronounced in a Presidential Address to the Royal Institute of Chemistry that those over 60 should not be relied upon to run chemistry departments and in keeping with that view, he retired from the Chair in 1983, while continuing research and other service part time. He died at his home in Tyninghame, Scotland, in 1998, survived by his wife, three children, and several grandchildren.

A few representative papers are listed here (*4–10*).

References

(*1*) Schwarz, P. Obituary: Professor Charles Kemball. *The Independent (London)*, **1998**.

(*2*) Donovan, R. J.; Charles Kemball: C.B.E., MA, ScD(Cantab), DSC(H-W,QUB), C Chem, FRSC, MRIA, FRS. *Royal Society of Edinburgh, Obituaries*, Available at: http://www.royalsoced.org.uk/cms/files/fellows/obits_alpha/kemball_charles.pdf (accessed June 12, 2014).

(*3*) Roberts, M. W.; Charles Kemball: C.B.E., 27 March 1923–4 September 1998: Elected F.R.S. 1965. *Biogr. Mems Fell. R. Soc.* **2000**, *46*, 285–298.

(*4*) Kemball, C. The Reaction of Ammonia and Deuterium on Evaporated Nickel Catalysts. *Trans. Faraday Soc.* **1952**, *48*, 254–258.

(*5*) Kemball, C. Contact Catalysis and Surface Chemistry. *Annu. Rev. Physicl Chem.* **1953**, *4*, 303–322.

(*6*) Kemball, C. The Deuteration and Exchange of Ethylene on Evaporated Metal Catalysts at Low Temperatures. *J. Chem. Soc.* **1956**, 735–743.

(*7*) Kemball, C. Exchange Reactions on Oxide Catalysts. *Ann. N. Y. Acad. Sci.* **1973**, *213*, 90–104.

(*8*) Wong, T. C.; Brown, L. F.; Haller, G. L.; Kemball, C. Hydrogenolysis and Hydrogenation of Hydrocarbons on Supported Rh-Ir Bimetallic Catalysts. *J. Chem Soc. Faraday Trans. 1* **1981**, *77*, 519–533.

(*9*) Brown, R.; Kemball, C. Exchange Reactions of Alkanes Containing Quaternary Carbon Atoms over Supported Metal Catalysts. *J. Chem Soc. Faraday Trans. 1* **1989**, *85*, 2159–2172.

(*10*) Pope, G. C.; Kemball, C.; McDougall, G. S. Combined Deuterium Nuclear Magnetic Resonance and Mass-Spectrometric Studies of the Exchange Reactions of Ketones over Supported Metal Catalysts. *Trans. Faraday Soc.* **1990**, *86*, 747–752.

Keith A. Nier
Independent Historian, Madison, NJ, USA

Frederick W. Lampe

b. January 5, 1927, Long Island, NY
d. November 8, 2000, State College, PA

Professor Frederick W. Lampe, former head of the Department of Chemistry at Penn State University, was an American chemist known for the development of mass spectrometric techniques to probe the fundamental chemistry of atomic and molecular systems that had been subjected to the influence of excitation and ionization fields. He made numerous contributions to the knowledge and understanding of the chemical nature and mechanistic roles of reactive intermediates in the fields of photochemistry and radiation chemistry.

Lampe was born on Long Island, NY on January 5, 1927. He served in the US Navy during World War II but, due to contracting chicken pox, did not accompany the rest of his basic training group on the ill-fated voyage of the USS Indianapolis. Following the war, he earned a bachelor's degree in chemical engineering at Michigan State College in 1950 and his masters and Ph.D. in chemistry at Columbia University in 1951 and 1953 respectively. Following graduation from Columbia, he served as a research chemist at Humble Oil and Refining Company for seven years in the group led by Joe Franklin (see this volume, *Joe Louis Franklin, Jr.*) in Baytown, TX. The studies in ion molecule chemistry done during that period served as a base for much of the future work in that field. He joined the Penn State faculty as an Associate Professor in 1960, was promoted to Professor in 1965, and became assistant head of the department in 1978. He served as head of the Chemistry Department from 1983 to 1988 and retired as Professor Emeritus in 1992. Lampe died in State College, PA, on November 8, 2000.

Fred Lampe held visiting professorships at the University of New Hampshire (1963), the Institute of Physics University of Freiburg (1966–67), and the Hahn–Meitner Institute for Nuclear Research, Berlin (1973–74 and 1984). He was awarded an NSF Senior Postdoctoral Fellowship (1966–67), elected a Fellow of the American Physical Society (1970), awarded a U.S. Senior Scientist Award of the Alexander von Humboldt Foundation Germany (1973), and named a Robert A. Welch Foundation Lecturer (1982). Lampe was a member of the American Chemical Society, the American Physical Society and the American Society for Mass Spectrometry. He served on the ASMS Board of Directors from 1981–1983. He published more than 160 research papers and held five patents. At the time of his death he was working on the third edition of the book he co-authored with Harry Allcock, also of the Penn State Department of Chemistry, *Contemporary Polymer Chemistry*.

Lampe himself noted three aspects of his work that he considered particularly significant. His earliest, and, in his words, perhaps most important work was conducted in the mid 1950s with coworkers F. H. Field and J. L. Franklin (see this volume, *Frank Henry Field*). This seminal research demonstrated the extreme rapidity and copiousness of ion-molecule reactions in hydrocarbons (*1*) and he showed for the first time (*2*) that such ionic reactions played a key role in the radiation chemistry of gases. These and subsequent papers played a major role in initiating the extensive activity in the study of ion–molecule reactions that has taken place in the last 45 years; indeed, the term 'ion-molecule reactions' was coined in these early endeavors. A singularly important individual contribution in these early years was the discovery of hydride ion abstraction by positive ions (*3*) – its universal occurrence in the mass spectra of hydrocarbons and in any chemical system in which positive ions and chemically bonded hydrogen are present. Lampe later showed how these ion-molecule reactions played a central role in the radiation-induced polymerization of isobutylene in the liquid phase at $-80\ ^\circ C$.

From the mid-1960s through the 1970s, Lampe and his students used pulsed mass spectrometric techniques to elucidate the mechanisms of the associative ionization reactions of excited atoms. They were the first to use pulsing techniques to determine both the average and specific reaction rates of the general process: $X^* + Y \rightarrow XY ++ e$. Working with Hotop and Niehaus, Lampe demonstrated that such reactions were not limited to atomic reactants but also occurred with excited molecules, particularly H_2^* (*4*).

A third area to which Lampe made significant contributions was the ionic chemistry of volatile silicon and germanium compounds and the lifetimes of their ion–molecule collision complexes. This work began in 1966 and continued until his retirement.

He and his students discovered the existence of SiH_5^+ and GeH_5^+ and measured the proton affinities of SiH_4 and GeH_4. With his coworkers, he demonstrated the enormous ease with which hydrogen compounds of silicon donate hydride ions to attacking positive ions. These studies were also used to demonstrate that the termination of silyl radical reactions in the 10–60 Torr range was a disproportionation to SiH_4 and SiH_2, with the latter persisting to form disilane (*5*). This observation is in distinct contrast to the termination of methyl radicals that yield ethane directly.

After retirement he returned to his study of the piano, an interest that had been interrupted by a broken arm in childhood. He was a talented athlete and, while he gave up ice hockey and squash in his mid-50s, he become an avid golfer.

From a personal perspective there were two very remarkable aspects to Fred Lampe's personality. First, he was extraordinarily patient as a teacher and mentor. He was talented in finding ways to explain difficult topics, and he was patient and supportive in situations in which the complexities of older generation mass spectrometers misbehaved. One of his favorite sayings was that "mass spectrometers worked best when you put nothing into them". Second, Fred had an exceptional talent for identifying the basic physical principles behind phenomena and finding mathematical expressions for them.

References

(*1*) Field, F. H.; Franklin, J. L.; Lampe, F. W. Gaseous-Ion Reactions in Methane and Ethylene. *J. Am. Chem. Soc.* **1956**, *78*, 5697–5698.

(*2*) Lampe, F. W. High-Energy Electron Irradiation of Methane. *J. Am. Chem. Soc.* **1957**, *79*, 1055–1058.

(*3*) Field, F. H.; Lampe, F. W. Reactions of Gaseous Ions VI. Hydride Ion Transfer Reactions. *J. Am. Chem. Soc.* **1958**, *80*, 5587–5592.

(*4*) Hotop, H.; Lampe, F. W.; Niehaus, A. Collision Reactions of Excited Hydrogen Molecules. *J. Chem. Phys.* **1969**, *51*, 593–597.

(*5*) Austin, E. R.; Lampe, F. W. Hydrogen-atom Sensitized Decomposition of Monosilane. *J. Phys. Chem.* **1976**, *80*, 2811–2817.

Alfred L. Yergey
(Lampe Ph.D. student 1963–67)
National Institute of Child Health and Human Development, NIH, Bethesda, MD, USA

Chava Lifshitz

b. March 26, 1936
d. March 1, 2005

Photo available at J. Mass Spectrom. and Ion Proc. ***1997****, 167/168, xiii.*

Chava Lifshitz (nee Eva Wolf) was an Austrian-born Israeli chemist best known for her work in ion molecule chemistry, particularly her ability to blend theory and experiment in support of the quaisi-equilibrium theory (QET). By demonstrating that the dissociations (including breakdown patterns and appearance potentials) of positive ions trapped in the space charge of a concentrated electron beam could be followed in real time, Lifshitz demonstrated that the kinetic shift was not only real but could be overcome (*1,2*).

Lifshitz was born on March 26, 1936, in Austria, the daughter of a prominent Vienna attorney. The family fled Austria after the Anschluss, resettling in Eretz Yisrael, then a British Mandate area. Eva, thereafter known by the Hebrew version of her name, Chava, grew up in a cultured home that retained many of the old European traditions. A gifted pupil, she included in her studies piano lessons from the Israeli composer Paul Ben-Haim. Availing herself of a special program for gifted students that allowed her to combine compulsory military service with her education, she began studying chemistry at the Hebrew University of Jerusalem (Israel) in 1953, where she received an MSc in 1958 and a PhD in 1961. Her dissertation entitled 'Isotope Effects in the Radiation Chemistry of Aqueous Solutions' was supervised by Gabriel Stein. While at the Hebrew University, she met and married fellow chemist Assa Lifshitz with whom she had three children, a son and two daughters (*3*).

Following completion of her PhD, she spent the next two years in the US as a post-doctoral student of Franklin Asbury Long at Cornell, where she was introduced to a field of study pioneered by Long and Lewis Friedman: the use of a mass spectrometer to analyze reaction products in which stable isotopic labels have replaced radioactive tracers. It was an area in which she would find fertile ground for the rest of her career.

Lifshitz returned to Israel in 1963, where she joined the faculty of the Hebrew University. She rose to the rank of full professor in 1976, a position she retained until her death in 2005. She was a frequent visitor in the laboratories of her US collaborators, spending 1965 as a Visiting Scientist at Argonne National Laboratory (Argonne, IL) with W. A. Chupka and J. Berkowitz; 1969 as a Visiting Scientist at Cornell University with F. W. McLafferty (see this volume, *Fred W. McLafferty*); 1969–1970 and 1972 as a Visiting Scientist at Wright Patterson Air Force Base (Ohio) with T. O. Tiernan; and again with Tiernan during 1976–1977 as a Visiting Scientist, Wright State University (Ohio); and 1989 and 1995 again as a Visiting Scientist at Argonne National Laboratory with J. Berkowitz and K. R. Lykke.

As the more than 200 publications in her curriculum vitae show (*4*), Lifshitz' exploration of a subject never stopped with one paper. Rather, as she focused on a new topic, a series of publications would result, sometimes coming in rapid succession and sometimes over a period of years, each extending understanding of gas phase chemistry through experimentation combined with modeling and calculation. Early in her career the focus was on information about kinetics that isotopic labeling could provide (*5,6*) as well as the chemistry of fluorinated ethylenes (*7–11*), and the electron affinities that could be determined from endothermic negative ion charge transfer reactions (*12–14*). She continued to explore the kinetic shift through the 1970s and '80s and into the '90s, enhancing her ability to measure with new techniques and more sophisticated calculations (*15–19*). Understanding the chemistry of toluene occupied her attention for most of her career (*20–23*), whereas the study of clusters and peptides began toward the end (*24,25*). As Rob Dunbar and Tilman Märk note in the special issue in honor of Lifshitz's 60th birthday, "Although she has been primarily an experimentalist, her work has constantly been infused with the most current theoretical understanding. Few people achieve an interplay of theory and experiment that is so organic and stimulating" (*26*).

Lifshitz was a member of editorial and advisory boards for Mass Spectrometry Reviews, the International Journal of Mass Spectrometry and Ion Processes, Rapid Communications in Mass Spectrometry, and the Journal of Mass Spectrometry. From

1991–1996 she was President of the Israel Society for Mass Spectrometry and she led or helped in organizing many international meetings both in Jerusalem and Berlin.

Lifschitz's contributions to science were honored with many awards, among them the Kolthoff Prize (Technion 1985), the Max Planck Research Award (Alexander von Humboldt Foundation, 1991), which she shared with longtime collaborator Helmut Schwarz, (see this volume, *Helmut Schwarz*) and the Israel Chemical Society Prize (2003). On the occasion of her 60th birthday, she was honored with a special issue of the International Journal of Mass Spectrometry and Ion Processes.

In 2005, Lifschitz finally succumbed to the cancer against which she had been fighting for many years. At the memorial symposium held in her honor at Hebrew University on March 26, 2006, she was remembered by one of the scientists she had mentored as "not being 'like' anybody: she was an original. At once unfailingly polite in the best European tradition and bluntly outspoken (especially about politics) in a way that could startle even Israelis. At once a sworn feminist and a doting mother; a driven scientist and a "mother lioness" to all who worked for or with her; a political liberal in many ways yet simultaneously the most strident of right-wing nationalists; at once an uncompromisingly analytical thinker and loving beyond reason" (*3*).

References

(*1*) Lifshitz, C.; Peers, A. M.; Weiss, M.; Weiss, M. J. A Direct Measurement of the 'Kinetic Shift' in Benzene. *Adv. Mass Spectrom.* **1974**, *6*, 871–876.

(*2*) Klotz, C. E. Chava Lifshitz: An Appreciation. *J. Mass Spectrom. and Ion Proc.* **1997**, *167/168*, xi–xii.

(*3*) Anon. *Chava Lifshitz Memorial Symposium, Hebrew University,* Jerusalem, Israel. Mary 26, 2006. http://theochem.weizmann.ac.il/chava.html (accessed June 8, 2012).

(*4*) Lifshitz, C. Chava Lifshitz: Curriculum Vitae. *Int. J. Mass Spectrom. and Ion Proc.* **1997**, *167/168*, xiii–xii.

(*5*) Lifshitz, C. Isotope Effects in Neutral H_2O-D_2O Irradiated Solutions and the Nature of the Reducing Radical. *Can. J. Chem.* **1962**, *40*, 1903–1908.

(*6*) Lifshitz, C.; Shapiro, M. Isotope Effects on Metastable Transitions: C_3H_8 and C_3D_8. *J. Chem. Phys.* **1967**, *46*, 4912–4920.

(*7*) Lifshitz, C.; Long, F. A. Appearance Potentials and Mass Spectra of Fluorinated Ethylenes. I. Decomposition Mechanisms and their Energetic. *J. Phys. Chem.* **1963**, *67*, 2463–2468.

(*8*) Lifshitz, C.; Long, F. A. Appearance Potentials and Mass Spectra of Fluorinated Ethylenes. II. Heats of Formation of Fluorinated Species and their Positive Ions. *J. Phys. Chem.* **1965**, *69*, 3731–3736.

(*9*) Lifshitz, C.; Long, F. A. Appearance Potentials and Mass Spectra of Fluorinated Ethylenes. II. Calculations Based on the Statistical Theory of Mass Spectra. *J. Phys. Chem.* **1965**, *69*, 3737–3740.

(*10*) Lifshitz, C.; Long, F. A. Appearance Potentials Mass Spectra of C_3F_6, C_3F_5Cl, and Cyclic C_3F_6. *J. Phys. Chem.* **1965**, *69*, 3741–3746.

(*11*) Lifshitz, C.; Long, F. A. Some Observations Concerning the Positive Ion Decomposition of C_2F_6 and C_3H_8. *J. Phys. Chem.* **1965**, *69*, 3746–3751.

(*12*) Lifshitz, C.; Tierman, T. O. Dissociation of Cyclobutane Molecule Ions in Charge Exchange Collisions. *J. Chem. Phys.* **1971**, *55*, 3555–3561.

(*13*) Hughes, B. M.; Lifschitz, C.; Tiernan, T. O. Electron Affinities from Endoergic Negative-Ion Charge Transfer Reactions III. NO, NO_2, SO_2, CS_2, Cl_2, Br_2, I_2, and C_2H. *J. Chem. Phys.* **1973**, *59*, 3162–3183.

(*14*) Lifshitz, C. On the Reaction Mechanism of $O + H_2O \rightarrow OH^- + OH$ at Low Incident Ion Energies. *J. Phys. Chem.* **1982**, *86*, 3634–3637.

(*15*) Lifshitz, C.; Tzidony, E. Kinetic Energy Release Distributions for $C_3H_6O^+$ Ion Dissociations. A Further Test of the Applicability of the Energy Randomization Hypothesis to Unimolecular Fragmentations. *Int. J. Mass Spectrom. Ion Phys.* **1981**, *39*, 181–195.

(*16*) Lifshitz, C. Time Resolved Appearance Energies, Breakdown Graphs and Mass Spectra. The Elusive 'Kinetic Shift'. *Mass Spectrom. Rev.* **1982**, *1*, 309–348.

(*17*) Aviyente, V.; Shaked, M.; Feinmesser, A.; Grefen, S.; Lifshitz, C. Kinetic Energy Release Measurements with an Ion Trapping and Time-Resolved Ion Momentum Spectrometer. *Int. J. Mass Spectrom. and Ion Proc.* **1986**, *70*, 301–320.

(*18*) Aviyente, V.; Elam, M.; Ohmichi, N.; Lifshiz, C. Time Dependent Mass Spectra and Breakdown Graphs. 11. Time-Resolved Ion Momentum Spectrometry (TRIMS) of Anisole. *J. Phys. Chem.* **1986**, *92*, 6548–6553.

(*19*) Lifshitz, C. Energetics and Dynamics Through Time Resolved Measurements in Mass Spectrometry: Aromatic Hydrocarbons, Polycyclic Aromatic Hydrocarbons (PAHs) and Fullerenes. *Int. Rev. Phys. Chem.* **1997**, *16*, 113–139.

(*20*) Lifschitz, C. Toluene-Metastable-Ion Lifetimes. *J. Chem. Phys.* **1967**, *47*, 1870–1871.

(*21*) Lifshitz, C. Unimolecular Fragmentation of Toluene Ions Near Threshold. *J. Chem. Phys.* **1975**, *62*, 1602–1603.

(*22*) Lifshitz, C. Tropylium Ion Formation from Toluene; Solution of an Old Problem in Organic Mass Spectrometry. *Accts. Chem. Res.* **1994**, *27*, 138–144.

(*23*) Lifshitz, C.; Gotkis, Y.; Laskin, J.; Ioffe, A.; Shaik, S. Threshold Formation of Benzylium (Bz^+) and Tropylium (Tr^+) from Toluene. Non-Statistical Behavior in Franck Condon Gaps. *J. Phys. Chem.* **1993**, *97*, 12291–12295.

(*24*) Lifshitz, C. Unimolecular Collision-Induced Decomposition of Selected Cluster Ions. In: *Current Topics in Ion Chemistry and Physics: Clusters*; Baer, T., Ng, C. Y., Powis, I., Eds.; Wiley: New York, 1993; p 494.

(*25*) Cui, W.; Hu, Y.; Lifschitz, C. Time Resolved Photdissociation of Small Peptide Ions: Combining Laser Desorption with ion Trap/Reflectron TOF Mass Spectrometry. *Europ. J. Phys. D.* **2002**, *20*, 565–571.

(*26*) Dunbar, R. C.; Märk, T. D. Forward to Special Issue in Honor of Chava Lifshitz's 60th Birthday. *Int. J. Mass Spectrom.* **1997**, 167–168.

P. Jane Gale
Waters Corp., Milford, MA, USA

Frederick Pettit Lossing

b. August 10, 1915
d. May 22, 1998

Frederick Lossing was a Candian scientist whose active scientific career spanned some 50 years and produced some 140 publications. His greatest professional achievements lay in the design and construction of precision scientific apparatus that he used to measure significant thermochemical data and in the seeking and finding of valuable generalizations there from. In particular he constructed a double-hemispheric electron source that gave a mono-energetic beam of electrons with an energy spread of less than 25 mV. This enabled him to measure many important ionization and appearance energies with great precision.

Fred Lossing was born in Norwich, Ontario, Canada, in August, 1915. He obtained a B.A Degree from the University of Western Ontario in 1938 and an M.A. in 1940. After obtaining a PhD at McGill University in 1942 with a thesis on "The preparation and properties of S_2F_{10}," he worked in polymer chemistry at Shawinigan Chemicals until 1946, when he joined the National Research Council of Canada (NRC) in Ottawa. During his years at NRC, a number of distinguished Canadian scientists began their careers as Post-doctoral fellows in his laboratory, among them Keith Ingold, Alex Harrison and Paul Kebarle (see this volume, *Alexander George Harrison* and *Paul Kebarle*).

He worked at NRC until his retirement in 1980, whereupon he moved his apparatus to the laboratory of John Holmes at Ottawa University, where he continued actively in research for the next 15 years (see this volume, *John Leonard Holmes*). His long collaboration with Holmes began in 1974, when by happenstance, they discovered that they were working separately on the same molecules. That had to stop, they decided, and their resulting collaboration proved extremely rewarding. Fred gave freely of his time to many graduate students, for whom he often made energy measurements and to whom he gave much valuable advice.

His early research involved the first reported measurements of the ionization energies of free radicals produced by the pyrolysis of suitable precursor molecules. An extended series of no fewer than 46 publications were produced over a span of 24 years (*1,2*). This pioneering research led to his election to the Royal Society of Canada in 1956. His interest in the effects on ion thermochemistry of substitutions of functional groups within the ion, began with the description of $C_nH_{2n+1}O^+$ ions, showing clearly that (for example) in the ion $[CH_2OH]^+$, the charge is essentially located at the carbon atom and not, as chemical wisdom then suggested, at the oxygen (*3*).

It was at about the time of the publication of this latter paper that the collaboration with Holmes began and among their 33 joint publications is a series of papers that greatly extended understanding of the thermochemistry of organic ions and their neutral counterparts, particularly the relationships between structure and enthalpy of formation. The greater stability of enol ions over their keto isomers reverses the thermochemistry of the corresponding neutrals (*4*), and similar studies also permitted the evaluation of neutral heats of formation (*5–9*).

More general studies of ion thermochemistry followed. A search was made for generalizations that could relate ion structure to ion thermochemistry and several successful publications ensued (*10–13*). Some of these collaborative studies also involved Johan Terlouw (University of Utrecht and McMaster University, Hamilton) and the first experimental measurement of the heats of formation of distonic ions appeared in 1982 (*14*). This paper appeared back-to-back with the computational study by L. Radom, J. McLeod and W. J. Bouma (*15*). This work was extended soon afterwards in (*16*).

In 1995 The Canadian Society for Mass Spectrometry named its major award The Fred P. Lossing Award and awarded the first to Raymond March (see this volume, *Raymond E. March*).

Outside of chemistry Fred's chief hobby was astronomy, and he was very active in Ontario astronomy groups and the Royal Astronomy Society of Canada. He ground his own spherical mirrors for several large telescopes. He greatly enjoyed reading science fiction and played the cello in a local string groups. Working with Fred Lossing was always an unalloyed pleasure. Sitting in the laboratory we would discuss the current chemical problems and any other subject that cropped up. When the problems of the world had been dealt with to general satisfaction, talk returned anew to the baffling chemistry and its solution. For Fred there was no such thing as a silly question, because the answer might just show us what to do next. Fred died in Ottawa in May, 1998.

References

(1) Lossing, F. P.; Tickner, A. W. Part I. The Measurement of Methyl Radical Concentrations. *J. Chem. Phys.* **1952**, *20*, 907–914.

(2) Lossing, F. P.; Traeger, J. C. Free Radicals by Mass Spectrometry. Part XLVI. Heats of Formation of C_5H_7 and C_5H_9 Radicals and Cations. *Int. J. Mass Spectrom. Ion Phys.* **1976**, *19*, 9–22.

(3) Lossing, F. P. Heats of Formation of Some Isomeric $C_nH_{2n+1}O^+$ Ions. Substitutional Effects on Ion Stability. *J. Am. Chem. Soc.* **1977**, *99*, 7526–7530.

(4) Holmes, J. L.; Lossing, F. P. Gas Phase Heats of Formation of Keto and Enol Ions of Carbonyl Compounds. *J. Am. Chem. Soc.* **1980**, *102*, 1591–1595.

(5) Holmes, J. L.; Lossing, F. P. Heats of Formation of the Ionic and Neutral Enols of Acetaldehyde and Acetone. *J. Am. Chem. Soc.* **1982**, *104*, 2648–2649.

(6) Holmes, J. L.; Lossing, F. P. The Heats of Formation of Organic Radicals from Appearance Energies. *Int. J. Mass Spectrom. Ion Proc.* **1984**, *58*, 113–120.

(7) Holmes, J. L.; Lossing, F. P. Heats of Formation and Homolytic Bond Dissociation Energies in the Keto-Enol Tautomers C_2H_4O and C_3H_6O. *J. Am. Chem. Soc.* **1986**, *108*, 1086–1087.

(8) Holmes, J. L.; Lossing, F. P.; Maccoll, A. The Heats of Formation of Alkyl Radicals from Appearance Energies. *J. Am. Chem. Soc.* **1988**, *110*, 7339–7342.

(9) Holmes, J. L.; Lossing, F. P. Heats of Formation and Bond Dissociation Energies in Halogen Substituted Methyl and Ethyl Radicals. *J. Am. Chem. Soc.* **1988**, *110*, 7343–7345.

(10) Holmes, J. L.; Fingas, M.; Lossing, F. P. Towards a General Scheme for Estimating the Heats of Formation of Organic Ions in the Gas Phase, I, Odd Electron Cations. *Can. J. Chem.* **1981**, *59*, 80–93.

(11) Holmes, J. L.; Lossing, F. P. Towards a General Scheme for Estimating the Heats of Formation of Organic Ions in the Gas Phase, II, The Effects of Substitution at Charge Bearing Sites. *Can. J. Chem.* **1982**, *60*, 2365–2371.

(12) Lossing, F. P.; Holmes, J. L. Stabilization Energy and Ion Size in Carbocations in the Gas Phase. *J. Am. Chem. Soc.* **1984**, *106*, 6917–6920.

(13) Holmes, J. L.; Lossing, F. P.; Mayer, P. M. The Effects of Methyl Substitution on the Structure and Thermochemistry of the Cyanomethyl Radical and Cation. *Chem. Phys. Lett.* **1993**, *212*, 134–140, *[Lossing's last paper]*.

(14) Holmes, J. L.; Lossing, F. P.; Burgers, P. C.; Terlouw, J. K. The Radical Cation $CH_2OH_2^{+\bullet}$ and Related Gas Phase Ion-Dipole Complexes. *J. Am. Chem. Soc.* **1982**, *104*, 2931–2932.

(15) Bouma, W. J.; Radom, J. K; MacLeod, L. Experimental Evidence for the Existence of a Stable Isomer of CH3OH + .cntdot.: The Methylenoxonium Radical Cation, CH2OH2 + .cntdot. *J. Am. Chem. Soc.* **1982**, *104*, 2930–2931.

(16) Holmes, J. L.; Lossing, F. P.; Terlouw, J. K.; Burgers, P. C. Novel Gas Phase Ions. The Radical Cations $CH_2XH^{+\bullet}$ (X = F, Cl, Br, I, OH, NH_2 and SH) and $CH_2CH_2NH_3^{+\bullet}$. *Can. J. Chem.* **1983**, *61*, 2305–2309.

John L. Holmes
Chemistry Department, University of Ottawa, Ottawa, ON, Canada

Alexander A. Makarov

b. June 2, 1966, Irkutsk, Russia

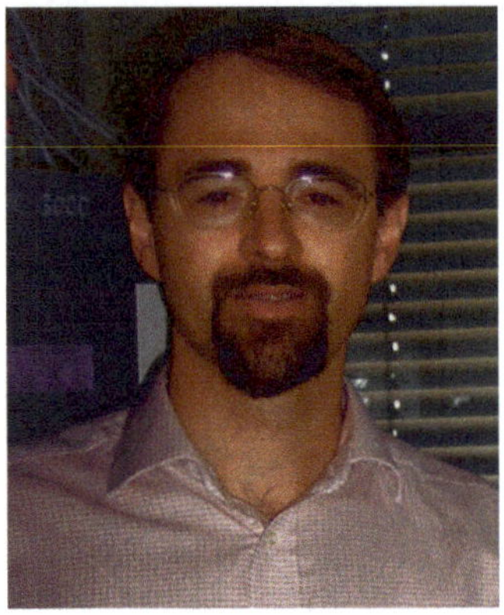

Makarov, Alexander A is a Russian physicist, now living in Germany, working mainly on the development of mass analyzers, with an emphasis on high-resolution, accurate-mass capabilities. I have been honored principally for the development of the Orbitrap mass analyzer.

I have started my work in mass spectrometry (time-of-flight at that time) in 1985 when I was still an undergraduate student at the Physics Engineering Institute (MPEI) in Moscow. I earned a MSc with honors in technical physics at the MPEI in 1989 and a PhD in physics and mathematics in 1992. After working as a scientist at the General Physics Institute of the Russian Academy of Sciences in Moscow (1992–1994) and a short time as a consultant for Shimadzu in the UK, I became a post-doc at the University of Warwick, Coventry, UK from 1994 to 1996.

In 1996, I joined a small high-tech company, HD Technologies in Manchester, UK, as a Research and Development Scientist. Though the company specialized in contract development of TOFMS, this is where I started to deviate from the beaten path and made the first steps toward the concept of the Orbitrap mass analyzer. Proof-of-principle results were presented at ASMS Conference in 1999.

Following acquisition of HD Technologies by Thermo Corporation in 2000, I became a Research Manager for Thermo Finnigan (Masslab Ltd), Manchester, UK. I concentrated my research on interfacing the analyzer to continuous rather than pulsed ion sources, with the practical goal of incorporating an electrospray ion source. Progress in this direction was accompanied by expansion of the research team as well as transfer of the project to the Thermo factory in Bremen, Germany, where I have held the positions of Senior Research Scientist FTMS, (2002–2006), Research Manager (2006–2007) and Director of Research in Life Sciences Mass Spectrometry, ThermoFisher Scientific, Bremen (since 2007). This work resulted in the commercial release of the new LTQ Orbitrap tandem mass spectrometer in 2005 and its enthusiastic acceptance in the mass spectrometry community, especially for proteomic and metabolic research.

I have continued to provide scientific guidance on further extensions of this instrument family as well as on other instruments including the Exactive single-stage mass spectrometer and, from 2011, the tandem mass spectrometer Q Exactive. In pursuit of the strategic goal of bringing Orbitrap-based mass spectrometry into new analytical applications, previously inaccessible for mass spectrometry, I continue to work on improvements in the technology, as illustrated by the introduction in 2011 of a new generation of Orbitrap analyzers and enhanced Fourier transform algorithms. Some my of significant publications are listed below (*1–5*).

I have received the Thomson Medal of the International Society for Mass Spectrometry (Kyoto, 2012), the Science and Technology award of the Human Proteome Organization (Geneve, 2011), the Distinguished Contribution to Mass Spectrometry Award of the American Society for Mass Spectrometry (Denver, 2008), the Gold medal of the Russian Mass Spectrometry Society (Moscow, 2007), the Gold Editors Award at Pittcon for the LTQ Orbitrap instrument (Orlando, 2006), and several other honors.

I have been married for 25 years, have two children, and live in Bremen. I enjoy traveling, mountain skiing, roller-blading, cycling, and volleyball.

References

(*1*) Makarov, A. Electrostatic Axially Harmonic Orbital Trapping: A High-Performance Technique of Mass Analysis. *Anal. Chem.* **2000**, *72*, 1156–1162.

(*2*) Hu, Q.; Noll, R.; Li, H.; Hardman, M.; Makarov, A.; Cooks, R. G. The Orbitrap: A New Mass Spectrometer. *J. Mass Spectrom.* **2005**, *40*, 430–443.

(*3*) Makarov, A.; Denisov, E.; Kholomeev, A., *et al.* Performance Evaluation of a Novel Hybrid Linear Ion Trap/Orbitrap Mass Spectrometer. *Anal. Chem.* **2006**, *78*, 2113–2120.

(*4*) Olsen, J.; Macek, B.; Lange, O.; Makarov, A.; Horning, S.; Mann, M. Higher-Energy C-Trap Dissociation for Peptide Modification Analysis. *Nat. Meth.* **2007**, *4*, 709–712.

(*5*) Michalski, A.; Damoc, E.; Hauschild, J.-P., *et al.* Mass Spectrometry-Based Proteomics Using Q Exactive, a High-Performance Benchtop Quadrupole Orbitrap Mass Spectrometer. *Mol. Cell. Prot.* 2011, http://dx.doi.org/10.1074/mcp.M111.011015, online.

Alexander A. Makarov
Thermo Fisher Scientific GmbH, Bremen, Germany

Boris Alexandrovich Mamyrin

b. May 25, 1919
d. March 5, 2007

Professor Boris Mamyrin was a Russian scientist best known for his invention of the electrostatic ion mirror mass spectrometer known as the reflectron (*1*). He published about 320 papers, was the author of 2 books, and was granted 35 patents. His last position was head of the mass spectrometry laboratory at Ioffe Physical Technical Institute of the Russian Academy of Sciences in St. Petersburg (formerly Leningrad).

Mamyrin was born May 25 1919 in Lipetsk, Russian Federation, in what was then the Soviet Union. After graduating from high school in 1937, he entered the Physical Mechanical Department of the Leningrad Polytechnic Institute. His master's degree was obtained just before World War II. Throughout the war he served in the army. Being demobilized from military service after a three-year delay, Mamyrin entered the Leningrad Physical Technical Research Institute (PTI) for diploma work at the invitation of Academician Abram Ioffe, the was PTI organizer and Director and who had been instrumental in getting Boris released from the army.

A major focus of the PTI at the time was isotope separation in the context of a project to develop atomic power, work that required the development of mass spectrometry methods. Mamyrin began working in this area and, in spite of six years post-university interruption, Boris' talent allowed him to refresh his knowledge of physics and in 1949 he wrote and successfully defended his Ph.D thesis on 'Modulating devices for uranium isotope separation by high frequency method.'

At the PTI Mamyrin had the good fortune to join the group headed by the famous mass spectrometrists Vladimir Dukelsky and Nikolay Ionov. He concentrated on mass spectrometers in which separation and mass measurement of ions are based mainly on time and on radio frequency fields. The first time-of-flight (TOF) mass spectrometer with compatible analytical characteristics was created by Mamyrin in the early 1950s. Several of these instruments were built in the institute workshops, and results obtained with them were published in 1953 in the *Soviet Journal of Technical Physics*. One of the instruments was demonstrated at the World Exhibition in Brussels in 1958 (see this volume, *The Development of Time-of-Flight (TOF) Mass Spectrometry*). It was the only TOF mass spectrometer shown there.

The desire to achieve significantly better resolution for TOF mass spectrometers brought Mamyrin to the idea of the reflectron-TOF with an electrostatic mirror to remove the energy spread of ions originating within the ion source zone. The reflectron is now the accepted standard for TOF mass spectrometry, produced by all commercial manufacturers world-wide (*2*). The reflectron received the Soviet Inventor's Certificate in 1967. The idea was published in *Soviet Physics JETP* in 1973 and Mamyrin obtained UK (1977), French (1978) and German (1980) patents. Mamyrin's Dr.Sc. thesis 'Investigations in the field of ion separation by time of flight' defended in 1966 already contained a chapter on the reflectron.

In the same period, Mamyrin created magnetic resonance mass spectrometers. An instrument of this type was first described in the Proceedings of the Academy of Sciences of the USSR in 1969. It had a resolution of 350,000, with absolute sensitivity of several thousand atoms in a sample and the ability to measure isotopic ratios of up to 10^{-11}. In 1965 Mamyrin performed the most accurate measurement (for that time) of the magnetic moment of proton in units of nuclear magneton (*3*). Later, similar instruments became available commercially from the Plant of Scientific Instrumentation of the Academy of Sciences of USSR.

Using this instrument, Mamyrin's group measured $^3He/^4He$ ratio for geophysical and geochemical goals. He and co-workers identified anomalous $^3He/^4He$ ratios, investigated its relation to heat fluxes in the Earth's crust (the 'helium isotope thermometer'), and helped launch helium isotope geology generally (*4*). The high sensitivity and dynamic range of the method allowed them to measure the half-life of tritium with high precision. For the first time the half-lives of tritium atoms, molecules and bare tritium nuclei were obtained and the influence of the orbital electrons on the tritium life time was demonstrated (*5*).

Mamyrin was a member of ASMS and the Soviet and Russian Academy of Sciences, a permanent member of the Committee on Data for Science and Technology (CODATA), and a permanent member of the editorial board of *Journal of Technical Physics* and *Letters to Journal of Experimental and Theoretical Physics*. He received Konstantinov's award of Presidium of Soviet Academy of Sciences. In 2000, the ASMS award for Distinguished Contribution in Mass Spectrometry and in 2005 the gold medal of the Russian Society for Mass Spectrometry (*6*).

References

(*1*) Mamyrin, B. A.; Karataev, V. I.; Shmikk, D. V.; Zagulin, V. A. The Mass-Reflectron, a New Nonmagnetic Time-of-Flight Mass Spectrometer with High Resolution. *Sov. Phys. – JETP* **1973**, 37–45.

(*2*) Mamyrin, B. A. Time-of-Flight Mass Spectrometry (Concepts, Achievements, and Prospects). *Int. J. Mass Spectrom.* **2001**, *206*, 251–266.

(*3*) Mamyrin, B. A.; Frantsuzov, A. Measurement of Magnetic Moment of Proton in Units of Nuclear Magneton. *J. Exp. Theor. Phys. (USSR)* **1965**, *48*, 416, English trans.: *Soviet Phys.-JETP,* **1965**, *21,* 274.

(*4*) Mamyrin, B. A.; Tolstikhin, I. N. *Helium Isotopes in Nature*; Elsevier: Amsterdam, 1984.

(*5*) Akulov, A.; Mamyrin, B. A. Difference between the Triton Beta Decay Constants in Atomic and Molecular Tritium Measured by the Helium Isotope Method. *Phys. Letts. B* **2004**, *600*, 41–47.

(*6*) Anon. Obituary: Boris Alexandrovich Mamyrin: 1919–2007. *Rapid Commun. Mass Spectrom.* **2007**, *21,* 1691.

Eugene Nikolaev
Laboratory of Ion and Molecular Physics, Institute for Energy Problems of Chemical Physics, Russian Academy of Sciences, Moscow, Russia; Laboratory of Mass Spectrometry of Biomacromolecules, Institute for Biochemical Physics, Russian Academy of Sciences, Moscow, Russia; and Institute Biomedical Chemistry, Russian Academy of Medical Sciences, Moscow, Russia

Raymond E. March

b. March 13, 1934

Raymond March is a Canadian chemist whose work in the field of mass spectrometry and gas-phase ion chemistry involved the development and application of mass spectrometric instruments, particularly quadrupole ion trap mass spectrometers (*1,2*) and hybrid mass spectrometers, for both fundamental studies and the formulation of analytical protocols for the determination of compounds of environmental interest.

Born in 1934 in Newcastle-upon-Tyne, England, Raymond E. March holds both British and Canadian citizenships. He obtained a BSc (Hons) in Chemistry from University in 1957; a Ph.D. from the University of Toronto in 1961 (supervised by Professor John C. Polanyi (*3*), Nobelist 1986; From 1954-57, Ray was a Cadet Pilot in the Leeds University Air Squadron (RAFVR) and, from 1958-63, a Flight Lieutenant in the RCAF (AUX). From 1960-61, he held a Canadian Industries Limited Research Fellowship. From 1962-63, he was a Post-Doctoral Fellow with Prof. H.I. Schiff at McGill University, and a Research Associate from 1963-65, during which time he lectured at McGill University and Loyola College. In 1965, he joined the faculty of Trent University where he has conducted independent research for some 48 years in gas-phase kinetics, optical spectroscopy, gaseous ion kinetics, analytical chemistry, nuclear magnetic resonance spectroscopy, and mass spectrometry. He is presently Professor Emeritus of Chemistry at Trent University in Peterborough, ON, Canada.

Dr. March is a co-author with Dr. Richard J. Hughes and Dr. John F.J. Todd of *Quadrupole Storage Mass Spectrometry*, published in 1989. A second edition of *Quadrupole Storage Mass Spectrometry*, co-authored by Dr. March and Dr. John F.J. Todd was published in 2005. Drs. March and Todd co-edited three volumes entitled *Practical Aspects of Ion Trap Mass Spectrometry*, published in 1995. Two further volumes, 4 and 5, in the series *Practical Aspects of Trapped Ion Mass Spectrometry* were published in 2010 and 2009, respectively. Dr. March is a co-author with Oscar V. Bustillos and André Sassine of *A Espectrometria de Massas Quadupolar*, published in Portugese in 2005.

As a member of the Water Quality Centre (www.trentu.ca/wqc/), of which he was a founding member, his principal research interest lies in the mass spectrometric and nuclear magnetic resonance spectroscopic investigation (*4*) of natural compounds that, having been formed by plants, may enter waterways and possibly the water table as well. Current research involves the study of flavonoids and flavonoid glycosides; such compounds are found often in those products that have become known as neutraceuticals. An important aspect of this research is the development of appropriate analytical protocols for flavonoid glycosides in water and in plant extracts. At present, Dr. March is carrying out a mass spectrometry/ion mobility spectrometry study of protein noncovalent complexes together with an investigation of volatile compounds formed by Ash trees in response to attack by the Emerald Ash Borer.

In 1975, Dr. March was an Exchange Fellow (NRC-CNRS) at Orsay, France, with Prof. Jean Durup (*5*); in 1983, an Exchange Fellow (NRC-Royal Society of London) Swansea, Wales, with Prof. J.H. Beynon; in 1989 and 1992, a Visiting Professor, Université de Provence, Marseille, France, with Prof. Fernande Vedel; in 1993 and 1995, a CNRS Visiting Professor, Université Pierre et Marie Curie, Paris, France, with Prof. Jean-Claude Tabet; and in 1999, a Visiting Professor, Université de Provence, Marseille, France, with Yves Zerega. Dr. March has enjoyed long-term collaborations with Professor John Todd (see this volume, *John Francis James Todd*), and with colleagues at l'Université de Provence and l'Université Pierre et Marie Curie (France), and with colleagues in Padova (Italy).

Professor March is a Fellow of the Chemical Institute of Canada and a member of the American, British, and Canadian Societies for Mass Spectrometry. In 2009, he received the Gerhard Herzberg Award of the Canadian Society for Analytical Sciences and Spectroscopy (CSASS). In 1995, he received the Distinguished Faculty Research Award from Trent University. He was twice honored by the Canadian Mass Spectrometry Society, first with the Recognition Award in 1995 and with the Distinguished Contribution Award in 1997. He received a a D.Sc. from Leeds University in 2000 and an honorary doctorate (D h.c.) from l'Université de Provence in 2008.

Ray and his wife Kathleen have been married for 55 years; they have three daughters, Jacqueline,

Roberta, and Sally with spouses Paul, Stuart, and Lauren, respectively, and nine grandchildren, Shawn, Jessica, Thomas, Daniel, Rebecca, Sara, James, Madeline, and Carson, in order of appearance.

References

(*1*) Doran, M. C.; Fulford, J. E.; Hughes, R. J.; Morita, Y.; Bonner, R. F.; March, R. E. Effects of Charge Exchange Reactions on the Motion of Ions in Three-Dimensional Quadrupole Electric Fields. III. A Two Ion Model. *Int. J. Mass Spectrom. Ion Phys.* **1980**, *33*, 139–158.

(*2*) Londry, F. A.; Wells, G. J.; March, R. E. High Mass-Resolution Chemical Analysis with Paul Traps. *Hyperfine Interact.* **1993**, *81*, 179–187.

(*3*) March, R. E.; Polanyi, J. C. Photolysis by Pulsed Illumination. *Proc. Roy. Soc. (London)* **1963**, *A273*, 360–371.

(*4*) March, R. E.; Burns, D. C.; Ellis, D. A. Empirically-Predicted ^{13}C NMR Chemical Shifts for 8-Hydroxyflavone Starting from 7,8,4′-Trihydroxyflavone and from 7,8-Dihydroxyflavone. *Magn. Reson. Chem.* **2008**, *46*, 680–682.

(*5*) Fournier, P.; Benoît, C.; Durup, J.; March, R. E. Spectroscopie de double transfert de charge: séparation des termes de multiplet de Xe^{+} et Xe^{++} produits par collison H^{+} + Xe. *C. R. Acad. Sci. Paris B* **1974**, *278*, 1039–1041.

Raymond E. March
Trent University, Peterborough, ON, Canada

Alan G. Marshall

b. May 26, 1944

Alan G. Marshall, Ph.D., is an American chemist best known for his co-invention and continued development of instrumentation and applications of Fourier transform ion cyclotron resonance mass spectrometry (FT-ICR MS). FT-ICR offers the highest broadband mass resolution and mass accuracy of any mass analyzer (see this volume, *A History of Ion Cyclotron Resonance (ICR) and Fourier Transform (FTICR) Mass Spectrometry*).

Marshall was born on May 26, 1944, in Bluffton, Ohio, USA. He received a B.A. with Honors in Chemistry from Northwestern University in 1965, and a Ph.D. in Physical Chemistry under the guidance of John Baldeschwieler from Stanford University in 1970. His Ph.D. research was split between NMR and ICR spectral lineshape theory. He was hired at the University of British Columbia as an instructor in 1969 and became an associate professor of chemistry in 1976. In 1980 he moved to Ohio State University as professor of chemistry and biochemistry, and Director of the Chemical Instrument Center. In 1993 he moved to Florida State University where he continues to work as professor of chemistry and biochemistry, and Director of the Ion Cyclotron Resonance Program and the National High Magnetic Field Lab (*1,2*).

Marshall learned Fourier transform NMR after arriving at the University of British Columbia in 1969, where he was joined two years later by Asst Prof Melvin B. Comisarow (see this volume, *Melvin Barnet Comisarow*). Together they developed FT-ICR starting in 1973. The first FT-ICR spectrum $[CH_4]^+$ was recorded on 17 Dec 1973 (*3,4*). The instrumentation was commercialized, and in 1983, after moving to Ohio State University, Marshall acquired a commercial FT-ICR instrument in 1983. He went on to develop the theory of FT-ICR excitation and detection and introduced stored-waveform inverse Fourier transform (SWIFT), the optimum excitation waveform for ICR and ion trap MS (*5*). He published a monograph on FT spectroscopy (*6*), and, after moving to Florida State University in 1993, an FT-ICR 'primer' that is the most highly cited paper in the field (*7*). He introduced external ion accumulation (*8*), enabling coupling of inherently continuous ion sources (e.g., electrospray) to inherently pulsed mass analyzers (e.g., ICR, orbitrap). He established and continues to extend world records for broadband mass resolving power and mass accuracy (*9–11*).

Apart from mass spectrometry, Marshall has published theory and experiments for other forms of spectroscopy: NMR, electron magnetic resonance, magnetic resonance imaging, uv–vis absorption/scattering/circular dichroism, Raman, infrared, x-ray crystallography, and perturbed γ-γ angular correlations.

Marshall pioneered applications of ultrahigh-resolution MS, notably the creation of the field of 'petroleomics' (correlation and prediction of the properties and behavior of petroleum crude oil and its products, based on unique determination of up to tens of thousands of elemental compositions from a single mass spectrum (*12*)). Apart from the ~800 FT-ICR MS systems installed to date worldwide, FT-ICR demonstration of the value of high resolution and accurate mass for biological applications (e.g., post-translational modifications of proteins) has stimulated improvement in mass resolution of other, lower-resolution mass analyzers (time-of-flight; orbitrap).

Marshall served as President of the American Society for Mass Spectrometry (2005–06). He has been elected to Fellow of the American Association for the Advancement of Science (1989), Fellow of the American Phys. Society (1989), Fellow (Inaugural Group) of the Society for Applied Spectroscopy (2004), Fellow (Inaugural Group) of the American Chemical Society (2009), and Fellow, American Academy of Arts & Sciences (2013). His honors include the Eastern Analytical Symposium Analytical Chemistry Award (1991), the American Chem. Society Franklin and Field Award (1995), the Spectroscopy Society of Pittsburgh's Maurice F. Hasler Award (1997), the New York Society for Applied Spectroscopy's Gold Medal (1998), the American Society for Mass Spectrometry's Award for Distinguished Contribution (1999), the International Mass Spectrometry Society's Thomson Medal (2000), the Pittsburgh Spectroscopy Award (2002), the Eastern Analytical Symposium's Award in Mass Spectrometry (2009), and William H. Nichols Medal, Amer. Chem. Soc. New York Section (2012), Association of Biomolecular Research Facilities Award (2012), and Analytical Chemistry Award from the Society for Analytical Chemists of Pittsburgh.

Marshall has been married to his wife Marilyn for 48 years. They have one daughter, Wendy and one son, Brian.

References

(*1*) Gross, M. L. Focus on FT-ICR MS Commemorating Distinguished Contribution Award to A. G. Marshall and M. B. Comisarow. *J. Am. Soc. Mass Spectrom.* **2001**, *12*, 243–303.

(*2*) Bowers, M. T. Special Issue in Honour of A. G. Marshall. *Int. J. Mass Spectrom.* **2004**, *234*(1–3), 261 pages, (Marshall 60th Birthday Issue).

(*3*) Comisarow, M. B.; Marshall, A. G. Fourier Transform Ion Cyclotron Resonance Spectroscopy. *Chem. Phys. Lett.* **1974**, *25*, 282–283, (717 citations through May, 2013).

(*4*) Comisarow, M. B.; Marshall, A. G. "Fourier Transform Ion Cyclotron Resonance Spectroscopy Method and Apparatus." U.S. Patent #3,937,955 (issued February 10, 1976).

(*5*) Marshall, A. G.; Wang, T.-C. L.; Ricca, T. L. Tailored Excitation for Fourier Transform Ion Cyclotron Resonance Mass Spectrometry. *J. Am. Chem. Soc.* **1985**, *107*, 7893–7897, (565 citations through May, 2013).

(*6*) Marshall, A. G.; Verdun, F. R. *Fourier Transforms in NMR, Optical, and Mass Spectrometry: A User's Handbook*; Elsevier: Amsterdam, 1990. 460 pp. (Hardcover: ISBN 0-444-87360-0; Paperback: ISBN 0-444-87412-7): (458 citations through November, 2010).

(*7*) Marshall, A. G.; Hendrickson, C. L.; Jackson, G. S. Fourier Transform Ion Cyclotron Resonance Mass Spectrometry: A Primer. *Mass Spectrom. Rev.* **1998**, *17*, 1–35, (977 citations through May, 2013).

(*8*) Senko, M. W.; Hendrickson, C. L.; Emmett, M. R.; Shi, S. D.-H.; Marshall, A. G. External Accumulation of Ions for Enhanced Electrospray Ionization Fourier Transform Ion Cyclotron Resonance Mass Spectrometry. *J. Am. Soc. Mass Spectrom.* **1997**, *8*, 970–976, (367 citations through May, 2013).

(*9*) Marshall, A. G. Fourier Transform Ion Cyclotron Mass Spectrometry. *Acc. Chem. Res.* **1985**, *18*, 316–322.

(*10*) Marshall, A. G.; Grosshans, P. B. Fourier Transform Ion Cyclotron Resonance Mass Spectrometry: The Teenage Years. *Anal. Chem.* **1991**, *63*, 215A–229A.

(*11*) Marshall, A. G. Ion Cyclotron Resonance Mass Spectrometry: A Brief History. In: *Encyclopedia of Nuclear Magnetic Resonance*; Grant, D. M., Harris, R. K., Eds.; Wiley: London, 1996; vol. 1, pp 486–489.

(*12*) Marshall, A. G.; Rodgers, R. P. Petroleomics: Chemistry of the Underworld. *Proc. Natl. Acad. Sci. USA* **2008**, *105*, 18090–18095.

Alan G. Marshall

National High Magnetic Field Laboratory, Florida State University, Tallahassee, FL, USA

Hisashi Matsuda

b. March 2, 1924
d. November 13, 2011

Hisashi Matsuda was a Japanese physicist noted for work in ion optics calculation and the development of mass spectrometers. He is credited particularly for the design and construction of sector-type mass spectrometers.

Dr. Matsuda was born on March 2, 1924, in Osaka, Japan. He studied physics and mathematics at Osaka University and obtained his BS degree from the Department of Physics in 1950. On finishing his undergraduate program, he started his MS-related research work with Professor Koreichi Ogata, an atomic and isotope physicist. Together, they published papers in *Physical Review* in 1953 describing a modified Bainbridge-Jordan-type of mass spectrometer with a resolving power of 60,000 (*1*,*2*). He finished the department's regular graduate course in 1955 and studied theoretical and practical ion optics related to the construction of high resolution mass spectrometers as a PhD student under Dr. Ogata. He received a PhD from Osaka University in 1957 for his thesis on the properties of a large mass spectrometer of what is now called the "Ogata-Matsuda type" with a resolving power of 900,000 (*3*). The work was published in the *Journal of the Physical Society of Japan* in 1956.

Dr. Matsuda devoted his career to the design and development of mass spectrometers. He developed mass spectrometers with novel ion optics and modified existing instruments (*4–16*). In 1963 he made a magnetic-sector-type instrument with resolving power of 1,200,000 by introducing the concept of a very large dispersion magnetic field. In 1974 he designed a new instrument, now known as the "Matsuda type double-focusing mass spectrometer" named CQH, for measuring large biomolecules. In 1988, he constructed an ultra-large double-focusing modified CQH (9 tons in weight and 7 m in length) at Osaka University. Dr. Matsuda was one of the first to realize the importance of high mass and high resolution studies in the analysis of peptides and proteins, through his collaborations with scientists in the Protein Research Institute at Osaka University.

Dr. Matsuda completed novel ion-optics software for calculating and representing the fringing fields of magnetic and electrostatic lenses in 1972. The software and his ideas have been widely used for making magnetic-sector-type instruments, including a double-focusing mass spectrometer JMS-D300 (JEOL) and a super high resolution ion microprobe (the SHRIMP, from Australian Scientific Instruments). The D300 was a best-seller worldwide, and the SHRIMP was used to determine the geological age of zircons in Canada's Acasta Gneiss, thereby identifying the oldest known rocks on Earth. Further, an electrostatic analyzer design from 1975 named the "Matsuda plate" has been employed in the design of magnetic sector type double-focusing mass spectrometers, a multi-turn time-of-flight analyzer MULTUM II (Osaka University), and a spiral type time-of-flight mass analyzer, JMS-S3000 (JEOL), with resolving power of 60 000 at m/z 2093.

Dr. Matsuda accepted a position as assistant professor in Osaka University in 1955, became associate professor in 1959, and was promoted to full professor in 1963. He retired from Osaka University and received the honor of being named emeritus professor in 1987, after which he worked as a guest professor of Kansai Medical University in Japan (1987–1991) and Justus-Liebig University in Germany (1987–88), where he had also served once before (1968–69).

Dr. Matsuda was awarded the Nishina Memorial Award for "the development of a large dispersion mass spectrometer for measuring accurate mass of atoms" (1969), the TORAY Science and Technology Award for 'the making of a high-resolution mass spectrometer and its application to molecular science" (1982), the International Mass Spectrometry Foundation's Thomson Medal (1991), and the Mass Spectrometry Society of Japan Award for "the development of a high performance mass spectrometer" (1997). He was elected President of the Mass Spectrometry Society of Japan in 1981, serving till 1985. In 1990 he founded the Matsuda Fund, a scholarship program for encouraging and promoting mass spectrometrists. Continuing the educational work he began with his students Takekiyo Matsuo and Itsuo Katakuse at Osaka University, Matsuda has been able to nourish the development of many mass spectrometrists through this fund.

Dr. Matsuda and his wife Yoko married in 1956. They had two children.

References

(*1*) Ogata, K.; Matsuda, H. Masses of Light Atoms. *Phys. Rev.* **1953**, *89*, 27–33.

(*2*) Ogata, K.; Matsuda, H. On the Substandards of Atomic Mass. In: *Mass Spectrometry in Physics Research*; Hipple, J. A., *et al.*, Eds.; National Bureau of Standards: Washington, DC, Circular 522, 1953; pp. 59–66.
(*3*) Matsuda, H. Focusing Properties of a Large Mass Spectrograph Newly Constructed at Osaka University. *J. Phys. Soc. Jpn.* **1956**, *11*, 183–191.
(*4*) Matsuda, H. Electric Analyzer with Variable Focal Length. *Rev. Sci. Instrum.* **1961**, *32*, 850–852.
(*5*) Matsuda, H.; Fukumoto, S.; Kuroda, Y. A New Mass Spectrograph with Very Large Dispersion. *Z. Naturforschg.* **1966**, *21a*, 25–33.
(*6*) Matsuda, H.; Matsuo, T. Atomic Masses of ^{1}H, ^{16}O and ^{32}S. *J. Phys. Soc. Jpn.* **1968**, *25*, 950–952.
(*7*) Matsuda, H. The Influence of a Toroidal Electric Fringing Field on the Trajectories of Charged Particles in a Third Order Approximation. *Nucl. Instrum. Methods* **1971**, *91*, 637–647.
(*8*) Matsuda, H.; Wollnik, H. Third Order Transfer Matrices for the Fringing Field of Magnetic and Electrostatic Quadrupole Lenses. *Nucl. Instrum. Methods* **1972**, *103*, 117–124.
(*9*) Matsuda, H. Double Focusing Mass Spectrometers of Second Order. *Int. J. Mass Spectrom. Ion Phys.* **1974**, *14*, 219–233.
(*10*) Matsuda, H. Ion Optics of Parallel Plane Condensers. *Int. J. Mass Spectrom. Ion Phys.* **1975**, *18*, 367–377.
(*11*) Matsuda, H. Mass Spectrometers of High Transmission and High Resolving Power. *Nucl. Instum. Methods* **1981**, *187*, 127–136.
(*12*) Matsuda, H. High-Resolution High-Sensitivity Mass Spectrometers. *Mass Spectrom. Rev.* **1983**, *2*, 299–325.
(*13*) Matsuda, H. High-Resolution High-Transmission Mass Spectrometer. *Int. J. Mass Spectrom.* **1985**, *66*, 209–215.
(*14*) Matsuda, H.; Matsuo, T.; Fujita, Y.; Sakurai, T.; Katakuse, I. A New Mass Spectrometer at Osaka University. *Int. J. Mass Spectrom. Ion Processes* **1989**, *91*, 1–10.
(*15*) Matsuda, H. A New Mass Spectrograph for the Analysis of Dissociation Fragments. *Int. J. Mass Spectrom. Ion Processes* **1989**, *91*, 11–17.
(*16*) Matsuo, T. High Performance Sector Mass Spectrometer: Past and Present. *Mass Spectrom. Rev.* **1989**, *8*, 203–236.

Mitsuo Takayama
Yokohama City University, Mass Spectrometry Laboratory, Yokohama, Kanagawa, Japan

Josef Heinrich Elisabeth Mattauch

b. November 21, 1895
d. August 10, 1976

Josef Mattauch was an Austrian-born nuclear physicist and chemist who spent most of his career in Germany. He is primarily known for the design of the double-focusing sector Mattauch-Herzog geometry for mass spectrographs, and for determinations of isotope abundance ratios and isotope stabilities. He also lobbied enthusiastically among physicists for the introduction of the ^{12}C-mass reference scale proposed by A. O. C. Nier.

Josef Mattauch was born in 1895 in the industrial and coal mining city of Mährisch-Ostrau (Moravská Ostrava), in the what is now the Czech Republic, then part of the Austrian-Hungarian Empire.

He received his high school diploma in Wien (Vienna) in 1913. During several summer vacations he had become irresistibly attracted to chemistry and physics through informal conversations with a professor at the Prager Deutschen Technischen Hochschule who was a friend of his father. Mattauch was particularly fascinated by the attempts of Kamerlingh Onnes to liquefy hydrogen and helium gas. At the end of 1913, he enrolled into the Technische Hochschule in Wien; however, his studies were cut short in 1915 by three years military service in World War I. Upon his return, Mattauch enrolled in the School of Education at the Universität Wien and received a doctorate degree with teaching credentials in mathematics and physics in 1920. He remained affiliated with the physical institute of the university as a researcher and, after a seminal 2-year research stipend from the Rockefeller Foundation to study mass spectrography in the laboratories of Nobel Laureate R. A. Millikan in California, was promoted to full professor for physics in 1928. Intrigued by the precision of mass separation in the instruments developed by Aston (England) and Dempster (USA) and by the Einstein relation of mass and energy, Mattauch began a systematic investigation of isotope ratios in elements and their mass defects (see this volume, *Francis William Aston* and *Arthur Jeffrey Dempster*). He became one of the leading mass spectroscopists in Europe, and he was the first to develop a generalized rule for the stability of isotopes. The 'isobar rule' was formulated in 1934 and states that there are no stable isobars with neighboring atomic numbers and that there are no stable isobars with odd mass numbers. (*1,2*) In connection with this work, the $^{87}Rb^{87}Sr$ radioactive decay was developed into a method for geological dating by Mattauch in collaboration with Otto Hahn at the Kaiser Wilhelm Institute (KWI) für Chemie in Berlin-Dahlem, Germany. (*3*) The isobar rule along with several specific limiting conditions was important in the advancement of nuclear physics and atomic energy. Aston had shown that a measurable relativistic mass defect corresponding to the binding energies between protons and neutrons existed. With the capability of measurements of mass accuracies of up to 5 ppm, Mattauch was able to show that ionic species with the same nominal mass such as $^{12}CD_4$, $^{14}ND_3$, and $^{16}OD_2$ could be separated based on their mass defects representing the respective energy gain upon incorporation of deuterons (Fig. 1) (*4*).

Mattauch contributed substantially to the development of mass spectrographic instrumentation. To improve velocity focusing of ions, Mattauch and W. R. Smythe developed a mass spectrometer based exclusively on alternating electric fields without any magnetic field in 1932. (*5*) However, the problem of persistent contamination of instruments by mercury from the diffusion pumps was not solved and instruments of this kind did not become successful.

Mattauch and his student and colleague R. Herzog introduced an important advance in the technique of double-focusing mass spectrometry (see this volume, *Magnetic and Electrostatic Analyzers before 1960* and *Richard Franz Kark Herzog*). Mattauch was not the first to publish a double-sector design and in the middle and later 1930s other double-focusing instruments were built by Dempster, Bainbridge and Jordan, and Asada. (*6–8*) However, Herzog and Mattauch designed and built an instrument with an improved double focusing configuration. (*9–11*) The beam of ions was first parallelized in an electric field (velocity focusing) before the ions were separated by their respective mass-to-charge ratio in the magnetic field (directional focusing). The dispersion and line widths were directly proportional to $\sqrt{m/z}$. The instrument was able to measure over a wide mass range ratio of 36:1, which made measurements of lithium up to uranium in one mass spectrum possible. The instruments had exceptional sensitivity, and the

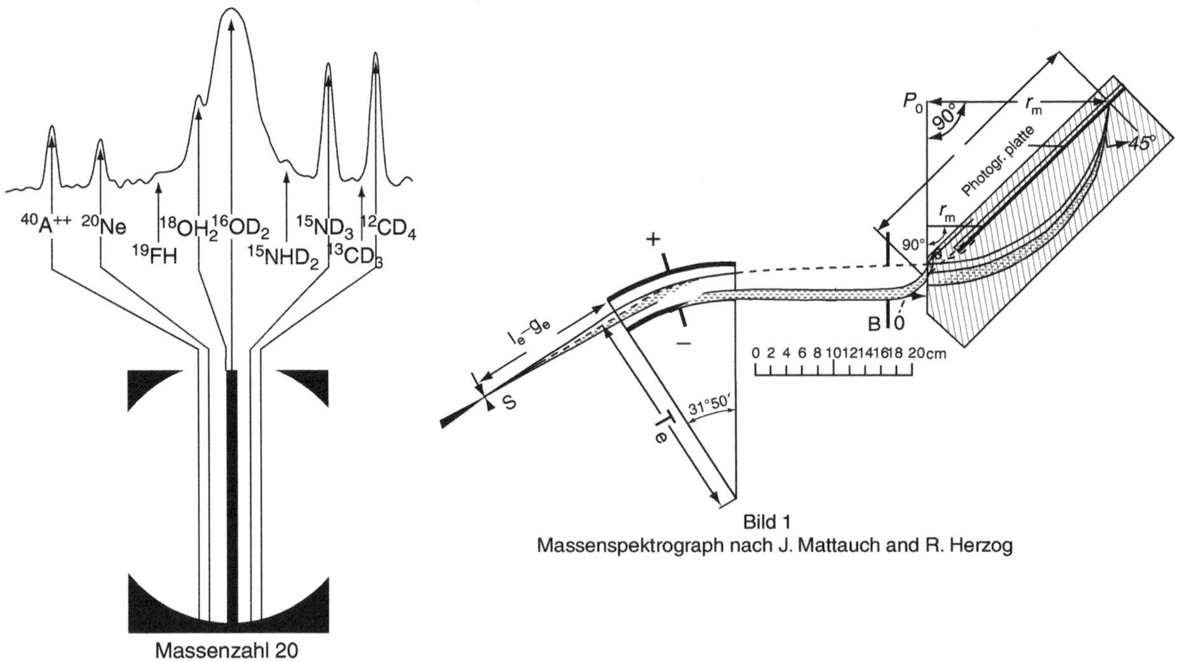

Figure 1
Left: Photometric detection of ions with mass number 20 to demonstrate the mass defect (energy gain) observed upon incorporation of one or two deuterons S. Hirzel Verlag, reproduced with permission from Mattauch, J. Substandards in Mass Spectrography, *Physik. Z.*, **1938,** *39*, 892–896; right: The double-focusing Mattauch-Herzog geometry for mass spectrographs. John Wiley & Sons, Ltd., reproduced with permission from Mattauch, J., Stabile Isotope, ihre Messung und ihre Verwendung, *Angew. Chem.*, **1947,** *59*(2), 37–42.

photographic plates allowed the detection of impurities at the ppm-level in a few minutes and at the ppb-level in a matter of a few hours exposure time. These features were particularly useful for the determination of isotopic compositions of elements. The Mattauch–Herzog configuration was incorporated in several commercial instruments. Interestingly, in a post-World War II publication, Mattauch lamented the success of mass spectrometers (electrometric detection) based on the directional focusing without energy focusing over mass spectrographs (*12*). Although the mass spectrometers had proven advantageous in quantitative isotope ratio measurements and were widely popular, the sensitivity of the photometric detection in mass spectrographs, on which he had relied, was still superior at that time.

With an initial donation from Carl Bosch (20,000 Reichs Mark) in 1935, Otto Hahn had initiated a major development of nuclear science in the Kaiser Wilhelm Institute (KWI) in Berlin. Lise Meitner, who was Jewish, lost her position as director of the physics institute of the KWI in 1938 and went to Sweden shortly before the 'Reichkristallnacht' and the acceleration of the anti-Jewish drive in Germany. In 1938 Mattauch, in German-annexed Austria, had refused twice to join Hitler's NSDAP (Nazi Party) but amid increasing political pressure, at the request of Otto Hahn, in 1939 he left Vienna and succeeded Meitner, who herself had supported his appointment. Mattauch quickly worked out plans for a large particle accelerator for the beginning of the 'Minerva Project', the German analog of the American 'Manhattan Project.' In 1941, he became a full member of the KWI and was promoted to vice director of the institute. Through the war years, the project suffered several setbacks including a British air strike against the accelerator. It was continued after the war as a smaller scale research project in the Max Planck Institute for Chemistry in Mainz. Mattauch was appointed director in 1947, a task of monumental proportions in the difficult years following WWII. The institute formally represented the continuation of the KWI and holds a wealth of historical documents in its archives today. (*13*)

Mattauch maintained good relations with many other international scientists, among them a close friendship, and friendly competition, with the American physicist Alfred O. C. Nier (see this volume, *Alfred Otto Carl Nier*). In 1957, Nier suggested to Mattauch that the different mass scales used in physical and chemical measurements should be unified on the basis of defining the mass of ^{12}C as exactly 12. (*14–16*) (A. Ölander independently noted that this could be a basis for the unification.) Mattauch, as a member of the international committee charged with working on the issue, was convinced and lobbied successfully for this new unified mass scale among physicists, while fellow member Edward Wichers did so among chemists. (*17*) The ^{12}C atomic mass scale is now the accepted standard.

In the years after WWII, Germany's democratic government began gradual political maneuvers toward the acquisition of atomic weapons, much to the dismay of many contemporary nuclear scientists. On 12 April 1957, Josef Mattauch and many others (the 'Göttinger Eighteen') signed the Göttinger Erklärung (Göttingen Manifesto), taking a very public and influential stand against Germany possessing nuclear weapons.

Mattauch remained prominent in nuclear studies until his retirement from the directorship of the Institute in Mainz on 21 November 1965. (*18*) He died on 10 August 1976 in Klosterneuburg, Austria. Shortly before his death, on the occasion of the 200-year anniversary of the USA, a Viking spacecraft landed a Mattauch-Herzog mass spectrometer on the surface of Mars and detected for the first time nitrogen in the Martian atmosphere (see this volume, *The Development of Mass Spectrometry in the Earth and Planetary Sciences*).

References

(*1*) Mattauch, J. The Systematics of Isotopes. *Z. Physik* **1934**, *91*, 361–371.
(*2*) Mattauch, J. Methods and Results of Research on Isotopes. *Physik. Z.* **1934**, *35*, 567–621.
(*3*) Mattauch, J. The Pair Rb87-Sr87 and the Isobar Rule. *Naturwissenschaften* **1937**, *25*, 189–191.
(*4*) Mattauch, J. Substandards in Mass Spectrography. *Physik. Z.* **1938**, *39*, 892–896.
(*5*) Smythe, W. R.; Mattauch, J. A New Mass Spectrometer. *Phys. Rev.* **1932**, *40*, 429–433.
(*6*) Dempster, A. J. New Methods in Mass Spectroscopy. *Proc. Am. Phil. Soc.* **1935**, *75*, 755–767.
(*7*) Bainbridge, K. T.; Jordan, E. B. Mass-Spectrum Analysis. 1. The Mass Spectrograph. 2. The Existence of Isobars of Adjacent Elements. *Phys. Rev.* **1936**, *50*, 282–296.
(*8*) Asada, T.; Okuda, T.; Ogata, K.; Yosimoto, S. Isotopic Weights of C12 and N14 by the Doublet Method. *Proc. Phys. Math. Soc. Japan* **1940**, *22*, 41–45.
(*9*) Mattauch, J.; Herzog, R. Mass Spectrograph. *Z. Physik* **1934**, *89*, 786–795.
(*10*) Mattauch, J. A Double-focusing Mass Spectrograph and the Masses of N15 and O18. *Phys. Rev.* **1936**, *50*, 617–623.
(*11*) Mattauch, J. Erratum: A Double-Focusing Mass Spectrograph and the Masses of N15 and O18. *Phys. Rev.* **1936**, *50*, 1089.
(*12*) Mattauch, J. Stable Isotopes, Their Measurement and Application. *Angew. Chem.* **1947**, *A59*, 37–42.
(*13*) Otto-Warburg-Haus, Boltzmannstraße 14, D-14195 Berlin-Dahlem, Germany. http://www.archiv-berlin.mpg.de (accessed October 24, 2014).
(*14*) Kohman, T. P.; Mattauch, J. H. E.; Wapstra, A. H. Carbon-12 as a Basis for a Common Scale of Nuclide Masses and Atomic Weights. *Naturwissenschaften* **1958**, *45*, 174–175.
(*15*) Mattauch, J. Mass Units for Atomic Weights and Nuclidic Masses. *Z. Naturforschung* **1958**, *13a*, 572–596.
(*16*) Mattauch, J. Rational Choice of a Unified Scale for Atomic Weights and Nuclidic Masses. *J. Am. Chem. Soc.* **1958**, *80*, 4125–4126.
(*17*) Holden, N. E. Atomic Weights and the International Committee: A Brief Historical Review. *Chem. International* **2004**, *26*, 4–7.
(*18*) Mattauch, J. Precision Measurements of Atomic Masses and Some of Their Implications on Nuclear Structure and Synthesis. In: *Advances in Mass Spectrometry*; Mead, W. L., Ed.; The Institute of Petroleum: London, 1966; vol. 3, pp 1–19.

Andreas H. Franz
Department of Chemistry, University of the Pacific, Stockton, CA, USA

James A. McCloskey

b. June 25, 1936

James McCloskey is an American chemist known for his work on the mass spectrometry (MS) of nucleic acids, for using MS to elucidate the structure and function of post-transcriptionally-modified nucleosides in RNA, for mapping these modifications in the ribosomal RNA of bacterial thermophiles for phylogenetic comparison with rRNA's from Archaea and Eucarya.

Jim McCloskey was born on June 6, 1936, in San Antonio, TX. He received a B.S. in chemistry in 1957 from Trinity University in San Antonio and a Ph.D. from M.I.T. in 1963, where he was one of the very early graduate students of Klaus Biemann (*1* and see this volume, *Klaus Biemann*). After completing his doctoral work he spent a year as an NIH Postdoctoral Fellow at the CNRS, Gif-sur-Yvette, France with Edgar Lederer. Between the time of completing his B.S. and Ph.D. he served as a Chemist in the United States Army Chemical Corps, U.S. Army Biological Laboratories, Fort Detrick, Maryland, 1959–61.

Dr. McCloskey began his academic career at the Institute for Lipid Research and Department of Biochemistry, Baylor College of Medicine in 1964 as an Assistant Professor, attaining the rank of Professor in 1971. He left Baylor in 1974 and became Professor of Biomedical Chemistry, Department of Medicinal Chemistry, University of Utah, Salt Lake City, UT, where he has remained, becoming Research Professor of Medicinal Chemistry in 2003. In addition he was a Visiting Professor at Tokyo University in Japan, from September 1971 to March 1972 and Visiting Investigator at the National Cancer Center Research Institute in Tokyo from 1971 to 1992.

In 1968, McCloskey and colleagues demonstrated the utility of trimethylsilyl (TMS) derivatization of the four major nucleosides and nucleotides of RNA and DNA. TMS-derivatization of these highly polar analytes was shown to be facile, quantitative, and afforded thermally stable products of sufficient volatility leading to successful analyses by GC-MS (*2*). McCloskey and colleagues also demonstrated the utility of perdeutero-TMS derivatization to facilitate the interpretation of mass spectra of trimethylsilyl compounds (*3*). Of the 107 known nucleosides that occur in RNA (*4*), nearly all have relied on mass spectrometry as the principal technology in structural elucidation. McCloskey and colleagues are responsible for determining 36 of the known nucleoside structures (*4–8*).

Dr. McCloskey's research interests have spanned two interrelated areas: the mass spectrometry of nucleic acids, and its use in structural studies and the discovery, structure determination and function of post-transcriptionally-modified nucleosides in RNA. There are nearly 100 different modified nucleosides known in RNA. They play a diverse but poorly understood variety of functional roles, including influence on secondary and tertiary structure and stabilization, and enhancement of fidelity of intermolecular interactions in protein synthesis. In his work McCloskey seeks a clearer understanding of modification – functional relationships through knowledge of the structures and sequence locations of these modified species, from a phylogenetically diverse source of organisms. Recent developments in mass spectrometry methodology offer significant new approaches to structural studies in this field. In particular his current interests include the chemistry of collision-induced dissociation of oligonucleotides.

McCloskey has been a long-time member of the American Society for Mass Spectrometry (ASMS), serving as secretary (1974–76), Vice-President for Programs (1976–78) and President (1978–80). He was made a Fellow of the Japan Society for the Promotion of Science in 1972 and an Honorary Professor at Kurume University, Kurume, Japan, in 1979. In 1989 he received the University of Utah Distinguished Research Award, and in 2005 the ASMS Distinguished Contribution in Mass Spectrometry for his work in the mass spectrometry of nucleic acids. In a special issue of the *Journal of the American Society for Mass Spectrometry* (JASMS) in honor of McCloskey, Guest Editors Pat Limbach and Dave Muddiman note in introducing McCloskey's achievements for the issue: "It is rare to readily identify a single individual as the undisputed leader in the practice of mass spectrometry for a particular field of study. Even more rare is the individual who has maintained an unparalleled level of success in his or her chosen field for over 35 years. Professor Jim McCloskey is such an individual. Jim's singular achievement has been to pioneer and continually advance the frontiers of mass spectrometry in the analysis of nucleic acid constituents" (*9*).

References

(1) Biemann, K.; McCloskey, J. A. Application of Mass Spectrometry to Structure Problems VI. Nucleosides. *J. Am. Chem. Soc.* **1962**, *84*, 2005–2007.

(2) McCloskey, J. A.; Lawson, A. M.; Tsuboyama, K.; Krueger, P. M.; Stillwell, R. N. Mass Spectrometry of Nucleic Acid Components. Trimethylsilyl Derivatives of Nucleotides, Nucleosides, and Bases. *J. Am. Chem. Soc.* **1968**, *90*, 4182–4184.

(3) McCloskey, J. A.; Stillwell, R. N.; Lawson, A. M. Use of Deuterium-Labeled Trimethylsilyl Derivatives in Mass Spectrometry. *Anal. Chem.* **1968**, *40*, 233–236.

(4) Cantara, W. A.; Crain, P. F.; Rozenski, J., *et al.* The RNA Modification Database, RNAMDB: 2011 Update. *Nucleic Acids Res.* **2011**, *39*, D195–D201.

(5) Sauerwald, D.; Sitaramiah, J. A.; McCloskey, D.; Söll; Crain, P. F. N6-Acetyladenosine: A New Modified Nucleoside from Methanopyrus Kandleri tRNA. *FEBS Lett.* **2005**, *579*, 2807–2810.

(6) Del Campo, M.; Recinos, C.; Yanez, G., *et al.* Number, Position and Significance of the Pseudourdines in the Large Subunit Ribosomal RNA of *Haloarcula marismortui* and *Deinococcus radiodurans. RNA* **2005**, *11*, 210–219.

(7) Pomerantz, S. C.; McCloskey, J. A. Detection of the Common RNA Nucleoside pseudouridine in mixtures of oligonucleotides by mass spectrometry. *Anal. Chem.* **2005**, *77*, 4687–4697.

(8) Wagner, T.; Nair, V.; Guymon, R., *et al.* A Novel Method for Sequence Placement of Modified Nucleotides in Mixtures of Transfer RNA. *Nucleic Acids Symp. Ser.* **2004**, *48*, 263–264.

(9) Limbach, P.; Muddiman, D. Editorial for Focus in Honor of James A. McCloskey, Recipient of the 2005 ASMS Award for a Distinguished Contribution in Mass Spectrometry. *J. Am. Soc. Mass Spectrom.* **2006**, *17*, I1–I2.

Alfred L. Yergey

National Institute of Child Health and Human Development, NIH, Bethesda, MD, USA

Fred W. McLafferty

b. May 11, 1923

Fred McLafferty is an American mass spectrometrist known for the breadth of his contributions to the field in a career spanning more than six decades. His early work focused on fundamental studies of electron ionization fragmentation pathways (one such mechanism now bears his name: the McLafferty rearrangement and is the only organic chemistry reaction named for a mass spectrometrist), the development and refinement of GC/MS interfaces, and pivotal efforts to create and mine mass spectral data bases, as well as to use computers in the interpretation of such spectra (see this volume, *The Beginnings of Organic Mass Spectrometry* and *A History of the Development of Mass Spectral Databases* and see also Chapter 4 (Volume 4): *McLafferty and Related Rearrangements*). His more recent activities have involved the use of FTICR instruments and studies of electron capture fragmentation of intact protein ions formed by electrospray ionization, so called top-down proteomics (see this volume, *Mass Spectrometry of Peptides and Proteins up to Proteomics*).

Fred W. McLafferty was born in Evanston Illinois on May 11th, 1923, and grew up in Nebraska. His high school teacher Mr. Dorsey engendered in him an interest in chemistry that lead him to major in the subject at the University of Nebraska in 1940. Having obtained an accelerated bachelor's degree he enlisted in the Army, becoming a member of a rifle infantry company that shipped overseas on Thanksgiving Day, 1944. He was awarded the Purple Heart, Presidential Unit Citation, and five bronze star medals for valor.

McLafferty returned to the University of Nebraska for the fall term of 1945 and received a Masters Degree in Analytical Chemistry in 1947. He pursued his doctorate at Cornell, working for William T. Miller on perchlorofluoro- compounds, and then a postdoctorate at the University of Iowa under Ralph Shriner. Unable to find an academic appointment, he entered industry in 1950 with the Dow Chemical Company in Midland, Michigan. There McLafferty found the mass spectrometry lab the most interesting, although he had never before seen such an instrument.

Working with a Westinghouse 90° sector instrument, he became interested in the interpretation of mass spectra. Dow had a ready supply of reference compounds that had been used in their pioneering development of the infrared spectrometer as an analytical tool. McLafferty had electron ionization (EI) mass spectra of these pure compounds measured whenever the mass spectrometers were free. He studied these spectra to build a body of interpretative information, as well as a library of mass spectra. He and his associate Roland Gohlke developed the first combined gas chromatograph/mass spectrometer (GC/MS) instrument based on the Bendix time-of-flight mass spectrometry (*1*, and see this volume, *The Development of Time-of-Flight (TOF) Mass Spectrometry* and *Roland S. Gohlke*). This seminal work lead others to investigate a variety of other techniques to interface the two instruments.

In 1956, convinced that mass spectrometry didn't offer much opportunity for advancement, McLafferty accepted Dow's offer to move to the East Coast and set up a laboratory for basic research in Framingham, Massachusetts. While there, he was able to spend time studying his mass spectral library that lead to the publication of his first book on the interpretation of mass spectra in 1963 (*2*). This formed the basis of his later (now classic) book *Interpretation of Mass Spectra* published in 1966 (*3*). The 4th edition with coauthor Frank Tureček was published in 1993. By 2009, the 9th Edition of the "Registry of Mass Spectral Data," which has its roots in the spectral compilations he had begun at Dow and Stenhagen and Abramsson had begun at Upsala, Sweden, had grown to more than 600,000 entries.

After successfully managing the start-up of the Eastern Research Laboratory for Dow, McLafferty moved in 1964 to Purdue University to help in the development of a modern program in analytical chemistry. It had by that time become obvious that mass spectrometry had more of a future than he had thought a decade earlier, so McLafferty dove into the field again with fundamental studies, analytical applications, and new instruments. In particular, he started to explore the analytical possibilities of using the mass spectrometer to measure product masses from the dissociation of a specific precursor mass, '2D-MS' or 'MS/MS'. McLafferty was intrigued by metastable ions in mass spectra and saw their potential to provide useful information regarding the structure of organic molecules (*4*). These ions had previously been considered an annoyance in mass spectra from single focusing magnetic sector instruments. He showed that MS/MS had far wider applicability using collisional activation to achieve dissociation, applying 'CAD' widely for the identification of organic compounds and then for mixture analysis (*5–7*). His group was the first to demonstrate the utility of MS/MS in the sequencing of peptides in a mixture, probably the largest current use of MS/MS (*8*).

McLafferty was wooed away from Purdue in 1968 by his alma mater Cornell, where he continued to pursue the use of computers in the acquisition, reduction, and interpretation of mass spectral data. By the mid-70s the application of GC/MS to the analysis of environmental pollutants had become a standard tool and the use of reliable library search routines was needed to deal with the massive volumes of data that could be generated in a single GC/MS analysis. This led to the development of Probability Based Matching (PBM), novel in its reverse search approach that is particularly powerful for the identification of compounds in the spectra of mixtures (*9*).

When liquid chromatography (LC) became recognized as a separation technique comparable in its importance to GC, its MS coupling became a primary problem that drew the attention of many researchers (see this volume, *A General Perspective on the Development of Liquid Chromatography Mass Spectrometry (LC/MS)*, and Chapter 3 [Volume 8]: *History of LC-MS Development and Interfacing*). McLafferty's approach depended primarily upon the use of chemical ionization, a technique popular in the 1970s (*10*). The serious resolving power handicaps of MS/MS were substantially alleviated by the McLafferty group's construction of the tandem double-focusing mass spectrometer (*11*). His research also used this instrument to establish the novel technique of Neutralization-Reionization MS (*12*). However, in the 1980s, the revolutionary new methods of electrospray ionization (ESI) and Fourier-transform (FT) MS would rapidly displace all other approaches to interfacing and MS/MS.

The Don Hunt and McLafferty groups first combined ESI with FTMS in 1989 began applying the unique resolving power and MS/MS capabilities of FTMS to large biomolecules (*13*, and see this volume, *Donald Frederick Hunt*). Efficient peptide and protein sequencing requires extensive MS/MS backbone fragmentation; electron capture dissociation, (ECD) was shown to be far more efficient than conventional dissociation methods such as collisional activation or infrared multiphoton dissociation (*14*). The McLafferty Group has been a leader in the approach to sequencing that introduces the whole protein, not peptides from its digestion, into the MS, so-called 'top-down' proteomics (*15*). This method has even been extended to the sequencing of >200 kDa proteins (*16*).

McLafferty is currently Cornell's Peter J. W. Debye Professor of Chemistry, Emeritus, and a member of the U. S. National Academy of Sciences, the American Academy of Arts and Sciences, and the Italian Academy of Sciences XL. His numerous awards include the American Chemical Society Awards in Chemical Instrumentation (1972), Analytical Chemistry (1981), and Mass Spectrometry (1989), the American Society for Mass Spectrometry Award for Distinguished Contribution to Mass Spectrometry (2003), International Mass Spectrometry Society's J. J. Thomson Medal (1985), Italian Chemical Society's G. Natta Medal (2000), Swedish Chemical Society's Torbern Bergman Medal (2001), Royal Society of Chemistry's Robert Boyle Medal (1992), and the French Chemistry Society's Lavoisier Medal (2004). McLafferty also received Honorary Degrees from the Universities of Nebraska (1983), Liege (1987), and Purdue (1995). McLafferty put some reflections on his own career in a memoir in 2011 (*17*).

References

(*1*) Gohlke, R. S.; McLafferty, F. W. Early Gas Chromatography/Mass Spectrometry. *J. Am. Soc. Mass Spectrom.* **1993**, *4*, 367–371.

(*2*) McLafferty, F. W.; *Mass Spectral Correlations*; American Chemical Society: Washington, 1963.

(*3*) McLafferty, F. W. *Interpretation of Mass Spectra; An Introduction*; W. A. Benjamin: New York, 1966.

(*4*) Shannon, T. W.; McLafferty, F. W. Identification of Gaseous Organic Ions by the Use of Metastable Peaks. *J. Am. Chem. Soc.* **1966**, *88*, 5021–5022.

(*5*) Haddon, W. F.; McLafferty, F. W. Metastable Ion Characteristics. VII. Collision Induced Metastables. *J. Am. Chem. Soc.* **1968**, *90*, 4745.

(*6*) McLafferty, F. W.; Kornfeld, R.; Haddon, W. F., *et al.* Application of Collisional Activation Spectra to the Elucidation of Organic Ion Structures. *J. Am. Chem. Soc.* **1973**, *95*, 3886.

(*7*) McLafferty, F. W.; Bockhoff, F. M. A Separation/Identification System for Complex Mixtures Utilizing Mass Separation and Mass Spectral Characterization. *Anal. Chem.* **1978**, *50*, 69–76.

(*8*) McLafferty, F. W.; Venkataraghavan, R.; Irving, P. Mass Spectrometric Studies of Peptides. IV. Determination of Amino Acid Sequences in Peptide Mixtures by Mass Spectrometry. *Biochemical and Biophysical Research Communications* **1970**, *39*, 274–278.

(*9*) McLafferty, F. W.; Hertel, R. H.; Villwock, R. D. Computer Identification of Mass Spectra. VI. Probability Based Matching of Mass Spectra. Rapid Identification of Specific Compounds in Mixtures. *Org. Mass Spectrom.* **1974**, *9*, 690–702.

(*10*) Baldwin, M. A.; McLafferty, F. W. Liquid Chromatography-Mass Spectrometry Interface. I. The Direct Introduction of Liquid Solutions Into a Chemical Ionization Mass Spectrometer. *Org. Mass Spectrom.* **1973**, *7*, 1111.

(*11*) McLafferty, F. W.; Todd, P. J.; McGilvery, D. C.; Baldwin, M. A. High-Resolution Tandem Mass Spectrometry (MS/MS) of Increased Sensitivity and Mass Range. *J. Am. Chem. Soc.* **1980**, *102*, 3360–3363.

(*12*) Danis, P. O.; Wesdemiotis, C.; McLafferty, F. W. Neutralization-Reionization Mass Spectrometry (NRMS). *J. Am. Chem. Soc.* **1983**, *105*, 7454–7456.

(*13*) Henry, K. D.; Williams, E. R.; Wang, B.-H.; McLafferty, F. W.; Shabanowitz, J.; Hunt, D. F. Fourier-Transform Mass Spectrometry of Large Molecules by Electrospray Ionization. *Proc. Natl. Acad. Sciences U.S.A.* **1989**, *86*, 9075–9078.

(*14*) Zubarev, R. A.; Kelleher, N. L.; McLafferty, F. W. Electron Capture Dissociation of Multiply Charged Protein Cations. A Nonergodic Process. *J. Am. Chem. Soc.* **1998**, *120*, 3265–3266.

(*15*) Kelleher, N. L.; Lin, H. Y.; Valaskovic, G. A.; Aaserud, D. J.; Fridriksson, E. K.; McLafferty, F. W. Top Down Versus Bottom up Protein Characterization by Tandem High-Resolution Mass Spectrometry. *J. Am. Chem. Soc.* **1999**, *121*, 806–812.

(*16*) Han, X.; Jin, M.; Breuker, K.; McLafferty, F. W. Extending Top-Down Mass Spectrometry to Proteins with Masses >200 kDa. *Science* **2006**, *314*, 109–112.

(*17*) McLafferty, F. W. A. Century of Progress in Molecular Mass Spectrometry. *Annu. Rev. Anal. Chem.* **2011**, *4*, 1–22.

Michael A. Grayson
Washington University in St. Louis, St. Louis, MO, USA (ret.)

Seymour Meyerson

b. December 4, 1916

Seymour (Sy) Meyerson is an American chemist known for his fundamental studies in mass spectrometry, concerning in particular but not exclusively the gas phase ion chemistry of aliphatic and aromatic hydrocarbons and their derivatives. He received for his outstanding achievements in mass spectrometry the Frank H. Field and Joe L. Franklin Award from the American Chemical Society in 1993 and was awarded an honorary doctorate in 1995 by Valparaiso University, Indiana.

Meyerson studied at the University of Chicago, where he obtained his bachelor's degree in chemistry in 1938 and subsequently took a few graduate courses in chemistry followed by several courses in social sciences – sociology, social psychology, and comparative religion. At that time jobs were very scarce, but in 1941 he obtained his first job as a chemist in a consulting laboratory in downtown Chicago. From there he went to a job as an inspector of powder and explosives at a plant near Joliet, Illinois, where he was employed by the army. In 1942 he enlisted in the Army Signal Reserve Corps for a training program in electronics, radio and radar. In the period of 1943–1946 he was in active service in the Army Signal Corps, from which he was detailed for about six months in the Army Corps of Engineers, with assignment to the Manhattan Engineering District – the atomic bomb project. At the end of 1945 he returned to the Signal Corps and went overseas for assignment in the Army of Occupation in Berlin, Germany. He returned to civilian life in August 1946 and later that year started to work as a chemist with Standard Oil Company (Indiana), which subsequently became American Oil Company and then Amoco Corporation, in Whiting, Indiana. Standard's research laboratories moved in 1970 to Naperville, Illinois. The first instrument he worked with was a Consolidated Engineering Corporation 21–101 mass spectrometer, which had been installed in 1943, the second commercially manufactured analytical mass spectrometer in the world (see this volume, *Magnetic and Electrostatic Analyzers before 1960* and *Mass Spectrometry Manufacturers and Markets from the Beginning to 1960*). He retired in 1984 as a research consultant, the company's highest professional job classification. Sy lives presently in Ashville, North Carolina with his wife Lotte. They have two daughters, Elana and Sheella, and one granddaughter, Jodi.

In the early stages of his career Meyerson began to correlate the electron-ionization mass spectra of alkylbenzenes with their structures (*1*). This helped to lay a basis for an understanding of the underlying chemistry that the ions undergo in the mass spectrometer, in effect a basis for the field of gas-phase ion chemistry, which has since been widely explored by mass spectrometrists and other chemists (see this volume, *The Beginnings of Organic Mass Spectrometry and Mass Spectrometric Methods for the Determination of Thermodynamic Data*). Highlights of his work in the 1950s are phenyl-cationized cyclopropane (*2*) and the tropylium ion (*3*). The phenyl-cationized cyclopropane structure was proposed for the (M-ethyl) cation of *t*-amylbenzene, which loses a molecule of ethylene with the carbon atoms of the side-chain participating equivalently as revealed by ^{13}C-labeling. This was the first suggested ion/molecule complex occurring in a unimolecular dissociation of an ion in the gas phase, a concept that was not explicitly recognized in the mass-spectrometry community until the eighties, after which it came to be invoked frequently in gas-phase ion chemistry.

The tropylium ion structure was proposed for the (M-H) cation of toluene because of the equivalent behavior of all its hydrogen atoms during the loss of acetylene, as shown by deuterium labeling and afterwards further confirmed by ^{13}C-labeling of the α-position, which demonstrated the equivalence of all carbon atoms in the ion. This ion structure is common for the $C_7H_7^+$ ions of *m*/*z* 91 in the electron-ionization mass spectra of alkylbenzenes (*4*).

Another important contribution from Meyerson's work concerns the chlorine isotopic abundances. In correcting the parent-peak intensities of unlabeled *p*-chlorobiphenyl for naturally-occurring isotopes he found that the then generally accepted value of 75.4% for the ^{35}Cl isotope found in reference tables did not match his data. Analysis of the mass spectra of several other chlorine-containing compounds led to an abundance value for the ^{35}Cl isotope of 75.8% (*5*), in excellent agreement with the 75.770_5 value redetermined soon afterwards and published by the National Bureau of Standards.

Five years before his retirement Meyerson and collaborators at the National Institute of Environmental Health Sciences, the University of North Carolina at Chapel Hill and the State University of New York at Stony Brook published work on the prominent peak at *m*/*z* 79 in the negative-ion chemical ionization mass spectra of some organophosphorus pesticides. Meyerson's longstanding collaboration with an organophosphorus chemist led him to suspect that the peak at *m/z* 79 was attributable to the monomeric metaphosphate anion PO_3^- (*5*). This was confirmed by accurate mass measurements and was of great interest for the following reason: the metaphosphate anion was first proposed in 1955 as an intermediate in the aqueous hydrolysis of phosphomonoesters, and phosphate chemists had been trying for about twenty-five years to establish its existence, while in the mass spectrometry literature much evidence was available for the metaphosphate anion and other metaphosphate species, albeit with no explicit reference to these species. As Meyerson stated: "Thus, organic chemists and mass spectrometrists interested in organophosphorus compounds had worked in isolation from each other all these years, quite oblivious to each others concern, even when they were working with similar or identical compounds."

During his career Meyerson noted many parallels between reactions in the mass spectrometer and those occurring in more familiar organic-chemical systems, an observation that led him to study thermally driven reactions of neutral molecules in collaboration with colleagues from university and government laboratories, as well as Standard Oil. As an industrial chemist he has shown by his fundamental research that mass spectrometry is not merely an instrumental technique but also a scientific discipline by which much can be learned about the chemistry of charged species in the gas phase! Meyerson's investigation of these phenomena has inspired many people to perform research in this area in later years.

The selected publications cited below are representative of the author's most significant contributions to the field (*1–5*).

Besides his scientific work, Seymour served as a member of the ASTM (American Society for Testing of Materials) Committee E-14 on Mass Spectrometry, which in 1969 was reorganized as the American Society for Mass Spectrometry. In addition, he served on the editorial boards of the journals Organic Mass Spectrometry and Mass Spectrometry Reviews.

References

(*1*) Meyerson, S. Correlations of Alkylbenzene Structures with Mass Spectra. *Appl. Spectrosc.* **1955**, *9*, 120–130.

(*2*) Rylander, P. N.; Meyerson, S. Organic Ions in the Gas Phase. I. The Cationated Cyclopropane Ring. *J. Am. Chem. Soc.* **1956**, *78*, 5799–5802.

(*3*) Rylander, P. N.; Meyerson, S.; Grubb, H. M. Organic Ions in the Gas Phase. II. The Tropylium Ion. *J. Am. Chem. Soc.* **1957**, *79*, 842–846.

(*4*) Meyerson, S. Reminiscences of the Early Days of Mass Spectrometry in the Petroleum Industry. *Org. Mass Spectrom.* **1986**, *21*, 197–208.

(*5*) Meyerson, S. Tropylium, Chlorine Isotopic Abundances, Monomeric Metaphosphate Anion, and Conestoga Wagon Theory. *J. Am. Soc. Mass Spectrom.* **1993**, *4*, 761–768.

N.M.M. Nibbering (d.)

Laser Lab Amsterdam, Vrije Universiteit, HV Amsterdam, The Netherlands

Fred Loomis Mohler

b. August 23, 1893
d. December 2, 1974

Fred Mohler was an American analytical chemist whose whole research career was with the National Bureau of Standards (NBS, now the National Institute of Standards and Technology, NIST). After extensive research work on atomic spectra and ionization, he turned to mass spectrometry following World War II and became head of the Bureau's section for mass spectrometry. He was a leading organizer of the movement that resulted in the creation of the national scientific society for mass spectrometry in the US.

Fred Mohler was born in 1893 in Wilbraham, Massachusetts. His father, John Frederick Mohler, was a physicist and mathematician who moved to Dickinson College, Carlisle, Pennsylvania, in 1896 as professor of physics. Fred Mohler graduated from Dickinson in 1914 and earned his Ph.D. at Johns Hopkins in 1917 under the direction of R. W. Wood. He was hired for the NBS pyrometry section but became part of the atomic physics section when it was created in the wake of World War I. For over two decades his research centered on topics intimately connected with the development of atomic physics and quantum mechanics. He headed the atomic physics section starting in 1927.

After distinguished service as analyst for the US Army Air Force in World War II and a short job as a consultant for declassifying Manhattan Project materials, Mohler resumed work at the NBS and was asked to set up a mass spectrometer for some work on measuring trace impurities. He became chief of the Bureau's mass spectrometry section in 1946, a position in which he continued until retirement.

Mohler quickly realized that a catalog of standardized mass spectra for pure substances would be very useful for anyone undertaking analyses of mixtures, and he was successful in getting this category added to the data collected and disseminated by Project 44, an on-going effort of the American Petroleum Institute, based at that time at the NBS, devoted to identifying characteristics of hydrocarbons and related substances. Mohler then dedicated a full-time staff position in his section to measuring and tabulating mass spectra of pure compounds (*1*,*2*, and see this volume, *A History of the Development of Mass Spectral Databases*).

Mohler's mass spectrometric research, much of it done together with Vernon Dibeler, was mostly oriented to determining the mass spectra of various types of hydrocarbons (see this volume, *Vernon Hamilton Dibeler*). Considerable attention was devoted to characterizing metastable ions. Some examples of his research in mass spectrometry are listed below (*3*–*9*).

Mohler helped organize and took part in a symposium on mass spectrometry in physics research as part of the NBS's semi-centennial celebrations in 1951. He also played a leading role in creating a scientific organization for mass spectrometry oriented primarily to chemistry, organizing a gathering of those working with mass spectrometers within the framework of the recently-created Pittsburgh Conference on Analytical Chemistry and Applied Spectroscopy (PITTCON). Mohler found that interest was strong enough to support creation of a formal group for mass spectrometry as a committee under the aegis of the American Society for Testing and Materials (ASTM). This committee (E-14) officially met first in 1952, during the third PITTCON; it organized annual meetings from 1953 on until the American Society for Mass Spectrometry was organized separately from ASTM in 1969 (*10* and see this volume, *Origins of Professional Mass Spectrometry Societies*).

Fred Mohler retired from the Bureau in 1960 though he continued as a consultant for some years. He died in Washington in 1974.

References

(*1*) Mohler, F. L. *Autobiographical Notes on Scientific Work*. Archives, National Institute of Standards and Technology: Gaithersburg, Maryland, US, 1963.

(*2*) Biographical Records for Fred L. Mohler. Archives, National Institute of Standards and Technology: Gaithersburg, Maryland, US. 1960

(*3*) Dibeler, V. H.; Wise, C. E. Jr.; Mohler, F. L. Metastable Ion Transitions in the Mass Spectra of Monocholoropropenes. *Phys. Rev.* **1947**, *71*, 381.

(*4*) Mohler, F. L. Mass Spectra of Hydrocarbons. *J. Wash. Acad. Sci.* **1948**, *38*, 193.

(*5*) Mohler, F. L.; Williamson, L.; Dean, H. M. Total Ionization of Hydrocarbons from Mass Spectral Data. *J. Res. NBS*. **1950**, *45*, 235.

(*6*) Mohler, F. L.; Dibeler, V. H.; Reese, R. M. Mass Spectra of Flourocarbons. *J. Res. NBS.* **1952**, *49*, 343.
(*7*) Mohler, F. L.; Dibeler, V. H.; Reese, R. M. Some Ions of High Kinetic Energy in Mass Spectra of Polyatomic Molecules. *J. Chem. Phys.* **1954**, *22*, 394.
(*8*) Bradt, P.; Mohler, F. L. Analysis of Flourinated Polyphenols by Mass Spectrometer. *Anal. Chem.* **1955**, *27*, 875.
(*9*) Mohler, F. L.; Bradt, P.; Dibeler, V. H. Mass Spectra of Aromatic Hydrocarbons Filtered from Smoky Air. *J. Res. NBS.* **1958**, *60*.
(*10*) Grayson, M. A., Ed. *Measuring Mass From Positive Rays to Proteins*; Chemical Heritage Press: Philadelphia, PA, 2002.

Keith A. Nier
Independent Historian, Madison, NJ, USA

Howard Redfern Morris

b. August 4, 1946

Howard R. Morris is a British biological chemist widely noted for his work in the field of biomolecular mass spectrometry. He has worked in the field of mass spectrometry research for 46 years and has made contributions that include structure elucidations of important natural products, new methods and strategies for biopolymer analysis and instrumentation development (*1–10*).

Morris was born on August 4, 1946, in Bolton, Lancashire, England. He received a B.Sc. Joint Honors in Mathematics and Chemistry from the University of Leeds in 1967 and a Ph.D. in Biophysics from the University's Astbury Department of Biophysics in 1970. From 1970–72 he served as SRC Post-Doctoral Fellow at the Cambridge University Chemical Laboratory, and from 1972–75 as a staff member of the Cambridge MRC Laboratory of Molecular Biology. In 1975 he moved to Imperial College in London, where he founded the Biomolecular Mass Spectrometry Group. He first served as a Lecturer in Biochemistry, later becoming Professor of Biological Chemistry (1980) and serving as Head of the Department of Biochemistry (1985–1988).

Morris's achievements in mass spectrometry include structure elucidations of a number of natural products, including characterisation of the active form of Vitamin D as 1,25 dihydroxycholecalciferol, the characterisation of Enkephalin (the first endorphin), Slow Reactive Substance of Anaphylaxis (SRS-A), the bronchoconstrictor released in the asthma crisis, and the inducer of malaria development in the mosquito. Morris is also noted for the first demonstration of protein sequence determination by MS (1968/9) including his introduction of the 'peptide mixture analysis' strategy for protein sequencing that eliminated the requirement for purified peptides in classical sequence analysis, the first development of glycopeptide sequencing strategies (1978), the development of a mass spectrometric method for disulphide bridge assignment (1985) and glycosylation analysis (1978/1980), and the introduction of the concept and method of 'peptide mapping' (fingerprinting) as a protein identity and post-translational modification test (1983).

Other contributions to the field have resulted from his proposals to instrument manufacturers for new developments to advance the field of Biopolymer research, including the first High Field Magnet mass spectrometer (mid 1970s) and the concept of a novel geometry Q-TOF mass spectrometer (early 1990s) for low collision energy ultra-high sensitivity MS/MS analysis that facilitated unambiguous spectral interpretation and helped drive the proteomics revolution. In 1979, Morris founded M-SCAN: Mass Spectrometry Consultants and Analysts to the Chemical, Petrochemical, Pharmaceutical and Biotechnology Industries to facilitate the transfer of technological expertise from academia to industry.

Morris has held several invited Visiting Professor positions in the USA, the former USSR, and Europe. In both 1996 and 2001 he received the Rector's Award for Excellence in Teaching at Imperial College. Morris also served as the Royal Society Representative on the Council of the Institute of Cancer Research, London, from 1994–2001 and as a Trustee of the Institute from 2001–2005.

Morris has won the Biochemical Society Gold Medal and Prize for Analytical Biochemistry (1978), the Royal Society of Chemistry Medal and Prize for Macromolecules and Polymers (1982), the University of Naples, Italy, Gold Medal for Distinguished Contributions to Biopolymer Structure Elucidation and Mass Spectrometry (1989), the DOW Distinguished Lecturer Award in Analytical Chemistry from the University of British Columbia, Canada (1990), the Dottore Honoris Causa Degree in Molecular Biotechnology from the University of Naples, Italy (2005), the Blaise Pascal Medal and Prize for Life Sciences and Medicine from the European Academy of Sciences (2010), and the Franklin Medal and Prize from the Institute of Physics for 'contributions to mass spectrometer design which revolutionized peptide sequencing and fuelled the proteomics revolution' (2012). He was elected a Fellow of the Royal Society (FRS) in 1988 for his work in Biomolecular Mass Spectrometry.

Morris has been married twice, first to Lene Jensen and currently to Maria Panico (also a mass spectroscopist working at Imperial College, London). He is the father of four children: Helen and Anna with Lene and Christina and Alexander with Maria. Since 2001 he has been Emeritus Professor of Biological Chemistry and Senior Research Investigator, Department of Molecular Biosciences, at Imperial College, London.

The publications cited below are representative of the author's most significant contributions to the field.

References

(*1*) Hughes, J.; Smith, T. W.; Kosterlitz, H. W.; Fothergill, L.; Morgan, B. A.; Morris, H. R. Identification of two Related Pentapeptides from the Brain with Potent Opiate Agonist Activity. *Nature* **1975**, *258*, 577–579.

(*2*) Morris, H. R.; Thompson, M. R.; Osuga, D. T., *et al.* Antifreeze Glycoproteins from the Blood of an Antarctic Fish: The Structure of the Proline-Containing Glycopeptides. *J. Biol. Chem.* **1978**, *253*, 5155–5161.

(*3*) Morris, H. R.; Taylor, G. W.; Piper, P. J.; Tippins, J. R.; Samhoun, M. N. Structure of Slow-Reacting Substance of Anaphylaxis (SRS-A) From Guinea-Pig Lung. *Nature* **1980**, *285*, 104–106.

(*4*) Morris, H. R.; Dell, A.; McDowell, R. A. Extended Performance Using a High-Field Magnet Mass Spectrometer. *Biomed Mass Spectrom.* **1981**, *8*, 463–473.

(*5*) Morris, H. R.; Panico, M.; Karplus, A.; Lloyd, P. E.; Riniker, B. Elucidation by FAB-MS of the Structure of a New Cardioactive Peptide from Aplysia. *Nature* **1982**, *300*, 643–645.

(*6*) McDowell, R. A.; Morris, H. R. Fast Atom Bombardment Mass-Spectrometry – Biological Analysis Using An Ion Gun. *Int J Mass Spectrom.* **1983**, *46*, 443–446.

(*7*) Beloff-Chain, A.; Morton, J.; Dunmore, S.; Taylor, G. W.; Morris, H. R. Evidence that the Insulin Secretagogue, Beta-Cell-Tropin, is ACTH22–39. *Nature* **1983**, *301*, 255–258.

(*8*) Billker, O.; Lindo, V.; Panico, M., *et al.* Identification of Xanthurenic Acid as the Putative Inducer of Malaria Development in the Mosquito. *Nature* **1998**, *392*, 289–292.

(*9*) Morris, H. R.; Paxton, T.; Dell, A., *et al.* High Sensitivity Collisionally-Activated Decomposition Tandem Mass Spectrometry on a Novel Quadrupole/Orthogonal-Acceleration Time-Of-Flight (Q-TOF) Mass Spectrometer. *Rapid Commun Mass Sp.* **1996**, *10*, 889–896.

(*10*) Morris, H. R.; Dell, A.; Easton, R. L., *et al.* Gender-Specific Glycosylation of Human Glycodelin Affects its Contraceptive Activity. *J Biol Chem.* **1996**, *271*, 32159–32167.

Howard R. Morris

Division of Molecular Biosciences, Imperial College, London, UK

James Douglas Morrison

b. November 9, 1924
d. February 1, 2013

James Morrison was born in Glasgow, Scotland, on November 9, 1924 and received his tertiary education at the University of Glasgow from 1942–1948 following schooling at the Morgan and Grove Academies, Dundee, and the Lenzie Academy, Lenzie, Scotland. He started out his scientific career intending to become a synthetic organic chemist but, being more interested in chemical structure determination, chose the research area of x-ray crystallography instead, studying for his PhD under the supervision of Professor J. Monteath Robertson.

Morrison entered the field of mass spectrometry in 1949 following an appointment as a research officer with the Australian Council for Scientific and Industrial Research. His first task was to explore the potential of a newly acquired CEC 211–02 mass spectrometer as an analytical chemical tool. This he did, although he very quickly recognized that mass spectrometry could also be used to study the mechanisms of ion formation and energy transfer, thereby providing valuable information about the structures and energetics of gas-phase ions.

Morrison's initial research involved the detailed analysis of ionization processes induced by electron ionization. His determination of experimental threshold laws for different types of ionization processes confirmed the theoretical ideas of Wigner and Wannier that were based on very simple statistical mechanical arguments. For the first time this made it possible to interpret ionization efficiency curves to yield meaningful ionization energies and hence obtain bond energy data for the ground and excited states of ions (*1*). He was able to extend these studies to photon induced ionization-fragmentation processes following his involvement in the construction of the first variable wavelength photoionization mass spectrometer in 1956 (*2*). This produced photoionization efficiency curves with sufficient resolution that autoionization and vibrational fine structure in excited states of ions could be distinguished. Much of this early work helped to lay the foundations for an understanding of the energetics of gas-phase ions that forms the basis of all mass spectrometry (see this volume, *Mass Spectrometric Methods for the Determination of Thermodynamic Data*).

During this period Morrison devoted considerable effort to the problem of electron energy spread removal associated with threshold ionization experiments, designing and building several different electron velocity selectors for the production of monoenergetic electrons (*3*). However, in 1959 he decided to take a different approach, devising an analytical process using Fourier transforms that he called 'deconvolution' to remove or unfold the smearing effect from raw ionization efficiency data (*4*). This he first carried out manually with Beevers-Lipson strips (a consequence of his x-ray crystallography training) and an adding machine, and later by using CSIRAC, one of the world's first digital computers. Although there was much scepticism at the time, deconvolution is now a widely accepted technique for improving instrumental resolution, being used in numerous fields other than just mass spectrometry.

Morrison was a pioneer in the use of computers for identification of unknown mass spectra from catalogs of mass spectra (*5*) and developed programs to carry out complete ab initio identification, rather than library searching (*6*). The advent of minicomputers in the 1960s allowed him to construct one of the first computer-controlled gas chromatograph-mass spectrometer combinations in 1968. An innovative feature of this magnetic sector instrument was the large laminated magnet designed to minimize eddy currents and help overcome the hysteresis effects typically associated with fast mass scans. These magnets are now routinely used in such instruments.

As part of his countless machine building activities, Morrison, together with his PhD student Don McGilvery, initiated the development of the industry standard ion optics program called SIMION, designed to calculate potential distributions for different electrode configurations and to model electron and ion trajectories. The significance of this software was acknowledged in 1998 by the American Society for Mass Spectrometry with a joint award for a Distinguished Contribution in Mass Spectrometry to McGilvery and David Dahl.

In 1974, Morrison's primary interest in fundamental ion chemistry led him to construct a novel triple quadrupole mass spectrometer (see this volume, *A Perspective on the Development of Tandem Mass Spectrometry* and *Christie George Enke*). Its combination with a pulsed laser and etalon, all computer-controlled, allowed photodissociation studies of gas-phase cations to be carried out and provided the measurement of a very detailed rovibrational spectrum for a mass analyzed ion (*7*). The same triple quadrupole mass spectrometer was used to explore its application to mass spectrometric analyses (*8*), resulting in a patent in 1980 for what is now a very widely used analytical instrument (*9*).

Morrison had always had a keen interest in herbal chemistry and in 1986, with the very enthusiastic collaboration of aboriginal tribes of the Northern Territory, he became involved in an Australian bicentennial aboriginal pharmocopoeia project. This involved mass spectral examination of a vast amount of material, resulting in the publication of a reference book in 1988 (*10*).

There have been numerous awards recognizing Morrison's long-standing contributions to mass spectrometry. Notably he has received both the Rennie Memorial Medal (1954) and the H. G. Smith Medal (1961) of the Royal Australian Chemical Institute, a medal of the University of Liege (1963) and the Queen's Jubilee medal (1977). In 1964 he was elected as a Fellow of the Australian Academy of Science and in 1985 as a Fellow of the Royal Society of Edinburgh. This was followed in 1990 by his appointment by Queen Elizabeth II as an Officer of the Order of Australia. There have been special issues of *Organic Mass Spectrometry* (*11*) and the *International Journal of Mass Spectrometry* (*12*) in his honor. In recognition of the role he played in the advancement of mass spectrometry in Australia, the Australian and New Zealand Society of Mass Spectrometry established in 1990 the Morrison Lectureship, which is the flagship plenary lecture at each of its biennial conferences. All of these attest to the significant impact Morrison's work has had in the development of mass spectrometry.

The high personal regard in which Morrison was held by his colleagues is reflected in Fred McLafferty's *Organic Mass Spectrometry* tribute, entitled, "Jim Morrison: a mass spectrometrists for all seasons." (*13*) Morrison's personal reminiscences on his career were published in that same volume (*14*).

References

(*1*) Morrison, J. D. Electron Impact Spectroscopy. *Rev. Pure Appl. Chem.* **1954**, *4*, 22–59.

(*2*) Hurzeler, H.; Inghram, M. G.; Morrison, J. D. Photon Impact Studies of Molecules using a Mass Spectrometer. *J. Chem. Phys.* **1958**, *28*, 76–82.

(*3*) Marmet, P.; Morrison, J. D. A Mass Spectrometer for Ionization Studies using an Electron Velocity Selector. *J. Chem. Phys.* **1961**, *35*, 746–747.

(*4*) Morrison, J. D. On the Optimum Use of Ionization Efficiency Data. *J. Chem. Phys.* **1963**, *39*, 200–207.

(*5*) Crawford, L. R.; Morrison, J. D. Computer Methods in Analytical Mass Spectrometry, Identification of an Unknown Compound in a Catalog. *Anal. Chem.* **1968**, *40*, 1464–1469.

(*6*) Crawford, L. R.; Morrison, J. D. Computer Methods in Analytical Mass Spectrometry, Development of Programs for Analysis of Low Resolution Mass Spectra. *Anal. Chem.* **1971**, *43*, 1790–1795.

(*7*) McGilvery, D. C.; Morrison, J. D. A Mass Spectrometer for the Study of Laser-Induced Photodissociation of Ions. *Int. J. Mass Spectrom. Ion Phys.* **1978**, *28*, 81–92.

(*8*) Yost, R. A.; Enke, C. G.; McGilvery, D. C.; Smith, D. L.; Morrison, J. D. High Efficiency Collision Induced Dissociation in an R.F. only Quadrupole. *Int. J. Mass Spectrom. Ion Phys.* **1979**, *30*, 127–136.

(*9*) Enke, C. G.; Yost, R. A.; Morrison, J. D. Tandem Quadrupole Mass Spectrometer for Selected Ion Fragmentation Studies and Low Energy Collision Induced Dissociatior Therefor. U.S. Patent 4,234,791, Nov 18, **1980**.

(*10*) Morrison, J. D.; Smith, J. F.; Bratspies, G. K. Composition of Essential Oils. In: *Traditional Bush Medicine: An Aboriginal Pharmacopoeia*; Barr, A., Ed.; Greenhouse Publications: Richmond, Australia, 1988; pp 232–254, (ISBN: 0–86436-167-X).

(*11*) Editorial (page 179) Derrick, P. J., Maccoll, A, and Traeger, J. C. eds. Issue in Honour of James D. Morrison, *Org. Mass Spec.* **1991**. 26, 179–349.

(*12*) Traeger, J. C. Ed. Issue in Honour of James D. Morrison. *Int. J. Mass Spectrom.* **2000**, *194*, 93–296.

(*13*) McLafferty, F. W. Jim Morrison – Mass Spectrometrist for All Seasons. *Org. Mass Spectrom.* **1991**, *26*, 181–182.

(*14*) Morrison, J. D. Personal Reminiscences of Forty Years of Mass Spectrometry in Australia. *Org. Mass Spectrom.* **1991**, *26*, 183–194.

John C. Traeger
La Trobe University, Victoria, Australia

Milam Stephen Burnaby Munson

b. March 10, 1933

Milam Stephen Burnaby Munson was born on March 10, 1933 in Wharton, TX. Known to the mass spectrometry community as Burnaby, he is the son, grandson, and great-grandson of Texas lawyers. He might have joined the family trade had not a 1950 chemistry course at tiny Tarleton State College (now University, Stephenville, Fort Worth, TX) captured his fancy. Later, while completing his BA at the University of Texas in Austin, he initiated what was to be a long connection with Prof. Robbin Anderson, starting with senior research investigating thermal reactions of acetylene in flames. The mentorship continued to a 1956 MA thesis on 'Particle Formation in the Pyrolysis of Acetylene,' and culminated (after a brief experiment with Madison winters in John Margrave's high-temperature chemistry laboratory at the University of Wisconsin) with a 1959 UT PhD thesis on 'Effects of Oxygen and Chlorine on the Pyrolysis of Acetylene.'

While completing his thesis work, he was introduced to mass spectrometry when a friend at Humble Oil and Refining in Baytown, TX, provided him with a low-energy electron ionization mass spectrum of one of his complex reaction mixtures – GC/MS was not yet available in the 1950s. He joined Humble (later Esso Research and Engineering, then Exxon) as a Research Chemist in 1959, 'rising through the ranks' of Senior Research Chemist, then Research Specialist. His initial assignment at Humble was in the area of separations in solution thermodynamics and engineering. His education in mass spectrometry began in 1961, when Fred Lampe's departure from Humble to Penn State created a vacancy in Joe Franklin's basic research group (see this volume, *Joe Louis Franklin, Jr.*). The opening provided an opportunity for Burnaby to join Franklin, Frank Field, and technician Wilburn Geiger, with the apparent expectation that Munson's expertise in kinetics would complement Franklin and Field's knowledge of ion-molecule reactions and Geiger's instrumental prowess (see this volume, *Frank Henry Field*). Munson began working with an instrument that Field had designed to operate at pressures of 0.1–0.3 Torr ('ultra-high pressure' in those days). It was a fertile collaboration, accounting for roughly 10% of Munson's 118 publications, including his first reviewed publication (*1*) and his most-cited, the seminal 1966 *JACS* paper introducing chemical ionization mass spectrometry (*2*). Franklin left Humble for a Welch Chair in chemistry at Rice University in Houston, TX, in 1963, so that, while Field was visiting Leeds in 1964, Munson and Geiger had the opportunity "to play with the instrument mostly on [their] own." (*3*) Munson counts the resulting 1965 *JACS* paper (*4*) as one of his 'favorites,' in part because it contested the B-strain theory of H.C. Brown. The work was included with that cited (in 1992) when Munson received the American Chemical Society Field and Franklin Award for Outstanding Achievement in Mass Spectrometry (see this volume, *A History of Chemical Ionization (CI) in Mass Spectrometry*).

When Field left Baytown for the main Esso Labs in New Jersey in 1966, he took the high pressure mass spectrometer with him. Perhaps with that prompting, Munson commenced the academic phase of his career about a year later at the University of Delaware, rising from Associate to Full Professor in 1972, and commencing service to the University Honors Program in 1976, service that continues to the date of this writing.

Burnaby served as President of the American Society for Mass Spectrometry from 1980–82. In addition to the Franklin and Field Award previously cited, he shared the American Society for Mass Spectrometry Award for Distinguished Contribution in Mass Spectrometry with Frank Field in 1996 for their invention of chemical ionization. He received the Eastern Analytical Symposium Award for Outstanding Achievements in the Fields of Analytical Chemistry in 2007, over the course of his long career at the University of Delaware, he has been honored with numerous University and regional teaching and research awards.

References

(*1*) Munson, M. S. B.; Field, F. H.; Franklin, J. L. High Pressure Mass Spectrometric Study of Reactions of Rare Gases with N_2 and CO. *J. Chem. Phys.* **1962**, *37*, 1790–1799.

(*2*) Munson, M. S. B.; Field, F. H. Chemical Ionization Mass Spectrometry. I. General Introduction. *J. Am. Chem. Soc.* **1966**, *88*, 2621–2630.
(*3*) Munson, B. A Pseudo-Scientific Autobiography. *Eur. J. Mass Spectrom.* **2004**, *10*, 137–140.
(*4*) Munson, M. S. B. Proton Affinities and the Methyl Inductive Effect. *J. Am. Chem. Soc.* **1965**, *87*, 2332–2336.

Kelsey D. Cook
Department of Chemistry, University of Tennessee, Knoxville, TN, USA

Nicolaas Martinus Maria Nibbering

b. May 29, 1938
d. August 25, 2014

Nico (as he was generally known) Nibbering was a Dutch physical organic chemist known for his fundamental studies of the unimolecular and bimolecular chemistry of both positive and negative ions in the gas phase.

Nico was born on May 29, 1938 in Zaandam, The Netherlands. He received his bachelor's, master's, and PhD degrees all *cum laude* at the University of Amsterdam, The Netherlands. His PhD thesis, entitled, 'Mass spectrometry of some aralkyl compounds with a functional group in the side-chain' was supervised by the late Thymen J. de Boer. In 1967, a year before he completed his degree, he was appointed a permanent staff member in charge of mass spectrometry for the organic chemistry laboratory of the University of Amsterdam. He became associate professor of organic mass spectrometry in 1975 and full professor in 1980. When the Institute of Mass Spectrometry was founded at the University of Amsterdam in 1988, the name of his chair was changed to Chemical Mass Spectrometry and he became its scientific director, a position he held until his (early) retirement in October, 2001. Since his retirement he has been a guest professor at the Vrije Universiteit in Amsterdam and at the University of Twente in Enschede, The Netherlands. Nico lived in Abcoude near Amsterdam, with his wife Tini de Waart. He had four children Erik, Hans, Vincent, and Karin and two granddaughters and five grandsons.

From the beginning of his career, Nico has had interest not only in the background of the chemical behavior of ions in the gas phase but also in instrumental/methodological developments and analytical applications of mass spectrometry. His principal contributions to the field include

- the elucidation of reaction mechanisms of unimolecular reactions of organic gas-phase ions using stable isotopic labeling and electron ionization, chemical ionization and field ionization mass spectrometry, in particular time-resolved mass spectrometry (field ionization kinetics) in combination with collision-induced dissociations.
- negative ion mass spectrometry, including the development of negative ion field ionization and for analytical applications of negative ion field desorption mass spectrometry.
- the elucidation of reaction mechanisms of ion/molecule reactions (both positive and negative ions) using a drift-cell and later on an in-house designed and constructed broad-band Fourier transform ion cyclotron resonance mass spectrometer
- the design, construction and coupling of an orthogonal time-of-flight mass spectrometer to a reversed geometry double-focusing sector mass spectrometer in his laboratory to photodissociate a mass selected and decelerated ion beam by interaction with a laser beam.

Over the course of his long career Nico was a popular visitor in other laboratories. He was spent a year as a visiting faculty member of Cornell University in Ithaca, New York, USA, with Professor Fred W. McLafferty in 1974. He was a guest professor at the University of Colorado, Boulder, USA, with Professor Charles H. DePuy in 1980, where he presented a summer course in advanced mass spectrometry. In 2002, he was a guest professor at the 'La Sapienza' University of Rome, Italy with Professor Maurizio Speranza.

Nico served the international mass spectrometry community in a number of capacities. He was chair of the Dutch mass spectrometry discussion group (1973–1975) and chaired the 12th International Mass Spectrometry Conference in Amsterdam, The Netherlands, in 1991. He also served as president of the European Society for Mass Spectrometry (1997–2000). He was a founding member of the International Mass Spectrometry Society (renamed the International Mass Spectrometry Foundation in 2006) and served as its president from 2000–2003. He was elected as member of the Royal Netherlands Academy of Arts and Sciences in 1989.

Nico's work has been widely recognized with awards and prizes. He won the Unilever chemistry prize in 1964 and the Shell research chemistry prize in 1968. He was honored for his work in mass spectrometry by the International Committee of the 12th International Mass Spectrometry Conference with the J.J. Thomson Medal award in 1991 and by the Czechoslowak Spectroscopic Society with the Joannes Marcus Marci award in 1992. A representative sampling of his work is given in the references (*1–24*), in the last of which he recounts the pleasures of his career in the field in a piece entitled, 'Four decades of joy in mass spectrometry.'

References

(*1*) Dawson, J. H. J.; Noest, A. J.; Nibbering, N. M. M. 1,1 and 1,3 Eliminations of Water from the Reaction Complex of $O^{\bullet-}$ with 1,1,1-Trideuteroacetone. *Int. J. Mass Spectrom. Ion Phys.* **1979**, *30*, 189–191.

(*2*) De Koning, L. J.; Nibbering, N. M. M. Formation of the Long-Lived $H_2O^{\bullet-}$ Ion in the Gas Phase. *J. Am. Chem. Soc.* **1984**, *106*, 7971–7973.

(*3*) De Koning, L. J.; Fokkens, R. H.; Pinkse, F. A.; Nibbering, N. M. M. High Front- and Back-End Resolution MS/MS in Fourier Transform Ion Cyclotron Resonance Mass Spectrometry. *Int. J. Mass Spectrom Ion Processes* **1987**, *77*, 95–105.

(*4*) De Lange, W.; Nibbering, N. M. M. Evidence for Stepwise Processes in the Charge Reversal of Negative Ions. *Int. J. Mass Spectrom. Ion Processes* **1986**, *68*, 111–125.

(*5*) De Visser, S. P.; de Koning, L. J.; Nibbering, N. M. M. Chemical and Thermodynamic Properties of Methyl Chloride Dimer Radical Cations in the Gas Phase. *J. Am. Chem. Soc* **1998**, *120*, 1517–1522.

(*6*) Freriks, I. L.; de Koning, L. J.; Nibbering, N. M. M. Gas-Phase Ambident Reactivity of Acyclic Enolate Anions. *J. Am. Chem. Soc.* **1991**, *113*, 9119–9124.

(*7*) Gur, E. H.; de Koning, L. J.; Nibbering, N. M. M. The Bimolecular Hydrogen-Deuterium Exchange Behavior of Protonated Alkyl Dipeptides in the Gas Phase. *J. Am. Soc. Mass Spectrom.* **1995**, *6*, 466–477.

(*8*) Heck, A. J. R.; de Koning, L. J.; Pinkse, F. A.; Nibbering, N. M. M. Mass-Specific Selection of Ions in Fourier-Transform Ion Cyclotron Resonance Mass Spectrometry. *Rapid Commun. Mass Spectrom.* **1991**, *5*, 406–414.

(*9*) Kleingeld, J. C.; Nibbering, N. M. M. A Fourier Transform Ion Cyclotron Resonance Study of the Mechanism of Formation of Protonated Dimethyl Ether from Methanol by use of Naturally Occurring ^{18}O. *Org. Mass Spectrom.* **1982**, *17*, 136–139.

(*10*) Kleingeld, J. C.; Nibbering, N. M. M. The Long-Lived H_3O^- Ion in the Gas Phase: Its Formation, Structure and Reactions. *Int. J. Mass Spectrom. Ion Phys.* **1983**, *49*, 311–318.

(*11*) Kleingeld, J. C.; Ingemann, S.; Jalonen, J. E.; Nibbering, N. M. M. Formation of the NH_4^- Ion in the Gas Phase. *J. Am. Chem. Soc.* **1983**, *105*, 2474–2475.

(*12*) Maas, W. P. M.; Nibbering, N. M. M. Formation of Doubly Charged Negative Ions in the Gas Phase by Collionally-Induced Ion-Pair Formation from Singly Charged Negative Ions. *Int. J. Mass Spectrom. Ion Processes* **1989**, *88*, 257–266.

(*13*) Matimba, H. E. K.; Crabbendam, A. M.; Ingemann, S.; Nibbering, N. M. M. The Gas-Phase Acidity of the Phenyl Radical and Some $C_6H_4X^{\cdot}$ (X=F, Cl, CN) radicals. *J. Chem. Soc. Chem. Commun.* **1991**, *9*, 644–645.

(*14*) Molenaar–Langeveld, T. A.; Fokkens, R. H.; Nibbering, N. M. M. An Unusual Pathway for the Elimination of Hydrogen Cyanide from Benzonitrile upon Electron Impact as revealed by ^{13}C and ^{15}N Labelling. *Org. Mass Spectrom.* **1986**, *21*, 15–21.

(*15*) Nibbering, N. M. M. Mechanistic Studies by Field Ionization Kinetics. *Mass Spectrom. Rev.* **1984**, *3*, 445–477.

(*16*) Noest, A. J.; Nibbering, N. M. M. Homoconjugation vs. Charge Dipole Interaction Effects in the Stabilization of Carbanions in the Gas Phase. *J. Am. Chem. Soc.* **1980**, *102*, 6427–6429.

(*17*) Ott, K. H.; Röllgen, F. W.; Zwinselman, J. J.; Fokkens, R. H.; Nibbering, N. M. M. Negative ion Field Desorption Mass Spectra of Some Inorganic and Organic Compounds. *Org. Mass Spectrom.* **1980**, *15*, 419–422.

(*18*) Russell, D. H.; Gross, M. L.; van der Greef, J.; Nibbering, N. M. M. Decomposition of the 1,3-Butadiene Radical Cation on the Microsecond to Picosecond Time Scale. *J. Am. Chem. Soc.* **1979**, *101*, 2086–2090.

(*19*) Theissling, C. B.; Nibbering, N. M. M. An Ion Cyclotron Resonance Study of the Formation of the $[C_6H_6O]^{+\bullet}$ Ions from the Molecular Ions of 2-Phenoxyethyl Halides. *Advan. Mass Spectrom.* **1978**, *7B*, 1287–1295.

(*20*) Van der Greef, J.; Nibbering, N. M. M. The Detection of Unimolecular Gas-Phase Decompositions of Negative Ions with Lifetimes of 10^{-11}–10^{-5} s. *Int. J. Mass Spectrom. Ion Phys.* **1979**, *31*, 71–75.

(*21*) Von Köding, H.; Nibbering, N. M. M. A Fourier Transform Ion Cyclotron Resonance Study of Two- and One-Electron Capture Reactions Between Doubly Charged Rare Gas Ions and Rare Gas Atoms. *Int. J. Mass Spectrom* **1999**, *185/186/187*, 281–289.

(*22*) Nibbering, N. M. M. Gas-Phase Ion/Molecule Reactions as Studied by Fourier Transform Ion Cyclotron Resonance. *Acc. Chem. Res.* **1990**, *23*, 279–285.

(*23*) Nibbering, N. M. M. The McLafferty Rearrangement. A Personal Recollection. *J. Am. Soc. Mass Spectrom.* **2004**, *15*, 956–958.

(*24*) Nibbering, N. M. M. Four Decades of Joy in Mass Spectrometry. *Mass Spectrom. Rev.* **2006**, *25*, 962–1017.

N.M.M. Nibbering (d.)
Laser Lab Amsterdam, Vrije Universiteit, HV Amsterdam, The Netherlands

Alfred Otto Carl Nier

b. May 28, 1911
d. May 16, 1994

Source: www.nap.edu

Alfred Otto Carl Nier, generally known as Al, was an American physicist whose designs and improvements of mass spectrometers in the course of diverse investigations transformed the world of mass spectrometry (*1–4*). His research spread through the areas of nuclear science and earth and planetary science. His work on instrumentation usually was undertaken in order to address specific questions in those areas. He pioneered or significantly advanced the application of mass spectrometry to field after field throughout his career. The importance of his work was acknowledged by many major scientific honors.

Al Nier's parents each immigrated to the United States from Germany. Al was born in St. Paul, Minnesota, where his father operated a dry cleaning business. Al had only one sibling, a sister eleven years older who soon enough had a family of her own, so Al was almost an only child for much of his youth. For awhile he aspired to having his own machine shop but after graduating from high school at sixteen, in 1927 he entered the electrical engineering program at the University of Minnesota. The Great Depression significantly changed his career path since it eliminated clear job opportunities as he finished his bachelor's and then his master's degrees in electrical engineering. He therefore took advantage of an offer to become an assistant and a Ph.D. student in the physics department and to pursue a life in science rather than industry.

John Tate, then the most prominent member of the physics faculty at Minnesota, became his adviser. Al Nier applied his engineering background to significantly improve a mass spectrometer, based on designs of Dempster, that Tate no longer used (see this volume, *Arthur Jeffrey Dempster*). He responded to Tate's advice to find a research topic in the then new area of nuclear physics by taking up the study of isotopes. His first publication reported one way to make mass spectrometers work better; the second announced the discovery of ^{40}K, the potassium isotope of mass 40 (*5,6*). This 1935 combination, in which instrumental improvements led to opening new areas of study, characterized much of the rest of Al Nier's life (*7,8*).

His early success brought an offer of National Research Council support and a post-doctoral position with Ken Bainbridge. Newly married, he went to Harvard University in 1936, built a better mass spectrometer, and carried out research of epoch-making significance in several fields. In the two years before he returned to Minnesota, Nier made mass spectrometry crucial for the future of atomic weight determination and he pioneered the study of the isotopic variations of lead which allow determination of the ages of the earth and the solar system. In addition, he and Earl Gulbransen demonstrated that the ratio of stable carbon isotopes varies in the natural world, opening another future area of isotope geochemistry (*9,10* and see this volume, *The Development of Mass Spectrometry in the Earth and Planetary Sciences*).

Al Nier joined the physics faculty at the University of Minnesota in 1938 and continued that connection for the remainder of his life, though officially he was on leave during the last two years of World War II. Upon his return to Minnesota, his work proceeded in two different instrumental directions and several different topic areas. In terms of instrumentation, he built another 180° magnetic analyzer mass spectrometer, similar to the type he worked with before going to and when he was at Harvard, and at the same time he created a second, significantly modified, type of instrument.

The fundamental features of his prior instruments were adopted at this time in the design of the first successful commercial mass spectrometer, soon to be produced by Consolidated Engineering Corp. and marketed primarily for analysis in the oil industry (see this volume, *Mass Spectrometry Manufacturers and Markets from the Beginning to 1960*). Nier's further research with this 180° analyzer type of instrument was possibly even more influential than the industrial developments, for he achieved the first isolation of ^{235}U, the uranium isotope of mass 235, collecting tiny samples that were just enough to use in demonstrating that this is the highly fissionable isotope. The demonstration was a significant contribution to the growing understanding of nuclear phenomena; the isolation was a crucial step in the eventual development of nuclear weapons and nuclear power (see this volume, *A History of Mass Spectrometry in Nuclear Science*).

The second, quite new, type of instrument begun in 1939 was the 60° magnetic sector design that made it possible to introduce mass spectrometry to more and more fields of science. Unlike the 180° magnetic analyzer he had worked with before, the sector design placed the source and the detector outside the region of high field strength; it cost far less to build; was easier, safer, and less expensive to operate; and was as or more accurate. This design, published and publicized in 1940, made it more convenient to try changes in various parts of the system and far easier for others to plan research efforts in diverse fields that called for mass spectrometric analysis. Nier soon worked to demonstrate this with cooperative investigations of various questions in biochemistry employing carbon 13 as a tracer (*11–13* and see this volume, *Magnetic and Electrostatic Analyzers before 1960*).

During World War II, Nier was drawn into the Manhattan Project and the job of concentrating ^{235}U. He designed and helped build new sector mass spectrometers to analyze the isotope ratios of the outputs of the electromagnetic, centrifuge, and gaseous diffusion processes being developed (see this volume, *Mass Spectrometry and the Separation and Enrichment of Isotopes*). Then he produced the prototypes for the hundreds of sector instruments that checked every bit of the gaseous diffusion plants for leaks as they were built and those that monitored every stage of their operation. For decades after this, the overwhelming majority of mass spectrometers in the world were based on his 60° magnetic sector designs (*14,15*).

After the war, Nier was able to publish the essentials of the designs produced during it, including the summation of his improvements to the electron impact ionization source which thereafter became a much-replicated bit of apparatus on its own. He returned to studies of isotopic masses and abundances, both improving measurements and helping to collect and codify the best available information (*16*). In the process, he and L. T. Aldrich laid the foundation for development of the potassium-argon dating method for geochronology.

With the start of the 1950s, Al Nier became a somewhat more widely known authority. Having been elected to the National Academy of Sciences before turning forty, he was asked to author articles for both popular and technical reference works. He wrote several reports for scientists and engineers in other fields to publicize the capabilities of mass spectrometers. He was involved in the formation of the predecessor to the American Society for Mass Spectrometry and was one of the principle organizers of one of the earliest international conferences for those concerned with these instruments (*17*). And he also created a new instrument that pioneered the introduction of mass spectrometers in hospitals and biomedical studies.

More far-reaching consequences flowed from his design of a new double-focusing mass spectrometer in 1949. Using a 90° electrostatic field to focus ions differing in energy and a 60° magnetic sector for directional focusing, along with new control systems, made significantly more accurate mass measurements possible for atoms and larger molecules, and allowed more refined discrimination among components of a mixture. These capabilities have been fundamental for determination of universal standards of nuclidic, atomic, and molecular masses. He was also a pioneer in the application of mass spectrometry to the study of meteorites. Nier's promotion of the capabilities of this design spurred John Beynon's pioneering extension of mass spectrometry into the domain of organic chemistry – a movement he observed with interest but did not join (*18* and see this volume, *The Beginnings of Organic Mass Spectrometry* and *John Herbert Beynon*).

For decades Al Nier taught introductory, experimental, and theoretical physics to large numbers of undergraduate and graduate students at Minnesota, and welcomed many post-doctoral associates as well. He eagerly shared techniques and insights and even built apparatus for other labs to facilitate their research. On campus and at scientific meetings, he was happy to report his results, happier to discuss methods and devices, and happiest to get other scientists, particularly younger ones, to talk about their projects and plans. His first marriage broke up after thirty years; his second marriage lasted from 1969 to his death.

Al Nier's ostensible retirement in 1980 ended his classroom and committee duties but his research career changed very little, save for the diversion of some energy to retrospective articles. While much of the world of mass spectrometry had moved toward organic molecular and biological subjects, and toward other types of sources and analyzers, from the dawn of the space age Nier's research had been oriented more toward rockets and the solar system (*19*). In this he continued to rely on EI sources and sector analyzers. In addition to continuing an interest in meteorites, he was among the pioneers who made mass spectrometers sufficiently small, rugged, and reliable to function despite the extreme stresses of rocket launches and the environments of the upper atmosphere, space, and other planets. He made extensive and intensive studies of the composition and structure of the upper regions of the earth's atmosphere. He went on to lead the Viking Project team that analyzed the atmosphere of Mars and, as the only scientist named to two of Viking's science teams, to design the mass spectrometers used with gas chromatographs to investigate the Martian soil and search for products of living organisms. He subsequently joined the investigation of the atmosphere of Venus, and in his nominal retirement he undertook extensive studies of how the distribution of noble gases illuminates the nature and history of lunar and interplanetary dust particles. Several publications appeared in 1994, some posthumously. Nier's work thus was an integral part of two inter -related, large-scale, and simultaneous developments in the character of science: one is the transformation (and at least partial merger) of several scientific fields into the conglomerate area known as the earth and planetary sciences, and the other is the on-going spread of mass spectrometry through that vast area of science (*20*).

Nier received many scientific honors beyond his 1950 election to the National Academy of Sciences. His colleagues in mass spectrometry bestowed the Thomson Medal and the Field and Franklin Award. He received the Pittsburgh Spectroscopy Award, the Day Medal from the Geological Society of America, the Goldschmidt Medal from the Geochemical Society, the Bowie Medal from the American Geophysical Union, plus awards from the Atomic Energy Commission, from NASA, and more.

In light of the above it is no wonder that Alfred O. C. Nier has been called the father of modern mass spectrometry (*4*).

References

(*1*) Anonymous [Nier, A. O. C.], Nier, Alfred Otto Carl, in *McGraw-Hill Modern Scientists and Engineers*; McGraw-Hill: New York, 1980; Vol. 2, pp. 361–363.

(*2*) Grayson, M. A. Professor Al Nier and His Influence on Mass Spectrometry. *J. Am. Soc. Mass Spectrom* **1992**, *3*, 685–694.

(*3*) Reynolds, J. H. Alfred Otto Carl Nier. in *Biographical Memoirs of the National Academy of Sciences*; National Academy Press: Washington, DC, 1999; vol. 76, pp. 18–34.

(*4*) Delmore, J. E., Bowers, M. T. Eds. Special Issue: Honour Biography: Alfred O. C. Nier. *Int. J. Mass Spectrom. Ion Proc.* **1995**, *146/147*.

(*5*) Nier, A. O. C. Device to Compensate for Magnetic Field Fluctuations in a Mass Spectrograph. *Rev. Sci. Instrum.* **1935**, *6*, 254–255.

(*6*) Nier, A. O. C. Evidence for the Existence of an Isotope of Potassium of Mass 40. *Phys. Rev.* **1935**, *48*, 283–284.

(*7*) Nier, A. O. C. Some Reflections on the Early Days of Mass Spectrometry at the University of Minnesota. *Int. J. Mass Spectrom. Ion Proc.* **1990**, *100*, 1–13.

(*8*) Nier, A. O. C. Some Reminiscences of Isotopes, Geochronology, and Mass Spectrometry. *Annu. Rev. Earth Planet. Sci.* **1981**, *9*, 1–17.

(*9*) Birch, F. Presentation of the Day Medal to A. O. C. Nier: Citation. *Geol. Soc. Am. Proc. for 1956* **1957**, 77–78.

(*10*) Craig, H. Introduction of Alfred O. C. Nier for the V. M. Goldschmidt Medal 1984. *Geochim. Cosmochim. Acta* **1985**, *49*, 1661–1665.

(*11*) Nier, A. O. C. A Mass Spectrometer for Routine Isotope Abundance Measurements. *Rev. Sci. Instrum.* **1940**, *11*, 212–216.

(*12*) Nier, A. O. C. A Mass Spectrometer for Isotope and Gas Analysis. *Rev. Sci. Instrum.* **1947**, *18*, 398–411.

(*13*) De Laeter, J. R.; Kurz, M. D. Alfred Nier and the Sector Field Mass Spectrometer. *J. Mass Spectrom.* **2006**, *41*, 847–854.

(*14*) Groueff, S. *Manhattan Project: The Untold Story of the Making of the Atomic Bomb*; Little, Brown & Co: Boston, Toronto, 1967; pp. 187–194.

(*15*) Nier, A. O. C. Some Reminiscences of Mass Spectrometry and the Manhattan Project. *J. Chem. Educ.* **1989**, *66*, 385–388.

(*16*) De Laeter, J. R. Al Nier's Contribution to Atomic Weights. *Int. J. Mass Spectrom. Ion Proc.* **1988**, *178*, 1–7.

(*17*) Mass Spectrometry Symposium Committee (Hipple, J., chair), Eds.; *Mass Spectroscopy in Physics Research: Proceedings of the NBS Semicentennial Symposium; Held at the NBS on September 6–8, 1951*, National Bureau of Standards Circular 522, National Bureau of Standards: Washington, DC, 1953.

(*18*) Nier, A. O. C. The Development of a High Resolution Mass Spectrometer: A Reminiscence. *J. Am. Soc. Mass Spectrom.* **1991**, *2*, 447–452.

(*19*) Nier, A. O. C. Mass Spectrometry in Planetary Research. *Int. J. Mass Spectrom. Ion Proc.* **1985**, *66*, 55–73.

(*20*) Anderson, D. L. The William Bowie Medal, Alfred O. C. Nier: Citation. *EOS* **1992**, *73*, 273–274.

NB: The Chemical Heritage Foundation, Philadelphia, PA, has the transcript of a very extensive oral history interview of Alfred O. C. Nier by Michael A. Grayson.

Keith A. Nier
Independent Historian, Madison, NJ, USA

Clair Cameron Patterson

b. June 2, 1922
d. December 5, 1995

Clair Cameron Patterson (1922–1995) from Patrick N. Wyse Jackson, *The Chronologers' Quest: the Search for the Age of the Earth*. CUP, 2006 (courtesy of Laurie Patterson).

Clair Cameron (Pat) Patterson was a highly regarded and decorated American geochemist. His cutting-edge research was responsible for first determining accurately the age of the Earth as 4.55 billion years through analysis of primeval lead in meteorites. He also carried out important research on anthropogenic lead, warning that it was a serious contaminant in various environments, which led to the lowering of emissions of health-reducing pollutants.

Patterson was born on June 2, 1922, in Mitchellville, Iowa, USA. Following graduation from high school, Patterson entered Grinnell College, Iowa, where he studied chemistry, graduating A.B. in 1943. He then undertook graduate work at the University of Iowa under the supervision of George Glockler, gaining an M.A. in 1944 in molecular spectroscopy. He moved to the University of Chicago and then to Oak Ridge, Tennessee, to work on the Manhattan Project. In 1945 he returned to the University of Chicago where he attained his Ph.D. in 1951 for his studies on lead isotopic compositions of early granites.

Patterson's initial research using mass spectrometers was carried out at Oak Ridge prior to his doctoral studies. Under the supervision of Harrison Brown, for his doctoral studies Patterson devised methods to measure minute concentrations of lead isotopes in meteorites and Precambrian granites (see this volume, *Harrison Scott Brown*). From the latter he measured radiogenic lead in zircon crystals and together with George Tilton (who analyzed the uranium) realized that these crystals could be accurate geochronometers (see this volume, *George Robert Tilton*). The precision Patterson achieved in extracting lead was far in excess of what was previously possible and opened research strands on materials in which isotope concentrations had not been measured due to the coarseness of earlier experimental methodologies. Returning to meteorites during post-doctoral work, Patterson determined the isotopic composition of the lead in the Canyon Diablo iron meteorite. This lead was primordial in origin and yielded the age of the Earth. In a short but seminal paper published in 1956 he reported an age of 4.55 billion years, an estimate that is still the accepted value today. Analysis of additional meteorites provided coeval ages (*1–3* and see this volume, *The Development of Mass Spectrometry in the Earth and Planetary Sciences*).

During his doctoral studies Patterson had to overcome serious difficulties with anomalous results in his lead determinations due to contamination. Later in the 1950s he turned his attention to the source of these contaminants and quantified the volumes of lead in many terrestrial materials. Oceanic sediments and river waters were found to have elevated lead levels. He measured atmospheric lead in ice-cores from Greenland and documented the rise in lead levels over the previous two hundred and fifty years. He later extended this back to Greek and Roman times. An alarming acceleration in levels coincided with the use of lead to reduce 'knocking' in petrol (gasoline)-driven engines. Patterson voiced his concern that the high volumes of atmospheric lead generated from industrial and transport sources had implications for the health of humans. He later assessed lead levels in the bones of some early Peruvian civilizations and found that they contained about 1000 times less lead than do present day human bones; by the 1980s he was researching the lead levels found in foodstuff. Patterson's work in this area was to bring him into conflict with the automotive industry and various individuals for the next two and a half decades. Although Patterson served on the National Academy of Sciences Committee on Lead in the Human Environment, he disassociated himself from the report published in 1980, choosing instead to supply an alternative viewpoint. It wasn't until the mid-1980s that lead additives in petroleum were banned in the US (*4–7*).

Patterson was a member of the National Academy of Sciences and received the National Academy's Lawrence Smith Medal for meteoritical studies (1973), the Geochemical Society's V. M. Goldschmidt Award (1980), the University of Southern California's Tyler Prize for Environmental Achievement (1995), and honorary degrees from Grinnell College and the University of Paris VII. Patterson Peak in the Queen Maud Mountains, Antarctica, was named for him, as was the asteroid 2511 Patterson, discovered by Carolyn Shoemaker in 1980. The Clair C. Patterson Award of the Geochemical Society for innovation in environmental geochemistry was established after his death.

Patterson was married to Lorna Jean McCleary whom he met while both were students at Grinnell

College, Iowa. They had two sons and two daughters. His last position was Emeritus Professor of Geochemistry, California Institute of Technology, Pasadena. He died on December 5, 1995, at The Sea Ranch, California, USA.

References

(*1*) Patterson, C. C. The Pb^{207}/Pb^{206} Ages of Some Stone Meteorites. *Geochim. Cosmochim. Acta* **1955**, *7*, 151–153.

(*2*) Patterson, C. Age of Meteorites and the Earth. *Geochim. Cosmochim. Acta* **1956**, *10*, 230–237.

(*3*) Wyse Jackson, P. N. *The Chronologers' Quest: the Search for the Age of the Earth*; Cambridge University Press: Cambridge, 2006.

(*4*) Chow, T. J.; Patterson, C. C. The Occurrence and Significance of Lead Isotopes in Pelagic Sediments. *Geochim. Cosmochim. Acta* **1962**, *26*, 263–308.

(*5*) Patterson, C. C. Contaminated and natural environments of man. *Arch. Environ. Health* **1965**, *11*, 344–360.

(*6*) Patterson, C. C. An Alternative Perspective – Lead Pollution in the Human Environment: Origin, Extent, and Significance, *Lead in the Human Environment*; *Report prepared by the Committee on Lead in the Human Environment*; National Academy of Sciences: Washington, DC, 1980; pp. 265–349.

(*7*) McGrayne, S. B. *Prometheans in the Lab: Chemistry and the Making of the Modern World*; McGraw-Hill: New York, 2001.

Patrick N. Wyse Jackson
Department of Geology, Trinity College, Dublin 2, Ireland

Wolfgang Paul

b. August 10, 1913
d. December 7, 1993

Wolfgang Paul was a German nuclear physicist. He was most widely renowned for his leading role in the invention of the quadrupole mass spectrometer and the quadrupole ion trap. He shared the Nobel Prize for physics in 1989 for the ion trapping technique that makes it possible study a single particle with extreme precision.

Paul was born in Lorenzkirch, Saxony, but spent his childhood and school years in Munich. His father, Theodor Paul, who had been a student of Wilhelm Ostwald (Nobel Prize in chemistry), became professor of pharmaceutical chemistry at the University of Munich but died in 1928 before Wolfgang completed his secondary education.

On the advice of Arnold Sommerfeld, professor of physics at the University (teacher of many Nobel laureates), he undertook an apprenticeship in mechanics as a preparation for studying physics at the university level. He then studied physics at the Technische Hochschule (TH) in Munich starting in 1932 and moved to the TH in Berlin in 1934 where he worked primarily with Hans Kopfermann, a student of James Franck (Nobel Prize in physics). After receiving his diploma in 1937 he followed Kopfermann to the University in Kiel. His graduate work was interrupted by his induction into the German air force. He was able to complete his Ph.D. at the Berlin TH during a leave of absence from the military in 1940; shortly thereafter he was exempted from military service and returned to Kiel where he worked with Wilhelm Walcher, Kopfermann's assistant, on isotope separation and mass spectrometry. Investigations in these areas were strongly concentrated on specifics related one way or another to the German effort to develop atomic weapons and energy. After Kopfermann and Walcher moved to the University in Göttingen, Paul followed and became a privatdozent there in 1944. He also began to work with scientists who had built a betatron accelerator at the Siemens lab in Munich, a machine that was moved to Göttingen after the end of World War Two. In the immediate post-war period Paul continued various scattering experiments while also getting involved in work in radiation therapeutics (*1*).

In 1952, Wolfgang Paul became professor of experimental physics and director of the physics institute at the University in Bonn, in which positions he remained, despite some time abroad, until his 1981 retirement. He and his assistants and students started work in Bonn on high energy electron physics, molecular beam studies, and mass spectrometry. They began construction of the first electron synchrotron in Europe to employ the principles of strong-focusing that had recently been developed. Paul (like a few others) realized that the principles involved in the electron and molecular beam work could be used in a non-magnetic mass spectrometer, and he also quickly produced the first working quadrupole mass spectrometer, patented the basic design, and effectively brought it to the attention of the scientific community (*2–5*). He and his collaborators (particularly 'house theoretician' Helmut Steinwedel) also worked out the characteristics of modifications of the basic design, including the possibility of arranging the combination of electrostatic and radio-frequency fields to trap ions. Less than a decade after the initial publication, companies were offering quadrupole mass spectrometers (now commonly called quadrupole mass filters) for sale and, particularly when coupled with gas chromatography, quadrupoles fairly soon became the most widely used basic type of mass spectrometer. The development of the quadrupole ion trap as a mass spectrometer followed, but only many years later (see this volume, *The Development of the Quadrupole Mass Filter and the Quadrupole Ion Trap* and *The Development of the Business of Mass Spectrometry from 1960 to 1975*).

Apart from his seminal role in mass spectrometry, Wolfgang Paul worked primarily in the areas of experimental nuclear physics, particularly on neutron studies. He was able to follow the early synchrontron built at Bonn with one in 1965 that had five times greater energy. He worked at times at the Institute Laue-Langevin in Grenoble, France, was director of the nuclear physics division at CERN in Geneva for two years, and helped start the large German National Laboratory DESY (an electron synchrotron) at Hamburg. Much work in his later career was directed to trapping neutrons in magnetic storage rings (*6–8*).

Paul served in many major roles in scientific organizations and in 1957 he also was one of the Göttingener 18, a group of prominent German nuclear scientists who made a politically significant public declaration opposing allowing German rearmament to include tactical nuclear weapons. He retired in 1981 but remained active. His marriage to Liselotte Hirsche lasted until she died after 36 years. They had four children; the two sons each also became professors of physics while the daughters went into art history and pharmacy. In 1979 he married Doris Walch-Paul, professor of literature at Bonn. He died in Bonn in 1993.

References

(*1*) Paul, W. Autobiography. In: *Les Prix Nobel: The Nobel Prizes 1989*; Frängsmyr, T., Ed.; Nobel Foundation: Stockholm, 1990; and at: 'Wolfgang Paul - Autobiography'. Nobelprize.org. (http://130.242.18.21/nobel_prizes/physics/laureates/1989/paul.html)

(*2*) Paul, W.; Steinwedel, H. Ein neues massenspektrometer ohne magnetfeld. *Z. Naturforsch. A* **1953**, *8*, 448–450.

(*3*) Paul, W.; Steinwedel, H. German patent No. 944,900; US patent 2,939,958, 1953.

(*4*) Paul, W.; Raether, M. The Electrical Mass Filter. *Z. Phys.* **1955**, *140*, 262–273.

(*5*) Paul, W.; Reinhard, H. P. Das Elektrische Massenfilter als Isotopentrenner. In: *Proceedings of the International Symposium on Isotope Separation Held in Amsterdam, April 23–27, 1957*; Kistemaker, J.; Bigeleisen, J.; Nier, A. O. C., Eds.; North-Holland Publishing Company: Amsterdam, 1958, pp. 640–652.

(*6*) Kügler, K. J.; Moritz, K.; Paul, W.; Trinks, U. Nestor – A Magnetic Storage Ring for Slow Neutrons. *Nucl. Instrum. Methods A* **1985**, *228*, 240–258.

(*7*) Paul, W.; Anton, F.; Paul, L.; Paul, S.; Mampe, W. Measurement of the Neutron Lifetime in a Magnetic Storage Ring. *Z. Phys. C* **1989**, *45*, 25–50.

(*8*) Paul, W. Electromagnetic Traps for Charged and Neutral Particles. *Rev. Mod. Phys.* **1990**, *62*, 531–540.

Keith A. Nier

Independent Historian, Madison, NJ, USA

Rowland Ivor Reed

b. 1924
d. 16 March 2007

Rowland Ivor Reed was a British chemist who significantly helped advance the use of mass spectrometers in research on organic molecules. He was a pioneer in the application of mass spectrometry to several classes of compounds, in instrumentation, and in educating scientists about mass spectrometry.

Reed was born in Penibroke Dock in South Wales in 1924. He earned his B.Sc. (1943) and Ph.D. (1946) at University College, London, the latter under C. K. Inglod and E. D. Hughes. After one year on the staff at University College he obtained a position at the University of Glasgow and spent the rest of his career there. He died in Glasgow in his 87th year (*1–3*).

In the middle of the 1950s, Reed was one of the few researchers willing to try applying mass spectrometry to organic chemistry and to encourage others to do the same. He initially used a Metropolitan-Vickers MS2. An early work on steroid and terpenoid chemistry has been called a 'landmark paper' (*4,5*). Reed also was a pioneer in the application of mass spectrometry to the investigation of carbohydrates and flavones. His was the first academic laboratory to get an MS9, the commercial version of the high performance double-focusing mass spectrometer initially ordered by John Beynon at Imperial Chemical Industries. Reed was also among the first to show that better mass spectra could be obtained for many substances by putting them on a probe that was inserted into the ionization source rather than by first volatilizing them outside the source (*6–11*, see this volume, *The Beginnings of Organic Mass Spectrometry*).

Beyond his teaching and his own research, Reed worked to promote and publicize the potentials for mass spectrometry in organic chemistry and beyond and to spur others to enter these areas of science. He wrote surveys and texts and he organized, led, and edited the published products of special NATO-sponsored institutes in mass spectrometry (*12–18*).

References

(*1*) [Anon.] Rowland Ivor REED: Obituary. Gloucestershire Media, 27 March 2007. Available at : http://www.iannounce.co.uk/United%1eKingdom/23?_fstatus=search;keywords=%2BRowland%20%2BIvor%20%2BREED (accessed December 3, 2012).
(*2*) Clayton, E. Rowland Ivor Reed: 1924–2007. *Rapid Comm. Mass Spectrom.* **2008**, *22*, 1344.
(*3*) [Anon.] Biographical Notes. *Org. Mass Spectrom.* **1969**, *2*, 864.
(*4*) de Mayo, P.; Reed, R. I. Application of the Mass Spectrometer to Steroid and Terpenoid Chemistry. *Chem. Ind., London* **1956**, 1481–1482.
(*5*) Maccoll, A. J.S. Shannon – an Appreciation. *Org. Mass Spectrom.* **1992**, *27*, 989–990.
(*6*) Reed, R. I. Electron Impact and Molecular Dissociation. I. Some steroids and triterpenoids. *J. Chem. Soc.* **1958**, 3432–3436.
(*7*) Finan, P. A.; Reed, R. I.; Snedden, W. Application of the Mass Spectrometer to Carbohydrate Chemistry. *Chem. Ind., London* **1958**, 1172.
(*8*) Hill, H. C.; Reed, R. I. The Volatilization of High Melting Solids in a Mass Spectrometer. *J. Sci. Instrum.* **1963**, *40*, 259.
(*9*) Scrivens, J. Industrial and Environmental Applications. In: *A History of European Mass Spectrometry*; Jennings, K. R., Ed.; IM Pubs.: Chichester, UK, 2012; pp 214–233.
(*10*) McLafferty, F. W. Vlada Hanus, a Pioneer of Molecular Mass Spectrometry. *Eur. J. Mass Spectrom.* **1998**, *4*, 303–304.
(*11*) Cotter, R. J. Mass Spectrometry of Nonvolatile Compounds: Desorption from Extended Probes. *Anal. Chem.* **1980**, *52*, 1589A–1606A.
(*12*) Reed, R. I. *Ion Production by Electron Impact*; Academic Press: London, New York, 1962.
(*13*) Quayle, A.; Reed, R. I. Interpretation of Mass Spectra. In: *Interpretation of Organic Spectra.*; Mathieson, D. W., Ed.; Academic Press: London, New York, 1965.
(*14*) Reed, R. I., Ed. *Mass Spectrometry: Proceedings of the 1st NATO Advanced Study Institute on Theory, Design, and Applications, Glasgow, Scotland, August 1964*; Academic Press: London, New York, 1965.
(*15*) Reed, R. I. The Mass Spectrometer in Organic Chemistry. *Q. Rev. Chem. Soc.* **1966**, *20*, 527–543.
(*16*) Reed, R. I. *Applications of Mass Spectrometry to Organic Chemistry*; Academic Press: New York, 1966.
(*17*) Reed, R. I., Ed. *Modern Aspect of Mass Spectrometry: Proceedings of the 2nd NATO Advanced Study Institute of Mass Spectrometry on Theory, Design, and Applications, July 1966*, University of Glasgow, Scotland; Plenum Press: New York, 1968.
(*18*) Reed, R. I., Ed. *Recent Topics in Mass Spectrometry: NATO Study Institute of Mass Spectrometry, Lisbon, August 1969.*; Academic Press: Gordon and Breach, New York, 1971.

Keith A. Nier
Independent Historian, Madison, NJ, USA

John H. Reynolds

b. April 3, 1923
d. November 4, 2000

The modern sciences of geochronology and nuclear cosmochronology grew in large part out of the work of American physicist John Reynolds and his students. He was the first to detect isotopic anomalies that provided evidence of the preservation in the meteorites of micron-size grains of stellar origin. In 1960 he detected the xenon isotope of mass 129 trapped in meteorites, and from that inferred that the extinct radioactive isotope iodine-129 was present when the meteorites formed, indicating that the meteorites appeared in the early history of the solar system. In later studies he and his collaborators showed that other short-lived species were present in the cloud of gas that turned into our solar system 4.6 billion years ago. He will be remembered as the 'father' of extinct radioactivities.

John Reynolds was born in Cambridge, Massachusetts, in 1923. His father, Horace Mason Reynolds, taught English in various colleges and at Brown University and his mother, Catherine Whitford, was a writer. Reynolds was predisposed toward the academic life. Although his strongest aptitudes in school were in math and physics, he enjoyed most subjects and took piano and organ lessons. He tinkered with electricity and radio, sang in choirs and in his college glee club. He graduated from Harvard *Summa cum Laude* in electronic physics in1943 and served as an ensign in the US Navy as an ordnance officer.

Reynolds did graduate study in physics at the University of Chicago. His selections of mass spectroscopy as a topic and of Mark Inghram as advisor were based mainly on his friendship with Joseph Hayden, who was doing research with Inghram at the time (see this volume, *Mark Gordon Inghram II*). Reynolds was captivated by stellar roster of geochemists and cosmochemists at Chicago – Harold Urey, Harrison Brown, Hans Suess, and young scientists Clair Patterson, George Tilton, and Sam Epstein (see this volume, *Harrison Scott Brown, Samuel Epstein, Clair Cameron Patterson, George Robert Tilton*, and *Harold Clayton Urey*). Like others, he was strongly influenced by Enrico Fermi; he audited two of Fermi's courses, never missing a lecture.

At Chicago, he and Inghram discovered the double beta-decay of ^{130}Te by way of ^{130}Xe production in tellurium ores. Reynolds also discovered ^{81}Kr, the long-lived isotope of krypton, which later became the basis of the most precise cosmic ray exposure dating method for meteorites and lunar rocks.

In 1950 he completed his Ph. D., married Genevieve Marshall, and accepted an Assistant Professorship at the University of California, Berkeley. Reynolds was able to set up his own laboratory for mass spectrometry and, in a few years, to design and construct the first static (non-pumped) all-glass mass spectrometer, since referred to as the Reynolds-type mass spectrometer. He used it for isotopic analysis of noble gases. A key feature was his incorporation of a bakeable ultrahigh-vacuum system, which had just been invented by Daniel Alpert, without which he could not have achieved the sensitivity he sought.

In the early 1950s, the decay of ^{40}K into ^{40}Ar had become a dating method for 10^8- to 10^9-year old potassium-rich rocks. In 1956, by virtue of the factor 10^2 higher sensitivity afforded by his static mass spectrometer, Reynolds and graduate students Joe Lipson (Physics), Garniss Curtis and Jack Evernden (Geology), were able to date rocks as young as $\sim 10^6$ years. This opened the door for determining the time-scale of geomagnetic reversals, which provided a quantitative proof of plate tectonics.

Harrison Brown had suggested in 1947 that meteorites could be used to determine the age of the elements if the daughter of an extinct natural radioactive nuclide could be found. Mass spectrometrists tried to detect ^{129}Xe, the decay product of ^{129}I, an isotope with a 16-million-year half-life. Reynolds' discovery of the evidence of extinct ^{129}I, namely observation of a large excess of ^{129}Xe in the Richardton chondrite in 1960, was the crowning achievement of his career. In a letter supporting promotion of Reynolds to full Professor that same year, Willard Libby (1960 Nobel laureate) said Reynolds had "revolutionized much of cosmological theory. His latest result is the most important single event in the whole field." A key to his success was his static all-glass mass spectrometer, which made it possible to pass the same rare gas atoms through the instrument many times in search of an isotopic anomaly.

In a sabbatical during 1963–64, Reynolds set up a complete K-Ar laboratory at the University of Sao Paolo, Brazil, with Umberto Cordani. During his stay, the group measured ages of Brazilian rocks that fitted age patterns on the coast of Africa. The agreement was strong evidence that South America separated from Africa some 10^8 years ago.

Xenon is particularly rich in stable isotopes that provide cosmochemical information. A second extinct radioactivity amenable to a mass spectrometric search is the spontaneous fission of ^{244}Pu with an 82 million year half-life. Its fission products should include ^{136}Xe, ^{134}Xe, ^{132}Xe, and ^{131}Xe, but the relative amounts were not known. In an experiment on ^{244}Pu produced at Oak Ridge, in 1971 Reynolds and colleagues demonstrated that the relative abundances of the xenon isotopes resulting from spontaneous fission of ^{244}Pu agreed with those found in various meteorites. From the ^{129}I and ^{244}Pu results they concluded that the meteorites formed within $\sim 10^7$ years of each other, some 10^8 years after cessation of nucleosynthesis of the material that formed the solar system.

His next big project, which occupied him for the better part of the decade 1978–88, was to design, construct, and do research with a new type of mass spectrometer which, along with electronics and computing facilities, was carried in a 25-foot trailer. With this facility, Reynolds and his group were able to do precise *in-situ* measurements of isotopes of rare gases from deep terrestrial fluids, mainly geothermal.

In parallel with that project, he and his group followed up on an important 1975 discovery by Ed Anders and co-workers of traces of gases with exotic isotopic composition that had been trapped in tiny extrasolar grains in the most primitive carbonaceous meteorites. Reynolds and his students offered the interpretation now seen to be correct – that the exotic gases had been trapped in presolar grains with a different nucleosynthetic history from the gases in our solar system. The study of such grains, made in stellar atmospheres and later incorporated into the material of which our meteorites were formed, is on the forefront of cosmochemistry today.

John Reynolds retired in 1993. During his career his honors included receipt of two Guggenheim Fellowships, election to the National Academy of Sciences, and award of the Academy's J. Lawrence Smith Medal and the Meteoritical Society's Leonard Medal. He died in Berkeley in 2000, survived by his second wife, Ann, and several children.

Much more about his life and work can be discovered through the small sample of references listed here (*1–20* and see this volume, *The Development of Mass Spectrometry in the Earth and Planetary Sciences*).

References

(*1*) Reynolds, J. H.; Inghram, M. G. Double Beta-Decay of Tellurium-130. *Phys. Rev.* **1950**, *78*, 822–823.

(*2*) Reynolds, J. H. A New Long-Lived Krypton Activity. *Phys. Rev.* **1950**, *79*, 886.

(*3*) Reynolds, J. H. High Sensitivity Mass Spectrometer for Rare Gas Analysis. *Rev. Sci. Instrum.* **1956**, *27*, 928–935.

(*4*) Folinsbee, R. E.; Lipson, J.; Reynolds, J. H. Potassium–Argon Dating. *Geochim. Cosmochim. Acta* **1956**, *10*, 60–68.

(*5*) Curtis, G. H; Reynolds, J. H. Notes on the Potassium-Argon Dating of Sedimentary Rocks. *Bull. Geol. Soc. Am.* **1958**, *69*, 151–160.

(*6*) Reynolds, J. H. Determination of the Age of the Elements. *Phys. Rev. Lett.* **1960**, *4*, 8–10.

(*7*) Reynolds, J. H. Isotopic Composition of Primordial Xenon. *Phys. Rev. Lett.* **1960**, *4*, 351–354.

(*8*) Reynolds, J. H. Xenology. *J. Geophys. Res.* **1963**, *68*, 2939–2956.

(*9*) Butler, W. A.; Jeffery, P. M.; Reynolds, J. H.; Wasserburg, G. J. Isotopic Variations in Terrestrial Xenon. *J. Geophys. Res.* **1963**, *68*, 3283–3291.

(*10*) Hohenberg, C. M.; Podosek, F. A.; Reynolds, J. H. Xenon–Iodine Dating; Sharp Isochronism in Chondrites. *Science* **1967**, *156*, 233–236.

(*11*) Reynolds, J. H. Isotopic Anomalies in the Solar System. *Annu. Rev. Nucl. Sci.* **1967**, *17*, 253–316.

(*12*) Amaral, G.; Bushee, J.; Cordani, U. G.; Kawashita, K.; Reynolds, J. H. Potassium–Argon Ages of Alkaline Rocks from Southern Brazil. *Geochim. Cosmochim. Acta* **1967**, *31*, 117–142.

(*13*) Reynolds, J. H. Plutonium-244 in the Early Solar System. *Nature* **1968**, *218*, 1024–1028.

(*14*) Reynolds, J. H.; Hohenberg, C. M.; Lewis, R. S.; Davis, P. K.; Kaiser, W. A. Isotopic Analysis of Rare Gases from Stepwise Heating of Lunar Fines and Rocks. *Science* **1970**, *167*, 545–548.

(*15*) Reynolds, J. H.; Alexander, E. C. Jr.; Davis, P. K.; Srinivasan, B. Studies of K–Ar Dating and Xenon from Extinct Radioactivities in Breccia 14318: Implications for Early Lunar History. *Geochim. Cosmochim. Acta* **1974**, *38*, 401–417.

(*16*) Reynolds, J. H.; Frick, U.; Neil, J. M.; Phinney, D. L. Rare-Gas-Rich Separates from Carbonaceous Chondrites. *Geochim. Cosmochim. Acta* **1978**, *42*, 1775–1797.

(*17*) Smith, S. P.; Reynolds, J. H. Excess ^{129}Xe in a Terrestrial Sample as Measured in a Pristine System. *Earth Planet. Sci. Lett.* **1981**, *54*, 236–238.

(*18*) Reynolds, J. H. Isotopic Anomalies in Meteorites Explained? *Nature* **1983**, *302*, 213–214.

(*19*) Kennedy, B. M.; Reynolds, J. H.; Smith, S. P. Noble Gas Geochemistry in Thermal Springs. *Geochim. Cosmochim. Acta* **1988**, *52*, 1919–1928.

(*20*) Price, P. B. John H. Reynolds: 1923–2000. In: *Biographical Memoirs Vol. 85; D.C*; National Academies Press: Washington, 2004; pp. 1–21.

P. Buford Price
Physics Department, University of California, Berkeley, CA, USA

David Rittenberg

b. November 11, 1906
d. January 24, 1970

David Rittenberg was a chemist who was internationally recognized as one of the fathers of stable isotope technology for the elucidation of the dynamic state of living systems. He was born in New York, NY, USA on November 11, 1906. He received a B.S. degree from the College of the City of New York and a Ph. D. from Columbia University in 1934 working with Harold Urey for his thesis 'Some Equilibria Involving Isotopes of Hydrogen.' He was a faculty member at Columbia University for 36 years and was chairman of the Department of Biochemistry from 1956–1970.

When Urey identified and isolated deuterium in the early 1930s, he became interested in promoting its use in biology (see this volume, *Harold Clayton Urey*). On finishing his degree, Rittenberg was given this task by Prof. Hans T. Clark, the chair of Biochemistry at that time. Soon thereafter Rittenberg met and began a long, fruitful collaboration with Prof. Rudolf Schoenheimer; together their experiments designed to understand intermediary metabolism through the use of stable isotopes literally revolutionized biochemistry (see this volume, *Rudolf Schoenheimer*). Early in this work, deuterium was employed to elucidate the dynamic state of lipids and later, in 1937, they used ^{15}N for similar studies of proteins. This work clearly demonstrated that the biology was not static (as many believed) but rather that it was in a stable but dynamic state of biosynthesis and breakdown. Later, ^{13}C and ^{18}O were incorporated into their studies of a wide variety of their metabolic studies.

Rittenberg's introduction to mass spectrometry occurred during his graduate work with Harold Urey when he was engaged with studies of deuterium. In 1937, when he began studies with ^{15}N, he constructed a mass spectrometer with an 180° magnetic analyzer for the measurement of this stable isotope in N_2 gas obtained from chemical digestion of biological samples. He was the author of many papers describing advancements in both the sample preparation and underlying mass spectrometry technology that allowed these paradigm-altering studies to be accomplished successfully.

Rittenberg's scientific contributions were recognized in 1941 when he was awarded the Eli Lilly Prize in biological chemistry given by the American Chemical Society. He was elected to the National Academy of Sciences in 1953. He was a member of the Board of Governors of the Weizmann Institute in Israel and was made a honorary fellow of that institution in 1967. Selected references here are from his biography prepared by the National Academy of Sciences in online format: http://www.nasonline.org/publications/biographical-memoirs/alphabetical-listing/memoirs-r.html

The publications cited are representative of the author's most significant contributions to the field (*1–5*).

References

(*1*) Schoenheimer, R.; Rittenberg, D. Deuterium as an Indicator in the Study of Intermediary Metabolism. *J. Biol. Chem.* **1935**, *111*, 163–168.

(*2*) Schoenheimer, R.; Rittenberg, D. Deuterium as an Indicator of Intermediary Metabolism. IX. The Conversion of Stearic Acid into Palmitic Acid in the Organism. *J. Biol. Chem.* **1937**, *120*, 155–165.

(*3*) Schoenheimer, R.; Rittenberg, D. The Application of Isotopes to the Study of Intermediary Metabolism. *Science* **1938**, *87*, 221–226.

(*4*) Bloch, K.; Rittenberg, D. On the Utilization of Acetic Acid for Cholesterol Formation. *J. Biol. Chem.* **1942**, *145*, 625–636.

(*5*) Schmin, D.; Rittenberg, D. The Life Span of the Human Red Blood Cell. *J. Biol. Chem.* **1946**, *166*, 627–636.

Richard M. Caprioli
Departments of Biochemistry, Chemistry, Pharmacology and Medicine and the Mass Spectrometry Research Center, Vanderbilt University School of Medicine, Vanderbilt University, Nashville, TN, USA

Carol Vivien Robinson

b. April 10, 1956

Carol Robinson (nee Bradley) is a British chemist best known for her pioneering use of mass spectrometry as an analytical tool in ground-breaking research into the 3D structure of proteins. She is the first female Professor of Chemistry at the University of Oxford and was previously the first female Professor of Chemistry at the University of Cambridge.

Carol V. Bradley was born in Beckenham, near London, UK, in 1956. She worked as a lab technician for Pfizer starting at age 16. She became a Graduate of the Royal Society of Chemistry through part-time study and then worked with John Beynon at Swansea in 1979–1980, earning her MSc, (*1*). Carol was a graduate student at Churchill College, Cambridge, under the direction of Dudley Williams, from 1980–1982, completing her PhD in two years (*2* and see this volume, *John Herbert Beynon* and *Dudley Howard Williams*).

Following an eight-year career break to begin raising her three children, Carol V. Robinson returned to research at Oxford. By 1999 she had become a titular professor. In 2001 she returned to Cambridge and was elected Professorial Fellow at Churchill College. She became a Royal Society Research Professor in 2006. In 2009 she was elected Doctor Lee's Professor of Chemistry at the University of Oxford.

Robinson's research has developed from studies of protein folding to the architecture and operations of protein complexes and included development of techniques to investigate weakly linked structures through electrospray time-of-flight mass spectrometry. This work is helping bring mass spectrometry into the field of structural biology. A sampling of her research is given below (*3–9*).

Robinson's research has earned her numerous honors, including the Royal Society of Chemistry's award for mass spectrometry and the Biemann Medal from the American Society for Mass Spectrometry. She was elected Fellow of the Royal Society and has received that Society's Rosalind Franklin and Davy awards. She was made Dame Commander of the Order of the British Empire in 2013.

References

(*1*) Bradley, C. V.; Howe, I.; Beynon, J. H. Analysis of Underivatised Peptide Mixtures by Collision Induced Dissociation of Negative Ions. *J. Chem. Soc., Chem. Comm.* **1980**, 562–564.

(*2*) Williams, D. H.; Bradley, C. V.; Bojesen, G.; Santikarn, S.; Taylor, L. C. E. Fast Atom Bombardment Mass Spectrometry: A Powerful Technique for the Study of Polar Molecules. *J. Am. Chem. Soc.* **1981**, *103*, 5700–5704.

(*3*) Miranker, A.; Robinson, C. V.; Radford, S. E.; Aplin, R. T.; Dobson, C. M. Detection of Transient Folding Populations by Mass Spectrometry. *Science* **1993**, *262*, 896–899.

(*4*) Rostom, A. A.; Robinson, C. V. Detection of the Intact GroEL Chaperonin Assembly by Mass Spectrometry. *J. Am. Chem. Soc.* **1999**, *121*, 4718–4719.

(*5*) Ilag, L. L.; Videler, H.; McKay, A. R.; Fucini, P.; Nierhaus, K. H.; Robinson, C. V. Heptameric (L12)6/L10 rather than Canonical Pentameric Complexes Found by Tandem MS of Intact Ribosomes from Thermophilic Bacteria. *Proc. Natl. Acad. Sci. USA* **2005**, *102*, 8192–8197.

(*6*) Ruotolo, B. T.; Giles, K.; Campuzano, I.; Sandercock, A. M.; Bateman, R. H.; Robinson, C. V. Evidence for Macromolecular Protein Rings in the Absence of Bulk Water. *Science* **2005**, *310*, 1658–1661.

(*7*) Barrera, N. P.; Di Bartolo, N.; Booth, P. J.; Robinson, C. V. Micelles Protect Membrane Complexes from Solution to Vacuum. *Science* **2008**, *321*, 243–246.

(*8*) Zhou, M.; Morgner, N.; Barrera, N. P., *et al.* Mass Spectrometry of Intact V-Type ATPases Reveals Bound Lipids and the Effects of Nucleotide Binding. *Science* **2011**, *334*, 380–385.

(*9*) Pagel, K.; Natan, E.; Hall, Z.; Fersht, A. R.; Robinson, C. V. Intrinsically Disordered p53 and Its Complexes Populate Compact Conformations in the Gas Phase. *Angew. Chem. Int. Ed.* **2012**, *10*, 1002.

Carol Vivien Robinson
Physical and Theoretical Chemistry Laboratory, Oxford, UK

Sibyl Martha Rock

b. August 1, 1909
d. November 17, 1981

Sibyl Rock was an American mathematician and engineer active in the early commercialization of mass spectrometry in the oil industry and the formative years of the computer industry. She pioneered in developing and spreading the methods for data analysis in the use of Consolidated Engineering Corporation's mass spectrometers in analyzing hydrocarbon mixtures.

Sibyl Rock was born in 1909 in Butte, Montana, US. Her early years were divided among Butte, Denver, and Los Angeles. She earned her BA in mathematics at the University of California, Los Angeles (UCLA) in 1931. Rock worked as a geophysical computer (in the era when 'computers' were people) for Rieber Laboratories in San Francisco before joining United Geophysical Co. in Pasadena, CA, in the same role in 1938. She soon moved into the research department of United's subsidiary company, Consolidated Engineering Corp (CEC) as it began to develop a mass spectrometer as a commercial product under the direction of Harold Washburn (see this volume, *Magnetic and Electrostatic Analyzers before 1960, Mass Spectrometry Manufacturers and Markets from the Beginning to 1960*, and *Harold Williams Washburn*). She officially moved into the sales division in 1947 and then into the newly-formed computer department in 1952. Rock stayed and became a sales-engineer as that department was turned into a separate company, ElectoData Corp., in 1953, and in turn became part of the Burroughs Corp. in 1956. Though she traveled extensively in her work, she remained in southern California until her death in 1981 in Los Angeles (*1–3*).

Rock's work was crucial in three developments (*4–12*). CEC had to show potential purchasers in the petroleum refining industry that use of a suitable mass spectrometer would allow them to obtain more and better analytical information and to do it much faster than by other methods. Rock's analysis was crucial to demonstrating these capabilities. The company also had to develop among the customers the competence needed to do the data analysis involved in the measurements. Rock led the way in this work. At the same time, CEC became a pioneer in developing computing equipment to ease and speed the data analysis, and Rock played a fundamental role in the design and development work that produced the early ElectroData computers. Beyond all this, Rock was crucial to customer relations as well, both in sales and service, for both CEC and ElectroData. Perhaps the clearest indication of how central her work was is the program of the 1952 Pittsburgh Conference (third of the series), where the first scientific society for mass spectrometry was organized (see this volume, *Origins of Professional Mass Spectrometry Societies*). Out of 121 presentations listed, 23 were on aspects of mass spectrometry, making up all or part of 5 of the 18 sessions. Sibyl Rock was the sole, lead, or coauthor of 3 of the 23 talks (*13–15*).

References

(*1*) Anon. Sibyl Rock, Put Digits to Work. *Control Engineering*. **1955**, *2*. July, 11.

(*2*) Sawyer, T. J. Sibyl M. Rock. Available at: http://tjsawyer.com/B205SMR.htm (accessed 7 June 2014).

(*3*) Hunter-Lascoskie, S. Chemical Heritage Foundation. Sibyl Rock and Innovation. Available at: http://www.chemheritage.org/research/policy-center/oral-history-program/projects/critical-mass/sibyl-rock-and-innovation.aspx (accessed 7 June 2014).

(*4*) Washburn, H. W.; Wiley, H. F.; Rock, S. M. The Mass Spectrometer as an Analytical Tool. *Ind. Eng. Chem. Anal. Ed.* **1943**, *15*, 541–547.

(*5*) Washburn, H. W.; Wiley, H. F.; Rock, S. M.; Berry, C. E. Mass Spectrometry. *Ind. Eng. Chem. Anal. Ed.* **1945**, *17*, 74–81.

(*6*) Rock, S. M. *Computing Manual: Analysis of Gas and Liquid Mixtures by Means of the Mass Spectrometer: CEC Report MC 25.51d*; Privately Published by Consolidated Engineering Corp.: Pasadena, CA, USA, 1946.

(*7*) Berry, C. E.; Wilcox, D. E.; Rock, S. M.; Washburn, H. W. Computer for Solving Linear Simultaneous Equations. *J. Appl. Phys.* **1946**, *17*, 262–272.

(*8*) Gifford, A. P.; Rock, S. M.; Comaford, D. J. Mass Spectrometer Analysis of Alcohols and Other Oxygenated Derivatives. *Anal. Chem.* **1949**, *21*, 1026–1032.

(*9*) Alfin-Slater, R. B.; Rock, S. M.; Swislocki, M. Determination of Isotope Ratios of Known Deuterium-Hydrogen Samples with a Mass Spectrometer. *Anal. Chem.* **1950**, *22*, 421–423.

(*10*) Berry, C. E.; Rock, S. M. High-Resolution Mass Spectrometry. *J. Chem. Phys.* **1951**, *19*, 1208–1209.

(*11*) Rock, S. M. Qualitative Analysis from Mass Spectra. *Anal. Chem.* **1951**, *23*, 261–268.
(*12*) Shepherd, M.; Rock, S. M.; Howard, R.; Stormes, J. Isolation, Identification, and Estimation of Gaseous Pollutants of Air. *Anal. Chem.* **1951**, *23*, 1431–1440.
(*13*) Gifford, A. P.; Rock, S. M.; Goodwin, P. S. A Dual-Purpose Mass Spectrometer. In: *The Pittsburgh Conference on Analytical Chemistry and Applied Spectroscopy, March 5–7, 1952*; Analytical Div. Pittsburgh Section, ACS, and Spectroscopy Society of Pittsburgh: Pittsburgh, PA, 1952; p. 18.
(*14*) Rock, S. M.; Berry, C. E.; Sink, R. L. Automatic Conversion of Mass Spectra to Digital Form. In: *The Pittsburgh Conference on Analytical Chemistry and Applied Spectroscopy, March 5–7, 1952.* Analytical Div. Pittsburgh Section, ACS, and Spectroscopy Society of Pittsburgh: Pittsburgh, PA, 1952; p 18.
(*15*) Rock, S. M. A Keysort File of Mass Spectra for Qualitative Analysis. In: *The Pittsburgh Conference on Analytical Chemistry and Applied Spectroscopy, March 5–7, 1952.* Analytical Div. Pittsburgh Section, ACS, and Spectroscopy Society of Pittsburgh: Pittsburgh, PA, 1952; p. 25.

Keith A. Nier
Independent Historian, Madison, NJ, USA

Henry Meyer Rosenstock

b. 15 February 1928
d. 14 September 1982

Henry Meyer Rosenstock developed a theory for the absolute rates of reaction for isolated molecular systems that had a major impact on the theory of mass spectra. He was a lifelong leader in applying theoretical statistical techniques to problems in dissociation of molecular ions to understand fragmentation processes in mass spectrometry.

Born in Mannheim, Germany, in 1928, Rosenstock emigrated with his family to the United States in 1936, where Henry later became a citizen. Following graduation from high school in New York City, he entered the City College of New York and was graduated *magna cum laude* in 1949 with a Bachelor's degree in Chemistry. He obtained his PhD at the University of Utah in 1952. He and his wife Shirley had three sons, Peter, Paul, and Stephen. A humorist who kept his friends and colleagues entertained with one-liners and scientific humor and an excellent amateur musician who had a viola constructed for his playing in local orchestras, Henry loved to entertain the NBS guests at dinner parties in his home or at Washington area restaurants. He also very much enjoyed the invitations he received to travel abroad and to collaborate; his untimely death on September 14, 1982, occurred while he was attending a mass spectrometry conference in Portugal.

Henry's PhD thesis, written under Henry Eyring's supervision, was titled *Absolute Rate Theory for Isolated Systems and the Mass Spectra of Large Molecules* (*1*) and was the first exposition of a theoretical basis for fragmentation processes in mass spectra. This theory, which became known as the quasi-equilibrium theory (QET), remains the basis of more modern descriptions of mass spectra (*2*, and see this volume, *The Beginnings of Organic Mass Spectrometry* and *Austin L. Wahrhaftig*).

Immediately after obtaining his PhD, Henry was employed in the private sector and continued to publish papers on a variety of topics. Before coming to the National Bureau of Standards (NBS) in 1960, Henry was a chemist at the Oak Ridge National Laboratory, Oak Ridge and worked for commercial companies in the Washington area. In this period, he developed an early theory of ion/molecule reaction kinetics, which, although never formally published, enjoyed considerable circulation and reference amongst his peers. When he joined NBS, he collaborated closely with the experimental group of Vernon Diebler and Ken McCulloh, who were developing mass spectrometric techniques with photoionization sources. Later he collaborated closely with Roger Stockbauer in developing threshold photoelectron and photoion coincidence spectroscopy (TPEPICO), a technique that gives detailed rate and energetic information on molecular fragmentation processes. Early in his career at NBS, Henry turned his attention to the generation of critically evaluated data on ion thermochemistry. This led to the publication of data he and his collaborators compiled called *Ionization Potentials, Appearance Potentials, and Heats of Formation of Gaseous Positive Ions* (*3*). This volume was followed by subsequent updates titled *Energetics of Gaseous Ions* (*4*) that were based upon the work of the Ion Energetics Data Center at NBS.

From the late 1960s and until the time of his death, Henry managed the Ion Energetics Data Center and was intimately involved in the TPEPICO work. He helped develop time delay techniques by which the interval between the detection of a threshold electron and the extraction of the ion from which it came could be varied. This technique provided accurate information on the appearance potentials of gaseous ions that Henry used to calculate accurate heats of formation of molecular ions. His strategy was to select key molecular systems for study that allowed for generalization of the results to estimate the heats of formation of families of molecular ion systems: The work he and his collaborators published on bromobenzene is a good example of this work (*5*). He consulted with many research groups throughout the world who were doing work in ion energetics and published jointly with many of his collaborators. His office at NBS was an international clearing house for information on ion energetics and he hosted many visitors to the labs at NBS.

Henry was chairman of the ASTM Committee E-14 on mass spectrometry in the late 1960s and was instrumental in helping form the American Society for Mass Spectroscopy. He served in various leadership positions in the Society in the early years of its formation and always looked forward to participating in the Society's annual meeting.

He also served on the Editorial Board of the *International Journal of Mass Spectrometry and Ion*

Processes. Volume 58 of that journal was dedicated to Henry and contains invited papers from his colleagues and collaborators throughout the world. He was a founding member of the United States Senior Executive Service at NBS and received other NBS and Department of Commerce Awards.

References

(*1*) Rosenstock, H.M., Absolute Rate Theory for Isolated Systems and the Mass Spectra of Large Molecules, PhD Theses, Department of Chemistry, University of Utah, 1952.

(*2*) Rosenstock, H. M.; Wallenstein, M. B.; Wahrhaftig, A. L.; Eyring, H. Absolute Rate Theory for Isolated Systems and the Mass Spectra of Polyatomic Molecules. *Proc. Natl. Acad. Sci.* **1952**, *38*, 667–678.

(*3*) Franklin, J. L.; Dillard, J. G.; Rosenstock, H. M.; Herron, J. T.; Draxl, K.; Field, F. H. *Ionization Potentials, Appearance Potentials, and Heats of Formation of Gaseous Positive Ions*, NSRDS-NBS 26, Superintendent of Documents, U.S. Government Printing Office: Washington, DC, 20402, 1969.

(*4*) Rosenstock, H. M.; Draxl, K.; Steiner, B. W.; Herron, J. T. Energetics of Gaseous Ions. *J. Phys. Chem. Ref. Data* **1977**, *6*, Supplement No. 1.

(*5*) Rosenstock, H. M.; Stockbauer, R.; Parr, A. C. Photoelectron-Photoion Coincidence Study of the Bromobenzene Ion. *J. Chem. Phys.* **1980**, *73*, 773.

Albert Parr
National Institute of Science and Technology, Washington, DC, USA

Roger Stockbauer
Louisiana State University, Baton Rouge, LA, USA

Ragnar Erik Ryhage

b. October 1, 1919
d. June 21, 1994

Ragnar Ryhage was a Swedish chemist who helped pioneer the use of mass spectrometers in the study of molecules of biochemical and biomedical interest. He played a leading role in linking mass spectrometers and gas chromatographs and linking the combination with computers. His work was crucial in the research that led to Sune Bergström and Bengt Samuelsson sharing the 1982 Nobel Prize for physiology or medicine.

Ragnar Ryhage was born in 1919 in the village of Fjällbacka on Sweden's western coast near the Norwegian border. After graduation from the Stockholm Technical Institute, he was hired in 1947 by Prof. Einar Hammarsten to help with the new mass spectrometer laboratory he was establishing at the Karolinska Institute, a medical university in Stockholm. While working, and being promoted, in the lab, Ryhage also studied at the Royal Institute of Technology in Stockholm and earned a doctorate in 1960 without having completed an undergraduate degree. Ragnar and Anna-Lisa Ryhage married in 1947; they had two children and four grandchildren. He was in charge of mass spectrometry at the Karolinska from 1950 until he retired in 1987. The Swedish Chemical Society inaugurated its Aarhenius Medal with its award to Ryhage in 1962. The Karolinska Institute awarded him an honorary MD in 1978. He died in Stockholm in 1994, survived by his wife and children (*1–3*).

Hammarsten wanted to analyze non-petroleum-based organic compounds with mass spectrometry and had arranged to have an instrument for his research built during 1944–1945. After considerable trouble and with the help of parts brought by visiting Americans, two functioning instruments, 90° and 60° magnetic sectors, were in service by late in 1947. However, plans to modify one or the other were begun almost instantly, a pattern that was continued for decades. At the same time, the mass spectrometry laboratory became a provider of mass spectral analyses for researchers in various departments at the Karolinska and at other institutions both in Sweden and in other countries. Along with this service role, the lab acquired another instrument, based on a 180° design, in the mid-1950s from the Stockholm College and which it put into service after considerable reconstruction.

In addition to keeping the lab running and working on modifications of the instruments, Ryhage began a research collaboration with Einar Stenhagen of Uppsala (see this volume, *Einar August Stenhagen*). In many papers with various collaborators, they demonstrated the power of mass spectrometry for analysis of particular organic materials and for mixtures, pioneering in its use not merely for identifications but for structural determinations (*4,5*). Their work with fatty acid esters was pioneering in itself and the example helped drive the adoption of MS in research on lipids, steroids, terpenes, alkaloids, and various other classes of substance of importance in natural product chemistry (see this volume, *The Beginnings of Organic Mass Spectrometry*). This collaboration ended with Stenhagen's move to the University of Gothenburg but similarly important work was done with Bergström (who followed Hammersten as chemistry professor at the Karolinska) and Samuelsson, among many others (*6–11*).

In addition to this ground-breaking work in organic analysis, Ryhage was a pioneer in the development of linked gas chromatographs and mass spectrometers (GC/MS), beginning in some of the research with Stenhagen (*1*). The basic problem at that time was the output of GC instruments – a gas flow about 100 times more than the maximum amount a mass spectrometer could handle – so ways were sought to eliminate most of the GC's carrier gas without losing too much of the analytes. Ryhage achieved some modest success in 1962 working with Erik von Sydow, of the Swedish Research Institute, Gothenburg, using a splitter to eliminate much of the carrier gas (*12*). Ryhage then developed one of the major ways in which the problem generally handled – the molecular jet separator, modifying a device recently developed in isotope separation work in Germany. This became a key feature of the first commercially marketed GC/MS instrument, the LKB 9000, which was designed and initially constructed in Ryhage's lab in partnership with the LKB company. Development of the instrument also involved creating a mass spectrometer that, unlike the Atlas MAT CH4 model that had been used the earlier experiments, was specifically designed with features like better vacuum pumping in order to work with the chromatograph (*13–16*, and see this volume, *A History of Gas Chromatography Mass Spectrometry [GC/MS]*). In the early years of GC/MS, before fused silica capillary columns largely eliminated the linkage problem, these instruments and their derivatives were the dominant type of machine in use.

As mass spectrometers were linked to gas chromatographs, the amount of information and the speed with which it came made increased computerization of data acquisition mandatory. Ryhage's lab was among the pioneers in this development as well, investigating several ways forward and providing the basis for LKB's subsequent products on these lines (*17,18*).

Throughout his career Ryhage demonstrated a great ability to manage a service lab while also working on new instrumentation and carrying out collaborative research of high quality with many other scientists.

References

(*1*) Ryhage, R. The Mass Spectrometry Laboratory at the Karolinska Institutet, 1944–1987. *Mass Spectrom. Rev.* **1993**, *12*, 1–49.
(*2*) Ryhage, L., personal communication, 6 Sep 2012.
(*3*) Sjövall, J., personal communication, 19 Aug 2012.
(*4*) Bergstrom, H.; Ryhage, R.; Stenhagen, E. Constituents of Tall Oil. I. The Nature of Carnauba Acid from Pitch Wood. *Sven. Papperstidn.* **1956**, *59*, 593–598.
(*5*) Ryhage, R.; Stenhagen, E. Mass Spectrometric Studies. VI. Methyl Esters of Normal Chain Oxo, Hydroxy, Methoxy, and Epoxy Acids. *Ark. Kemi* **1960**, *15*, 545–547.
(*6*) Bergström, S.; Dressler, F.; Krabisch, L.; Ryhage, R.; Sjövall, J. The Isolation and Structure of a Smooth Muscle Stimulating Factor in Normal Sheep and Pig Lungs. *Ark. Kemi.* **1962**, *20*, 63–66.
(*7*) Bergström, S.; Ryhage, R.; Samuelsson, B.; Sjövall, J. The Structures of Prostaglandin E, $F_{1\forall}$, and $F_{1\exists}$. *J. Biol. Chem.* **1963**, *238*, 3555–3564.
(*8*) Bergström, S. The Prostaglandins: From the Laboratory to the Clinic. In: *Nobel Lectures, Physiology or Medicine 1981–1990*; Frängsmyr, T., Lindsten, J., Eds.; World Scientific Publishing Co.: Singapore, 1993; pp 93–112.
(*9*) Bruun, T.; Hollis, D. P.; Ryhage, R. The Constitution of Fragilin. *Acta Chem. Scand.* **1965**, *19*, 839–844.
(*10*) Johansson, B.; Ryhage, R.; Westöö, G. Identification and Determination of Methylmercury Compounds in Fish Using Combination Gas Chromatograph-Mass Spectrometer. *Acta Chem. Scand.* **1970**, *24*, 2349–2354.
(*11*) Enzell, C. R.; Wahlberg, I.; Ryhage, R. Mass Spectra of Tobacco Isoprenoids. *Mass Spectrom. Rev.* **1984**, *3*, 395–438.
(*12*) Ryhage, R.; von Sydow, E. Mass Spectrometry of Terpenes. I. Monoterpene Hydrocarbons. *Acta Chem. Scand.* **1963**, *17*, 2025–2035.
(*13*) Ryhage, R. Use of a Mass Spectrometer as a Detector and Analyzer for Effluent Emerging from High Temperature Gas Liquid Chromatography Columns. *Anal. Chem.* **1964**, *36*, 759–764.
(*14*) Ryhage, R.; Wikstrom, S.; Waller, G. R. Mass Spectrometer Used as Detector and Analyzer for Effluent Emerging from a Capillary Gas Liquid Chromatography Column. *Anal. Chem.* **1965**, *37*, 435–436.
(*15*) Ryhage, R. Efficiency of Molecule Separators Used in Gas Chromatograph-Mass Spectrometer Applications. *Ark. Kemi* **1966**, *26*, 305–316.
(*16*) Ryhage, R. Integrated Gas Chromatography-Mass Spectrometry. *Quart. Rev. Biophys.* **1973**, *6*, 311–335.
(*17*) Pettersson, B.; Ryhage, R. Mass Spectral Data Processing: Identification of Aliphatic Hydrocarbons. *Anal. Chem.* **1967**, *39*, 790–793.
(*18*) Hedfjall, B.; Ryhage, R. Computerized Fast-scanning Gas Chromatograph Mass Spectrometer. *Anal. Chem.* **1975**, *47*, 666–670.

Keith A. Nier
Independent Historian, Madison, NJ, USA

Rudolf Schoenheimer

b. May 10, 1898
d. September 11, 1941

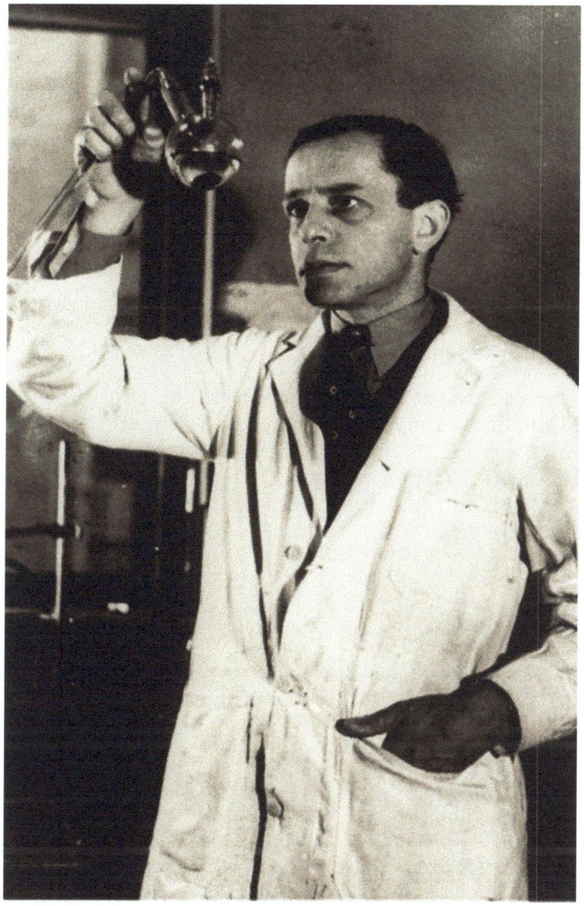

Rudolf Schoenheimer (circa 1937), courtesy of Dr. Salome Glücksohn Schoenheimer Waelsch.

Rudolf Schoenheimer was a German physician and biochemist known for his studies of cholesterol and fatty acid metabolism and for changing the fundamental way in which body constituents are understood by formulating the concept of the dynamic state (*1*). The importance of his work continued to be explicitly recognized with references to it in Nobel Prize announcements and lectures in 1943 (George de Hevesy), 1964 (Konrad Bloch), and 1985 (Michael Brown and Joseph Goldstein). He is also remembered more importantly in the context of this volume, for his metabolic studies using stable isotopes of nitrogen and for pioneering in the use of mass spectrometry in metabolic studies employing ^{15}N. This work was done in collaboration with the chemist David Rittenberg at the urging of Harold Urey, who was interested in seeing development of biochemical applications of stable isotopes (see this volume, *David Rittenberg* and *Harold Clayton Urey*).

Rudolf Schoenheimer was born in Berlin, Germany on May 10, 1898. He completed gymnasium studies in 1916 and entered the University of Berlin, receiving a medical degree in 1922. He served as a pathologist at the Moabit Hospital in Berlin for a year and it was during that time that he became interested in experimental aspects of atherosclerosis and cholesterol metabolism (*2*), interests that persisted through the rest of his career. Schoenheimer went to Leipzig in 1923 to work with Karl Thomas, one of the few biochemists in Germany at the time. While working in Leipzig, he applied for and received a fellowship from the Rockefeller Foundation. Schoenheimer moved to Freiburg in 1926, with the encouragement of Thomas, to study with Ludwig Aschoff, an expert on atherosclerosis. Aschoff, who was the Director of the Pathological Institute of the University of Freiburg, named Schoenheimer as an assistant in chemistry and pathology. Also working in Freiburg at the time was Adolf Windaus, an organic chemist who was awarded the Nobel Prize in 1928 (with Heinrich Wieland) for their work on cholesterol. Schoenheimer's publications during this period reflect the influence of Aschoff and Windaus and their studies in sterol chemistry (*3*).

In 1927 Schoenheimer was appointed acting head of the new department of pathological chemistry in Freiburg. After a fellowship year (1930) as a Research Associate in the Department of Surgery at Billings Hospital at the University of Chicago, Schoenheimer returned to Germany, where in 1931 he became head of the Department of Pathological Chemistry at the Pathological Institute. The work of Hans Adolf Krebs (1900–1981, Nobel Laureate 1953), who astonished the field of biochemistry in 1932 with his studies of nitrogen metabolism and the urea cycle, was an important influence on Schoenheimer's interests in his time in Freiburg. Another was physical chemist George de Hevesy (1885–1960, Nobel Laureate 1943), who had come to Freiburg in 1926 and who introduced Shoenheimer to the concept of isotopic indicators using radioactive tracers through de Hevesy's attempts to study the partition of lead between normal and tumor tissues (*4*). Schoenheimer produced forty-two papers on cholesterol and bile acid metabolism between 1924 and 1933.

From the start of 1933, Hitler's government in Germany increasingly restricted and oppressed people with any Jewish ancestry; in a few months this extended to their dismissal from all university positions (*5–7*). The Josiah Macy Foundation held a conference in New York in 1933 to organize research on atherosclerosis and cholesterol metabolism.

Schoenheimer was invited and arrived in New York on April 17 with his wife Salome (nee Glücksohn). On April 20, Schoenheimer received a telegram from Aschoff informing him that he had been dismissed from his position at the University of Freiburg (*8*). Through Aschoff's contacts in the United States, Schoenheimer found a new position, at first supported by The Macy Foundation, in the department of Hans T. Clarke, chairman of the Department of Biochemistry at the College of Physicians and Surgeons at Columbia University.

A fortuitous collaboration began in 1934 when David Rittenberg, a physical chemist who had trained with Urey, was unable to obtain a job in his field (see this volume, *David Rittenberg*). Urey, who had published the discovery of deuterium in 1932 and was eager to promote its use in biochemical research, encouraged his Columbia colleague Clarke to employ Rittenberg as a postdoctoral fellow. Clarke introduced him to his faculty and Schoenheimer decided to undertake research using deuterium tracers. Rittenberg was skilled in the production and analysis of deuterium and quickly added organic synthesis to his resume. In 1935 Schoenheimer and Rittenberg published the first of a series of seminal papers on the use of deuterium in biological studies of fatty acids (*9*). The analyses did not employ mass spectrometry. Instead, very precise measurements were performed either by refractive index (RI) or, in later studies, by careful measurement of the density of water using the 'falling drop' method (*10*). With these techniques they were able to measure deuterium enrichment with a precision of ±0.001 atom % (RI) or ±0.0001 atom % (water density).

The first reported use of stable isotopes and mass spectrometry for biochemical studies was in a letter to the editor of the *Journal of the American Chemical Society* in 1937 (*11*). In it, Schoenheimer, Rittenberg, and colleagues report biochemical studies using ^{15}N-enriched glycine and hippuric acid from ^{15}N-ammonia provided by Urey. No experimental details or results were given. However, in 1938 Schoenheimer and Rittenberg began to publish a second series of seminal studies on amino acid and protein metabolism using their homemade gas isotope ratio mass spectrometer, the design of which was described in a 1939 report (*12*). This instrument covered the mass spectrum from m/z 16 to m/z 75 and could detect ion currents as low as 10^{-14} A. At normal operating pressure, the ion current was 10^{-11} A, corresponding to a flux of 6×10^7 ions/sec. The device was able to achieve a precision of 0.368 atom % $\pm$ 0.003 atom % for the $^{15}N/^{14}N$ ratio measurement of natural nitrogen. Much of the credit for the implementation of the mass spectrometric measurements of ^{15}N belongs to David Rittenberg, whose name is firmly associated with the technique via the 'Rittenberg Tube' used to prepare nitrogen gas from ammonium sulfate. Schoenheimer's studies of nitrogen metabolism were strongly influenced by the earlier work of Krebs and consumed most of his attention for another three years until his death.

As Hans Clarke wrote, "Few men had more reason for desiring to live; his work gave him intense satisfaction, and its increasing importance was widely recognized" (*13*). Nevertheless, Shoenheimer had been troubled by depression and died through suicide in 1941.

Thorough accounts of Schoenheimer's work are available in several contemporary publications, notably the obituary written by Hans T. Clarke, a review by Henry Borsook of Schoenheimer's posthumous and seminal book, *The Dynamic State of Body Constituents*, and an account of the development of isotopic tracers given by George de Hevesy in his 1943 Nobel Prize lecture (*4,13–15*). Further context and detail can be found in the historical account by Kohler, the scientific perspective by Young and Ajami, and the book by Berthold (*8,16,17*).

Acknowledgments

The authors wish to acknowledge the contributions to this biography by Dr. Peter D. Klein (1927–2001), Jerry D. Eastman (1951–1997) and Dr. Heiner K. Berthold. They collected much original source material in 1990 during an interview with Schoenheimer's widow, Dr. Salome Glücksohn Schoenheimer Waelsch (1907–2007).

References

(*1*) Ratner, S. Applications of Stable Isotopes to the Study of Intermediary Metabolism: Historical Survey. In: *Biochemical Applications of Mass Spectrometry*; Waller, G. R., Ed.; John Wiley & Sons: New York, etc., 1972; pp 1–10.

(*2*) Schoenheimer, R. Über Die Experimentelle Cholesterinkrankheit Der Kaninchen. *Virchows Arch. Pathol. Anat. Physiol. Klin. Med.* **1924**, *249*, 1–42.

(*3*) Schoenheimer, R. New Contributions in Sterol Metabolism. *Science* **1931**, *74*, 579–584.

(*4*) George de Hevesy. Some Applications of Isotopic Indicators. Nobel Lecture, 1944. See http://nobelprize.org/nobel_prizes/chemistry/laureates/1943/index.html

(*5*) Bullock, A. *History: A Study in Tyranny*; Konecky and Konecky: New York, 1962.

(*6*) Salome Gluecksohn Schoenheimer Waelsch, interview by Peter D. Klein, New York, 23 October 1990.

(*7*) Kennedy, E. P. Hitler's Gift and the Era of Biosynthesis. *J. Biol. Chem.* **2001**, *276*, 42619–42631.

(*8*) Kohler, R. E. Jr. Rudolf Schoenheimer, Isotopic Tracers, and Biochemistry in the 1930s. *Hist. Studs. Phys. Sci.* **1997**, *8*, 257–298.

(*9*) Schoenheimer, R.; Rittenberg, D. Deuterium as an Indicator in the Study of Intermediary Metabolism. I. *J. Biol. Chem.* **1935**, *111*, 163–168.

(*10*) Schoenheimer, R.; Rittenberg, D. Deuterium as an Indicator in the study of Intermediary Metabolism. II. Methods. *J. Biol. Chem.* **1935**, *111*, 169–174.

(*11*) Schoenheimer, R.; Rittenberg, D.; Fox, M.; Keston, A. S.; Ratner, S. The Nitrogen Isotope (N^{15}) as a Tool in the Study of the Intermediary Metabolism of Nitrogenous Compounds. *J. Amer. Chem. Soc.* **1937**, *59*, 1768.
(*12*) Rittenberg, D.; Keston, A. S.; Rosebury, F.; Schoenheimer, R. Studies in Protein Metabolism. II. The Determination of Nitrogen Isotopes in Organic Compounds. *J. Biol. Chem.* **1939**, *127*, 291–299.
(*13*) Clarke, H. T. Rudolf Schoenheimer (1898–1941). *Science* **1941**, *94*, 553–554.
(*14*) Schoenheimer, R. *The Dynamic State of Body Constituents*; Harvard University Press: Cambridge, MA, US, 1946.
(*15*) Borsook, H. Review: Physiologic Chemistry, The Dynamic State of Body Constituents. *Science* **1943**, *98*, 63–65.
(*16*) Young, V. R.; Ajami, A. The Rudolf Schoenheimer Centenary Lecture. Isotopes in Nutrition Research. *Proc. Nutr. Soc.* **1999**, *58*, 15–32.
(*17*) Berthold, H. K. *Rudolf Schönheimer (1898–1941) – Leben und Werk*; Falk Foundation: Freiburg im Breisgau, Germany, 1998.

E. Roseland Klein
Baylor College of Medicine, Houston, TX, USA

David L. Hachey
Vanderbilt University, Nashville, TN, USA

Helmut Schwarz

b. August 6, 1943

Source: Schwarz, H. Angewante Chemie Author Profile. *Angew. Chem. Int. Ed.* **2012**, *51*, 5532–5534.

Helmut Schwarz is a German chemist best known for employing mass spectrometry to answer fundamental chemical questions not amenable to traditional chemical methods. Using the mass spectrometer to study molecules as chemists in the 19th Century used test tubes to study chemistry in bulk, his work provides fundamental insights into the step-by-step processes by which molecules react with each other. His approach is summarized succinctly in the title of his 1992 Schering Lecture, 'Chemie 'nackter' Moleküle – oder das Massenspektrometer als Laboratorium' ('The chemistry of 'naked' molecules – or the mass spectrometer as a laboratory') (*1*).

Schwarz was born in Nickenich, Rheinland-Pfalz, Germany, on August 6, 1943. He began his career as a Chemical Laboratory Assistant with Dynamit Nobel in Troisdorf (1960 to 1964) and afterward studied chemistry at the Technishe Universität (TU) in Berlin. He was awarded his Diplom-Chemiker in 1971, his Dr. rer. nat. in 1972 under F. Bohlmann, and his Habilitation in1974. Schwarz went on to post-doctoral appointments with J. Seibl at the ETH Zürich, Klaus Biemann at MIT, and Dudley Williams at Cambridge University. He was appointed Professor of Theory and Practice of Mass Spectrometry at the TU Berlin in 1978. He became Professor of Inorganic Chemistry at the TU Berlin in 1983, later adding responsibilities for Organic and Physical Chemistry. In January, 2008, he also became president of the Humboldt Foundation.

Chemistry first caught Schwarz's attention in high school when a teacher commented that "chemistry, in essence, is the only science that tries to cope with the permanent change of matter" (*2*). With that as a jumping off point for his scientific endeavors, it is no surprise that the most prominent characteristic of his achievements is their depth and breadth of vision. His laboratory is crammed with mass spectrometers that have been variously connected to reveal grand themes in the chemistry. Techniques and machines are characterized by names plucked from an alphabet soup – BEBE, MSMS, CAMS, NRMS, NIDD, NRMS-CA, *etc.* – generally names of increasing complexity as yet another piece is added; thus, wherever the signal may be diminished, somehow the detector sensitivity will increase.

In virtually every area to which he has turned, Schwarz has made important contributions. The form and sequence of his work follows a pattern (*3*): first, comes the initial vision; next, the highly complex experiments, invariably with computation to characterize the key features of the reaction hyper-surface that shapes the chemistry, and finally, a sense not just of the individual *tessera* but rather of its place in the mosaic as a whole. Schwarz reports his work in publications (numbering more than 950) that always include masterful exegesis and are compelling in their clarity. A few examples serve to illustrate.

He and his colleagues uncovered the first exception to the ergodic theorem (which states that for an excited chemical species, vibrational energy randomizes before bonds break) in the enol acetone radical cation (*4*). Other exceptions have since been found. He was the first to insert an atom into a cage, creating $[C_{60}\,He]^{+}$, a species bound by only ~6 eV but synthesized at 8,000 eV. This feat he accomplished with a very sophisticated tandem mass spectrometer that he created (*5,6* and see this volume, *A Perspective on the Development of Tandem Mass Spectrometry*). Using an electrospray ion source coupled to a tandem mass spectrometer, he completed a broad survey of different reactions involving a wide range of metals (see this volume, *The Development of Electrospray Ionization (ESI) for Mass Spectrometry*). This work, demonstrating that metals can act as leaving groups in S_N2 reactions, was reported in a paper that reviews the vast histories of organometallic chemistry and of nucleophilic displacement and puts this seemingly unfamiliar piece in the jigsaw of chemistry in its rightful place (*7*). Turning his attention to the synthesis and chemistry of non-existent and elusive neutral compounds (M), Schwartz used Neutralization-Reionization Mass Spectrometry, NRMS, to explore the chemistry of such species. Using charge transfer or collisional ionization of ionized counterparts, Schwarz created transient moities such as AuF, H_2CO_3, H_2SO_3, OCCO and even the of elusive neutral species CH_3O that lasted long enough in the ion source for their chemical nature to be probed (*8–11*). He has also explored the relationship between spin changes and reactivity, demonstrating that where one reactant has a low-lying excited state, minimum-energy pathways can involve two states, and crossing between them involves spin changes that affect the chemistry (*12,13*). This new principle is found to have a widespread relevance in such diverse areas as organometallic chemistry, oxidation catalysis, and the chemistry of cytochrome P-450 (*14*). In addition, Schwarz has used gas-phase ion chemistry with transition metal ions, both bare and complexed, to provide conceptual insights (*15*) and to probe the conversion of methane to both methanol and acetic acid for which no industrial catalytic processes has yet been found (*16*). This mechanistic insight has proved useful for the industrial synthesis of HCN from ammonia and methane (*17,18*). Schwarz has also been able to model the activation of carbon dioxide by carbonic anhydrase (30 kDa) in the gas phase by an ionic zinc complex, at 300 Da, one hundredth of the size (*19*).

Schwarz's remarkable scientific accomplishments should not, however, overshadow an essential element of the man himself: Helmut Schwarz is greatly and universally loved by friends and colleagues alike, and his enthusiasm for science is contagious among his colleagues and students. That trait is particularly apparent in his activities as President of the Alexander von Humboldt Foundation, where he is able to travel the world promoting and encouraging the achievements of others.

Schwarz is an avid cyclist: his preferred mode of transportation for his daily 20 km commute is by bicycle. He is keenly interested in music, literature, and the theatre. Indeed, his seat at the Berlin Opera is occupied at least once each week (*20*).

References

(*1*) Schwarz, H. 'Chemie 'nackter' Moleküle – oder das Massenspektrometer als Laboratorium' *Chemie in unswerer Zeit*, **1991**, *25*, 268–278.

(*2*) Pearson, R. Berlin chemist wins US award for reaction studies. http://news.bio-medicine.org/medicine-news-2/Berlin-chemist-wins-U-S–award-for-reaction-studies-9307-1/ (accessed August 28, 2012).

(*3*) Butschke, B.; Schwarz, H. "Rollover" Cyclometalation – Early History, Recent Developments, Mechanistic Insights and Application Aspects. *Chem. Sci.* **2012**, *3*, 308–326.

(*4*) Depke, G.; Lifshitz, C.; Schwarz, H.; Tzidony, E. Non-Ergodic Behavior of Excited Radical Cations in the Gas Phase. *Angew. Chem. Int. Ed.* **1981**, *20*, 792–793.

(*5*) Schwartz, H. Fullerene Eine Neue Dimensione in der Chemie? *Nova Acta Leopold.* **1993**, *69*, 167–181.

(*6*) Weiske, T.; Böhme, D. K.; Hrušák, J.; Krätschmer, W.; Schwarz, H. Endohedral Cluster Compounds: Inclusion of Helium within $C_{60}^{\bullet+}$ and $C_{70}^{\bullet+}$ through Collision Experiments. *Angew. Chem. Int. Ed.* **1991**, *30*, 884–887.

(*7*) Kretschmer, R.; Schlangen, M.; Kaupp, M.; Schwarz, H. Neutral Metal Atoms Acting as a Leaving Group in Gas-Phase S_N2 Reactions: $M(CH_3)^+ + NH_3 \rightarrow CH_3NH_3^+ + M$ (M=Zn, Cd, Hg). *Organometallics.* **2012**, *31*, 3816–3824.

(*8*) Srinivas, R.; Sützle, D.; Weiske, T.; Schwarz, H. Generation and Characterization of Neutral and Cationic 3-Sila-cyclopropenylidene in the Gas Phase: Description of a New BEBE Tandem Mass Spectrometer. *Int. J. Mass Spectrom. Ion Proc.* **1991**, *107*, 369–376.

(*9*) Goldberg, N.; Schwarz, H. Neutralization-Reionization Mass Spectrometry: A Powerful "Laboratory" to Generate and Probe Elusive Neutral Molecules. *Acc. Chem. Res.* **1994**, *27*, 347–352.

(*10*) Schröder, D.; Heinemann, D. C.; Schwarz, C. H., *et al.* Ethylenedione: An Intrinsically Short-Lived Molecule. *Chem. Eur. J.* **1998**, *4*, 2550–2557.

(*11*) Schröder, D.; Hrušák, J.; Tornieporth-Oetting, I. C.; Klapötke, T. M.; Schwarz, H. Neutral Gold(I) Fluoride Does Indeed Exist. *Angew. Chem. Int. Ed.* **1994**, *33*, 212–214.

(*12*) Schalley, C. A; Hornung, G.; Schröder, D.; Schwarz, H. Mass Spectrometric Approaches to the Reactivity of Transient Neutrals. *Chem. Soc. Rev.* **1998**, *27*, 91–104.

(*13*) Shaik, S.; Danovich, D.; Fiedler, A.; Schroder, D.; Schwarz, H. Two-State Reactivity in Organometallic Gas-Phase Ion Chemistry. *Helv. Chim. Acta.* **1995**, *78*, 1393–1407.

(*14*) Schröder, D.; Shaik, S.; Schwarz, H. Two-State Reactivity as a New Concept in Organometallic Chemistry. *Acc. Chem. Res.* **2000**, *33*, 139–145.

(*15*) Schwarz, H. On the Spin-Forbiddeness of Gas-Phase Ion–Molecule Reactions: A Fruitful Intersection of Experimental and Computational Studies. *Int. J. Mass Spectrom.* **2004**, *237*, 75–105.

(*16*) Böhme, D. K.; Schwarz, H. Gas-Phase Catalysis by Atomic and Cluster Metal Ions: The Ultimate Single-Site Catalysts. *Angew. Chem. Int. Ed.* **2005**, *44*, 2336–2354.

(*17*) Schwarz, H. Chemistry with Methane: Concepts rather than Recipes. *Angew. Chem. Int. Ed.* **2011**, *50*, 10096–10115.

(*18*) Diefenbach, M.; Brönstrup, M.; Aschi, M.; Schröder, D.; Schwarz, H. HCN Synthesis from Methane and Ammonia: Mechanisms of Pt^+-Mediated C–N Coupling. *J. Am. Chem. Soc.* **1999**, *121*, 10614–10625.

(*19*) Schröder, D.; Schwarz, H.; Schenk, S.; Anders, E. A Gas-Phase Reaction as a Functional Model for the Activation of Carbon Dioxide by Carbonic Anhydrase. *Angew. Chem. Int. Ed.* **2003**, *42*, 5087–5090.

(*20*) Everts, S. Interview with Helmut Schwarz, Jan 21, 2008, *86*, *C&E News.* p. 46.

P. Jane Gale

Waters Corp., Milford, MA, USA

Nicholas John Shackleton

b. June 23, 1937
d. January 24, 2006

Nicholas Shackleton (commonly known as Nick) was an English geophysicist who used mass spectrometry to determine the course of climate change over millions of years and to help resolve the question of the cause of ice ages. His outstanding scientific work was recognized with the 1995 Crafoord Prize and a knighthood, yet he himself suggested that he was better known as a player, collector, and historian of the clarinet (*1–3*).

Nick Shackleton was born in London in 1937. His father, Robert Millner Shackleton, FRS, was a renowned structural geologist and his mother, Gwen Isabel (nee Harland), was a professional musician. The couple divorced in 1949, the same year that Nick became a boarder pupil at Cranbrook School in Kent. After service in the army, where he became secretary of the regimental band, Nick entered Clare College in Cambridge University, where he earned a BA in physics in 1961.

As a graduate student at Cambridge, Shackleton took on the project of creating a lab to make mass spectrometric measurements of oxygen isotope ratios to determine paleotemperatures for the Sub-Department of Quaternary Research in the University's Botany Department. This was intended to follow upon the work of Cesare Emiliani. Harold Urey's team had created isotope paleoclimatology with their work on ratios in fossil carbonate shells and Emiliani had extended that to studies of foraminifera in oceanic sediments (see this volume, *The Development of Mass Spectrometry in the Earth and Planetary Sciences, Samuel Epstein*, and *Harold Clayton Urey*). Shackleton created a modified mass spectrometer capable of precisely measuring tiny isotope ratio variations with much smaller samples than Emiliani had required (*4*). He then was able to confirm the sequences of changes that Emiliani had reported but also demonstrated that they reflected more the changes in the volume of ice, and thus the extent of glaciation, than changes in sea-surface temperatures (*5*). His research brought him a Ph.D. (1967) and, before it was completed, official employment by the Sub-Department, starting in 1965. He never left, rising eventually to become director and staying as it was reborn the Godwin Institute of Quaternary Research. At the same time he held positions in his college. He retired in 2004 and less than a year and a half later died of leukemia at home in Cambridge.

Throughout his research career, Shackleton measured isotopic composition ratios in many samples, regularly seeking to correlate measurements of different types and from different locales, developing a global isotope stratigraphy and chronological scale for the more recent of geological times. He sought to have different sorts of studies and different data sets complement each other. He worked in ice cores and cores from deep sea drilling projects. He added carbon isotope measurements to his earlier focus on oxygen. He also brought changes in the earth's magnetic field into the overall picture. He entered a long and fruitful cooperation with researchers at the Lamont-Doherty Geological Observatory of Columbia University. He became a central figure in major international cooperative projects to chronologically and geographically map climate change. Surprising and fundamental results flowed from all this effort.

One major discovery was clear evidence that the climate changes marked by major times of glaciation fluctuated in complex long term cycles with the same periods as those variations in the earth's orbit that the mathematician Milutin Milanković had long before proposed as the causes of ice ages. The interrelations are complex but the analysis of the extensive data clearly showed the interconnection of these astronomical variations with terrestrial ice (*6,7*). Another major result of the studies by Shackleton and his colleagues was establishing that there have been many more glacial and interglacial periods than had been identified previously through continental evidence. In addition, he discovered through comparisons with his own measurements that many potassium-argon dates had to be corrected because significant argon loss had taken place. And, perhaps most salient, due to implications for present and future global policy issues, was the demonstration that increases in the atmospheric concentration of CO_2 precede rather than follow the major warming episodes. A few papers sampling his work are listed below (*8–11*).

From election as a Fellow of the Royal Society to his knighthood, Shackleton was widely honored for the quality and significance of his scientific work. Among many other awards, he received the

Wollaston Medal from the Geological Society of London, the Urey Medal from the European Geochemical Association, the Royal Medal from the Royal Society, and he shared the Crafoord Prize (commonly compared with the Nobel Prize for the fields it covers) with Willi Dansgaard in 1995 in recognition of their combined work revealing the history of global climate (see this volume, *Willi Dansgaard*).

Throughout his adult life Nick Shackleton also played and collected clarinets. He became one of the world's authorities on the history of this instrument. He bequeathed his collection, the largest in the world, to the Edinburgh University.

References

(*1*) McCave, I. N.; Elderfield, H. Sir Nicholas John Shackleton. 23 June 1937 – 24 January 2006. *Biogr. Mems Fell. R. Soc* **2011**, *57*, 435–462.

(*2*) Gibbard, P. Professor Sir Nicholas Shackleton. *Boreas* **2006**, *35*, 385–390.

(*3*) Ruddiman, W. F. Retrospective: Nicholas J. Shackleton (1937–2006). *Science* **2006**, *312*, 711.

(*4*) Shackleton, N. J. The High-Precision Isotopic Analysis of Oxygen and Carbon in Carbon Dioxide. *J. Scient. Instrum* **1965**, *42*, 689–692.

(*5*) Shackleton, N. J. Oxygen Isotope Analyses and Pleistocene Temperatures Re-Assessed. *Nature* **1967**, *215*, 15–17.

(*6*) Hays, J. D.; Imbrie, J.; Shackleton, N. J. Variations in the Earth's Orbit: Pacemaker of the Ice Ages. *Science* **1976**, *194*, 1121–1131.

(*7*) Imbrie, J.; Hays, J. D.; Martinson, D. G., *et al.* The Orbital Theory of Pleistocene Climate: Support from a Revised Chronology of the Marine $*^{18}$O Record. In: *Milankovitch and Climate, Part 1*; Berger, A. L., Imbrie, J., Hays, J. D., Kukla, G., Saltzman, B., Eds.; D. Reidel: Dordrecht, 1984; pp 269–305.

(*8*) Shackleton, N. J.; Hall, M. A.; Line, J.; Cang, S. Carbon Isotope Data in Core V19–V30 Confirm Reduced Carbon Dioxide Concentration in the Ice Age Atmosphere. *Nature* **1983**, *306*, 319–322.

(*9*) Shackleton, N. J.; Imbrie, J.; Pisais, N. The Evolution of Oceanic Oxygen Isotope Variability in the North Atlantic over the Past Three Million Years. *Philos. Trans. R. Soc. Lond. B Biol. Sci.* **1988**, *318*, 679–688.

(*10*) Shackleton, N. J.; Berger, A.; Peltier, W. R. An Alternative Astronomical Calibration of the Lower Pleistocene Timescale Based on ODP Site 677. *Trans. R. Soc. Edinb. Earth Sci.* **1990**, *81*, 251–261.

(*11*) Shackleton, N. J. The 100,000-Year Ice Age Cycle Identified and Found to Lag Temperature, Carbon Dioxide and Orbital Eccentricity. *Science* **2000**, *289*, 1897–1902.

Keith A. Nier
Independent Historian, Madison, NJ, USA

Andrew Gans Sharkey, Jr.

b. August 21, 1919
d. November 17, 2003

Andrew Gans 'Jack' Sharkey, Jr. was an American physicist whose career not only embodied high caliber research but is a sterling example of an individual's service to others. He excelled in three scientific arenas: industry, government and academe. In industry he contributed to development of mass spectrometers. In government he had a long and distinguished career in characterization of coal and in administration. In academe, he taught graduate courses in mass spectrometry and carried out basic research.

Sharkey was born on August 21, 1919, and raised in Mt. Lebanon, PA, USA, a suburb of Pittsburgh. He graduated from the College of Wooster (Ohio) with a B.A. in Physics in 1941 and received an M.S. degree from Case Institute of Technology in 1943. He did some additional graduate work at the University of Pittsburgh.

Sharkey began his professional career in 1943 at Westinghouse, working on the development of a commercial mass spectrometer. In 1946 Sharkey moved to the U. S. Bureau of Mines location in Bruceton, PA (also a Pittsburgh suburb) which ultimately transitioned to the Energy Research and Development Administration (ERDA) in 1974 and then the Department of Energy in 1977 Under the latter it was named the Pittsburgh Technology Center (PETC); Sharkey was its Deputy Director from 1978 until his retirement in 1985.

Because of his experience with development of modern scientific instrumentation at Westinghouse, Sharkey became an advocate for the application of spectroscopic instrumentation to the study of coal and coal liquids. His own research at PETC involved such studies on coal extracts and coal gasification and liquefaction products (*1–4*).

In addition to his full-time job at PETC, Sharkey made a significant impact on programs at the University of Pittsburgh. This began with a part-time adjunct appointment in the Department of Geology and Planetary Sciences begun in 1964. He developed two graduate-level courses in mass spectrometry: one on general mass spectrometry, and the other on mass spectral interpretation. These were among the first such courses in the country, and were taken by geology and chemistry students and local industrial chemists. Sharkey's effort in the department led to the development of organic geochemistry and isotope geochemistry programs. He also served the department as Assistant Chair and as Acting Chair when the Chairman was away.

On his retirement from PETC in 1985 he joined the Pitt Surface Science Center as Research Professor and established a program on the surface characterization of coal and how the oxidized surface properties of coal affected its end use. In these endeavors he used the laser microprobe mass spectrometer (LAMMA-1000) and time-of-flight secondary-ion mass spectrometry (*5*, and see this volume, *The Development of Time-of-Flight (TOF) Mass Spectrometry*). Additionally, he helped to modernize the chemistry department's mass spectrometry laboratory and served as Interim Assistant Chair of the department.

Sharkey was heavily involved in the support of professional societies. He worked tirelessly for the Pittsburgh Conference on Analytical Chemistry and Applied Spectroscopy (PittCon), serving annually on various of its committees and as its Chairman in 1961. Additionally he was a member of the Spectroscopy Society of Pittsburgh and its Chairman in 1953–54, he served on ASTM Committee E-14 on Mass Spectrometry and Allied Topics and as its Vice-President for Arrangements in 1963, and he was a charter member of the Board of the American Society for Mass Spectrometry in 1971 at the inception of that organization.

Sharkey received the highest awards from two federal agencies: Department of the Interior Distinguished Service Award (1970) and the Department of Energy Distinguished Career Service Award (1985). He also received recognition of his scientific accomplishments by two awards: Pittsburgh Award of the Pittsburgh Section of the American Chemical Society (1992) and as a Pioneer in Mass Spectrometry at the James L. Waters Symposium at PittCon (1997). The University of Pittsburgh honored him with the "Outstanding Service Award" from the Pitt American Chemical Society Student Affiliates (1992); and on his retirement from Pitt, by a two-day symposium, "Advances in Isotope Mass Spectrometry of the Earth and the Environment" in 1996.

Sharkey was devoted to family and community. He married his high school sweetheart ("Tooie") and they remained loyal fans of his high school football team, attending all home games as long as their health permitted. His last position was Research Professor, Department of Chemistry and Department of Geology and Planetary Science, University of Pittsburgh. He died on November 17, 2003.

References

(*1*) Sharkey, A. G.; Friedel, R. A.; Langer, S. H. Mass Spectra of Trimethylsilyl Derivatives. *Anal. Chem.* **1957**, *29*, 770–776.
(*2*) Friedel, R. A.; Schultz, J. L.; Sharkey, A. G. Mass Spectra of Alcohols. *Anal. Chem.* **1956**, *28*, 926–934.
(*3*) Sharkey, A. G. Jr.; Schultz, J. L.; Friedel, R. A. Mass Spectra of Pyrolyzates of Several Aromatic Structures Identified in Coal Extracts. *Carbon.* **1966**, *4*, 365–374.
(*4*) Langer, S. H.; Friedel, R. A.; Wender, I.; Sharkey, A. G. New Mass Spectrometric Method for Determining Alcohols and Water in Complex Mixtures. Fischer-Tropsch Product. *Anal. Chem.* **1958**, *30*, 1353–1356.
(*5*) Kubis, A. J.; Somayajula, K. V.; Sharkey, A. G.; Hercules, D. M. Laser Mass Spectrometric Analysis of Compounds Separated by Thin-Layer Chromatography. *Anal. Chem.* **1989**, *61*, 2516–2523.

David M. Hercules
Department of Chemistry, Vanderbilt University, Nashville, TN, USA

David Smith

b. November 26, 1935

David Smith is an English chemical physicist whose work has concentrated on ion physics and chemistry, be it in the interstellar medium or the interior of a clinical diagnostic instrument. He has made significant contributions in several areas, particularly in the development of the selected-ion flow tube (SIFT) and flowing afterglow/ Langmuir probe (FALP) techniques and their medical applications.

David Smith was born on 26 November, 1935, in the Potteries, in the heart of the Black Country (Staffordshire, England). The seventh of nine children from very modest circumstances, he worked as a laboratory technician after leaving high school to help support the family. There his interest in science was kindled and he was able to complete his secondary education through courses taken in night school. David became a member of the first class of the then newly founded University of Keele, graduating with a BA in physics and chemistry in 1959. He earned a Ph.D. in the department of electron physics at Birmingham University in 1962 under the supervision of Jim Sayers. Smith joined the department and remained as it evolved into the School of Physics and Space Research. It was here that his lifelong interest in the terrestrial atmosphere, the interstellar medium and the evolution of the cosmos began. In 1984 he was appointed to a personal chair in chemical physics. He became University Professor at the Institute for Ion Physics at the University of Innsbruck, Austria, in 1990. Returning to the University of Keele in 1995, he became professor in the Department of Biomedical Engineering and Medical Physics within its School of Medicine.

David Smith has furthered our understanding of reaction processes in ionized gases by developing two distinctly different instruments, the selected-ion flow tube (SIFT) and the flowing afterglow/Langmuir probe (FALP). These instruments straddle the disciplines of physics and chemistry in a fundamental way in that they provide the means by which the detailed parameters, measured for individual molecules in the mass spectrometer ion source, may be translated into their corresponding macroscopic counterparts. Circumventing the long-standing impasse to accomplishing this translation for chemical kinetics is the singular achievement of David Smith in his design of the SIFT instrumentation.

In the SIFT technique, a specific ion is formed and selected by mass, using traditional means. The resulting ion beam is then injected into an operating flow tube, typically against a pressure differential of six orders of magnitude and without significant diminution of the ion beam. Neutral reactants are added downstream. Both ions and neutrals are thermalized by repeated collisions with the buffer gas in the flow tube. When the plasma is then sampled by another mass spectrometer, the measured rate constant for the reaction between ions and neutrals is then a truly *thermal* rate constant for the temperature fixed by the temperature of the contents of the flow tube. Rate constants can be measured over a wide range of temperature and pressure, establishing SIFT as the workhorse for measuring the ion-neutral molecule rates of reactions. When the SIFT technique appeared in 1975, it attracted to Birmingham visitors from throughout the world: twenty years later, that same technique was being used in some 30 different laboratories worldwide.

In contrast, the more challenging FALP technique is used in perhaps six laboratories. It can be usefully viewed as a modification of the SIFT technique for those reactions involving free electrons, either as reactants or products. (Electron detachment and positive ion/electron combination are obvious examples.) Because the flow of ions and electrons in a flow tube differ, the electron concentration must be measured *in situ* using a Langmuir probe. The proper operation of a Langmuir probe is tricky and requires people with skill and experience to employ it properly.

These techniques have been used to explore elementary processes in plasmas, including ion-molecule, electron-molecule, electron-ion, and ion-ion interactions. They have been used in studies of the chemistry of stellar and interstellar regions, and they have been developed as a new way to carry out medical diagnosis. A few of Smith's many publications are listed below (*1–7*). A full publication list can be found in the issues of the International Journal of Mass Spectrometry honoring David Smith's accomplishments (*8*).

For nearly two decades, David Smith and his former student Patrik Španěl (now a senior scientist at the Physical Chemistry Institute of the Czech Academy of Sciences, Prague) have extended the SIFT and FALP techniques for the direct on-line monitoring of human breath. This project aims to measure quantitatively metabolites in exhaled human breath and in the headspace of biological fluids, the purpose being to support clinical diagnosis, therapeutic monitoring, and health and safety practices. In the techniques SIFT-MS and FALP-MS, derived from the SIFT and FALP techniques, a sample of the breath is ionized; each ion is identified by matching with standards using the SIFT technique; thus, components of the breath can quantified, e.g. molecular markers of disease in the clinical patient's breath, such as acetone for diabetes and ammonia for kidney failure.

David Smith was elected a Fellow of the Royal Society in 1983. He has received numerous prizes and medals, including the Schrödinger Prize (Austria), the Masaryk University Medal (Czech Republic), the Gold Medal of Comenius University (Slovak Republic). He has also received a number of honorary degrees including the first that the University of Keele ever awarded to one of its own graduates.

References

(*1*) Adams, N. G.; Smith, D. Selected Ion Flow Tube (SIFT) - Technique for Studying Ion-Neutral Reactions. *Int. J. Mass Spectrom. Ion Phys.* **1976**, *21*, 349–359.
(*2*) Smith, D.; Adams, N. G.; Alge, E. Attachment Coefficients for the Reactions of Electrons with CCl_4, CCl_3F, CCl_2F_2, $CHCl_3$, Cl_2 and SF_6 Determined between 200 K and 600 K Using the FALP Technique. *J. Phys. B Atom Mol. Opt. Phys* **1984**, *17*, 461–472.
(*3*) Smith, D. The Ion Chemistry of Interstellar Clouds. *Chem. Rev.* **1992**, *92*, 1473–1485.
(*4*) Smith, D.; Španěl, P. Studies of Electron-Attachment at Thermal Energies using the Flowing Afterglow Langmuir Probe technique. *Adv. Atom. Mol. Opt. Phys* **1994**, *32*, 307–343.
(*5*) Smith, D.; Španěl, P Selected Ion Flow Tube Mass Spectrometry (SIFT-MS) for On-Line Trace Gas Analysis. *Mass Spec. Rev.* **2005**, *24*, 661–700.
(*6*) Španěl, P.; Smith, D. Progress in SIFT-MS: Breath Analysis and Other Applications. *Mass Spec. Rev.* **2011**, *30*, 236–267.
(*7*) Henchman, M. David Smith. *Int. J. Mass Spectrom. Ion Phys.* **1995**, *149/150*, xiii–xv.
(*8*) Smith, David. Professor David Smith: List of Publications. *Int. J. Mass Spectrom. and Ion Processes.* **1995**, *149/150*. xvii-xxiv.

P. Jane Gale
Waters Corp., Milford, MA, USA

George C. Stafford, Jr.

b. October 2, 1950

George Stafford is an American chemist known for his skills as an experimentalist and for the development of a number of instruments particularly a practical Quadrupole Ion Trap mass filter. He was born on October 2, 1950, in Harrisonburg, VA and resides at present in San Jose, CA, USA. He is married and he and his wife have raised three children.

George received his B.S in chemistry in 1973 from Virginia Commonwealth University, Richmond, VA and his Ph.D. in chemistry in 1977 from the University of Virginia, Charlottesville, VA under the direction of Professor Donald Hunt (see this volume, *Donald Frederick Hunt*). While still a graduate student in Hunt's lab, where he was studying negative ion chemical ionization, George discovered that both positive and negative ion spectra could be recorded simultaneously by rapidly changing the polarity of the source (*1*). Pulsed Positive Ion Negative Ion Chemical Ionization (PPINICI), as the technique came to be called, was used extensively in the Hunt lab for a number of years, first in environmental and clinical applications and subsequently for peptide characterization (*2–4*).

George began his professional career as a Research Scientist at Finnigan Instruments, where his initial focus was on improving source design for chemical ionization and detection for negative ions. The former led to development of an interchangeable ion volume, known as the EI/CI source, and the latter to the conversion dynode addition to the electron multiplier. Both innovations were originally implemented commercially on the Finnigan 4000 GC/MS (*5*). During the late 1970s and early 1980s, from his position at Finnigan, George supported research and development in the Hunt lab on triple-stage quadrupole instruments, at the same time initiating a research program on ion trap mass spectrometry at Finnigan and collaborations with other academic collaborators, notably Graham Cooks at Purdue (see this volume, *R. Graham Cooks*).

In the 1983–84 academic year George served as a Visiting Professor at the University of Virginia, during which time he taught courses in analytical chemistry and instrumentation methods and continued research in the area of quadrupole ion traps under the sponsorship of Finnigan Instruments.

Returning to Sunnyvale in 1984 as Research Manager for Finnigan, George assumed responsibility for directing internal mass spectrometry instrumentation research and development. From 1993–1997 he was the Ion Trap Product Development Program Manager and was responsible for both new developments (*6–8*) including direct responsibility for the development of the LCQ series of instruments. In 1999 he became the R&D Program Director with responsibility for all internal mass spectrometry instrument research and development.

Upon joining Agilent Technologies in 2009 Stafford assumed responsibility for the single and triple quadrupole lines, as well as for Q-TOF instruments and API ion sources.

George Stafford received the 2001 ASMS Award for Distinguished Contribution in Mass Spectrometry for discovery of Ion Trap Mass-Selective Ejection Technology. The Association of Analytical Chemists has honored him with the 1995 Anal Chem award for his development of the conversion dynode electron multiplier negative ion detector and for development of ion trap technology.

References

(*1*) Hunt, D. F.; Stafford, G. C. Jr; Crow, F. W.; Russell, J. W. Pulsed Positive Negative Ion Chemical Ionization Mass Spectrometry. *Anal. Chem.* **1976**, *48*, 2098–2105.

(*2*) Hunt, D. F.; Stafford, G. C. Jr; Crow, F. W.; Russell, J. W. Pulsed Positive Negative Ion Chemical Ionization Mass Spectrometry. *Anal. Chem.* **1976**, *48*, 2098–2105.

(*3*) Hunt, D. F.; Shabanowitz, J.; Giordani, A. B. Collision Activated Decompositions of Negative Ions in Mixture Analysis with a Triple Quadrupole Mass Spectrometer. *Anal. Chem.* **1980**, *52*, 386–390.

(*4*) Hunt, D. F.; Bone, W. M.; Shabanowitz, J.; Rhodes, G.; Ballard, J. M. Sequence Analysis of Oligopeptides by Secondary Ion Collision Activated Dissociation Mass Spectrometry. *Anal. Chem.* **1981**, *53*, 1704–1706.

(*5*) Stafford, G. C. Instrumental Aspects of Positive and Negative Ion Chemical Ionization Mass Spectrometry. *Environ. Health Perspect.* **1980**, *36*, 85–96.

(*6*) Stafford, G. C. Jr; Kelley, P. E.; Syka, J. E. P.; Reynolds, W. E.; Todd, J. F. J. Recent Improvements in and Analytical Applications of Advanced Ion Trap

Technology. *Int. J. Mass Spectrom. Ion Processes* **1984**, *60*, 85–98.

(*7*) Louris, J. N.; Cooks, R. G.; Syka, J. E. P.; Kelley, P. E.; Stafford, G. C. Jr.; Todd, J. F. J. Instrumentation, Applications, and Energy Deposition in Quadrupole Ion-Trap Tandem Mass Spectrometry. *Anal. Chem.* **1987**, *59*, 1677–1685.

(*8*) Madsen, J. A.; Gardner, M. W.; Smith, S. I., *et al.* Top-Down Protein Fragmentation by Infrared Multiphoton Dissociation in a Dual Pressure Linear Ion Trap. *Anal. Chem.* **2009**, *81*, 8677–8686.

George C. Stafford Jr.
Agilent Technologies, Santa Clara, CA, USA

Kenneth Graham Standing

b. April 3, 1925

Ken Standing on right pictured with long time collaborator Werner Ens

Kenneth G. Standing is a Canadian physicist noted for the design and building of high-resolution time-of-flight mass spectrometers and for their use in measurements of the masses of large biomolecules, particularly proteins (*1,2*).

Ken Standing was born in Winnipeg, Manitoba, Canada on April 3, 1925. He is the son of an accountant who became the manager of a large wholesale drug company in Winnipeg. Ken attended public schools in Winnipeg and then the University of Manitoba, where he graduated in Mathematics and Physics in 1948. He proceeded to Princeton University for the A.M. (1950) and Ph.D. degrees, which he completed in nuclear physics, with Prof. Rubby Sherr as advisor, in 1955. He took up an appointment as Assistant Professor of Physics at the University of Manitoba, the University that would be his home for the remainder of his career, in 1953. There, he established a research program in nuclear physics, designed, built, and commissioned a 50 Mev Cyclotron, and served as Director of the Cyclotron Laboratory from 1959 to 1967 and from 1968 to 1974. During this period he spent leaves in 1958–9 at the University of Bristol (as a Nuffield Fellow); and in 1967–8 at the Université de Grenoble. Although Standing retired from the University of Manitoba with the title Professor Emeritus in 1995, his research program has continued unabated.

Standing became interested in biological problems when the Cyclotron Laboratory was approached to help develop a simpler way of measuring the nitrogen content of wheat to replace the cumbersome Kjeldahl wet chemistry method then in use. Although the method he devised was not adopted, the experience prompted Standing to look for other opportunities for a physicist to contribute to important biological problems. When he learned about the work being done by R. D. Macfarlane and D. J. Torgerson at Texas A & M University, using plasma desorption by fission fragments to produce unexpectedly large quasi-molecular ions from biological materials, which were then studied by time-of-flight (TOF) mass spectrometry, he determined to build such an instrument himself (see this volume, *The Development of Time-of-Flight (TOF) Mass Spectrometry* and *A History of ^{252}Cf-Plasma Desorption Mass Spectrometry*). A recently arrived post-doctoral fellow, Brian Chait – also trained as a nuclear physicist and also interested in biological applications – undertook the project, using a small grant from the University (see this volume, *Brian T. Chait*). A ^{252}Cf source for fission fragments was acquired but was never used; – instead, following on Alfred Benninghoven's success with secondary ion mass spectrometry (SIMS) and small organic molecules, Chait and Standing developed a novel ionization method that employed pulses of accelerated Cs^+ ions (see this volume, *The Early History of Secondary Ion Mass Spectrometry (SIMS)*, *The Development of SIMS for Imaging*, and *Alfred Benninghoven*). The TOF-MS instrument with these features was described in 1981 (*3*). Later, introduction of an electrostatic reflector into the ion flight path greatly improved the achievable mass resolution (*4*).

During the decade after 1979, there were intense worldwide efforts to develop desorption methods that would allow ionization of intact macromolecules. These efforts culminated in the invention of the two key methods now in use – matrix-assisted laser desorption ionization (MALDI) and electrospray ionization (ESI) (see this volume, *The Development of Matrix Assisted Laser Desorption Ionization (MALDI) Mass Spectrometry* and *The Development of Electrospray Ionization (ESI) for Mass Spectrometry*). Standing's group, which by now included his Ph.D. student and future colleague, Werner Ens, was soon able to fit the TOF instrument with a MALDI source. Using the matrix materials recently discovered by Karas & Hillenkamp, the first measurements with this MALDI-TOF instrument were reported in 1988 (*5* and see this volume, *Franz Hillenkamp* and *Michael Karas*).

Coupling ESI to time-of-flight mass measurement was technically a more difficult problem than MALDI coupling, since MALDI is a pulsed technique while ESI produces ions as a steady state. The solution was orthogonal injection – the ions were first brought into the starting region of the flight tube at right angles to the direction of flight, and the start signal was a voltage pulse (see this volume, *Michael*

Guilhaus). A young scientist whom Standing had met during a lecture tour of the Soviet Union, Anatoly Verentchikov, came to Manitoba as an NSERC International Fellow in 1991, and he, Standing and Ens built the group's first ESI-TOFMS instrument. This was described in 1994 (*6*). Verentchikov was the first of a series of Russian graduate students and post-doctoral fellows who passed through the Standing laboratory over the next decade and made important contributions, particularly to instrument development.

The first ESI/TOF instrument was used extensively to study ions arising from non-covalent macromolecular complexes (*7*). Introduction of a nanospray source allowed experiments with a wider variety of buffers, improving the stability of the non-covalent interactions (*8*). It was soon realized that orthogonal injection would also improve the precision of measurements on MALDI-derived ions, and a MALDI source incorporating orthogonal injection was invented (*9*). This was combined with an additional quadrupole where ion clusters were broken down and ion momentum reduced by collisions with gas molecules (*10*). With the powerful instrumentation now in hand, the Standing-Ens research group began a very productive period, in which aspects of a great variety of biological systems were investigated, working with collaborators from laboratories all over the world. Efforts were divided between further instrumental innovations and the development and use of methods for studying proteomics (*11–15*). The latter included development of off-line coupling of the MALDI-TOF instrument to HPLC, so that the effluent from the HPLC was effectively archived and could be accessed repeatedly as necessary; techniques to improve identification of peptides; and studies to help understand modifications that peptides undergo during preparation and in the MS instrument itself.

Standing received the Canadian Society for Mass Spectrometry's Award for Distinguished Contributions to Mass Spectrometry (1998); the Canadian Association of Physicists' Medal for Outstanding Achievement in Industrial and Applied Physics (2003); and the American Chemical Society's Frank H. Field and Joe L. Franklin Award for Outstanding Achievement in Mass Spectrometry (2004). In 2004 he was elected a Fellow of the Royal Society of Canada, and a Fellow of the American Physical Society. In 2007 Standing and nine collaborators received the Brockhouse Award of the Natural Sciences and Engineering Research Council (Canada) for collaborative research. The Ken Standing Award, established in 2007, is presented biennially at the International Symposium on Enabling Technologies for Proteomics. In 2009, he was awarded an honorary doctorate of science by the University of Manitoba, and in 2010, Standing and his long-time collaborator Ens shared the Encana Principal Award, the highest honour bestowed by the Ernest C. Manning Awards for their 'transformational innovations in mass spectrometry.'

References

(*1*) Standing, K. G. Timing the Flight of Biomolecules: A Personal Perspective. *Int. J. Mass Spectrom. Ion Processes* **2000**, *200*, 597–610.

(*2*) Duckworth,. H.W., personal reflections.

(*3*) Chait, B. T.; Standing, K. G. A Time-of-Flight Mass Spectrometer for Measurement of Secondary Ion Mass Spectra. *Int. J. Mass Spectrom. Ion Phys.* **1981**, *40*, 185–193.

(*4*) Tang, X.; Beavis, R.; Ens, W.; Lafortune, F.; Schueler, B.; Standing, K. G. A Secondary Ion Time-of-Flight Mass Spectrometer with an Ion Mirror. *Int. J. Mass Spectrom. Ion Processes* **1988**, *85*, 43–67.

(*5*) Tang, X.; Ens, W.; Standing, K. G.; Westmore, J. B. Daughter Ion Mass Spectra from Cationized Molecules of Small Oligopeptides in a Reflecting Time-of-Flight Mass Spectrometer. *Anal. Chem.* **1988**, *60*, 1791–1799.

(*6*) Verentchikov, A. N.; Ens, W.; Standing, K. G. Reflecting Time-of-Flight Mass Spectrometer with an Electrospray Ion Source and Orthogonal Extraction. *Anal. Chem.* **1994**, *66*(1), 126–133.

(*7*) Tang, X. L.; Brewer, C. F.; Saha, S.; Chernushevich, I.; Ens, W.; Standing, K. G. Investigation of Protein-Protein Noncovalent Interactions in Soybean Agglutinin by Electrospray Ionization Time-of-Flight Mass Spectrometry. *Rapid Commun. Mass Spectrom.* **1994**, *8*(9), 750–754.

(*8*) Donald, L. J.; Hosfield, D. J.; Cuvelier, S.; Ens, W.; Standing, K. G.; Duckworth, H. W. Mass Spectrometric Study of the Escherichia coli Repressor Proteins, IclR and GclR, and their Complexes with DNA. *Protein Sci.* **2001**, *10*(7), 1370–1380.

(*9*) Krutchinsky, A. N.; Loboda, A. V.; Spicer, V. L.; Dworkschak, R.; Ens, W.; Standing, K. G. A Orthogonal Injection of Matrix-Assisted Laser Desorption/Ionization Ions into a Time-of-Flight Spectrometer Through a Collisional Damping Interface. *Rapid Comm. Mass Spectrom.* **1998**, *12*, 508–518.

(*10*) Krutchinsky A.N., Loboda A.V., Spicer V.L., Ens W., & Standing K.G. Spectrometer Provided with Pulsed Ion Source and Transmission Device to Damp Ion Motion and Method of Use. US Patents: 6 331 702 B1, 18 December 2001; 6 680 475, 20 January 2004.

(*11*) Krokhin, O. V.; Ying, S.; Cortens, J. P., *et al.* Use of Peptide Retention Time Prediction for Protein Identification by Off-Line Reversed-Phase HPLC-MALDI MS/MS. *Anal. Chem.* **2006**, *78*(18), 6645–6650.

(*12*) Krokhin, O. V.; Ens, W.; Standing, K. G. MALDI QqTOF MS Combined with Off-Line HPLC for Characterization of Protein Primary Structure and Post-Translational Modifications. *J. Biomol. Tech.* **2005**, *16*(4), 429–440.

(*13*) She, Y-M; Krokhin, O.; Spicer, V.; Loboda, A.; Garland, G.; Ens, W.; Standing, K. G.; Westmore, J. B. "Formation of (b(n-1) + H(2)O) ions by collisional activation of MALDI-Formed peptide M + H(+) ions in a QqTOF mass spectrometer". *J Am Soc Mass Spectrom* **2007**, *18*(No. 6), 1024–1037.

(*14*) Vykova, N. V.; Rampitsch, C.; Krokhin, O.; Standing, K. G.; Ens, W. Determination and Characterization of Site-Specific N-Glycosylation Using MALDI-Qq-TOF Tandem Mass Spectrometry: Case Study with a Plant Protease. *Anal. Chem.* **2006**, *78*(4), 1093–1103.

(*15*) Krokhin, O. V.; Ens, W.; Standing, K. G. Characterizing Degradation Products of Peptides Containing N-Terminal Cys Residues by (Off-Line High-Performance Liquid Chromatography)/Matrix-Assisted Laser Desorption/Ionization Quadrupole Time-of-Flight Measurements. *Rapid Commun. Mass Spectrom.* **2003**, *17*(22), 2528–2534.

Harry W. Duckworth
Department of Chemistry, University of Manitoba, Winnipeg, MB, Canada

Einar August Stenhagen

b. April 6, 1911
d. December 14, 1973

Einar Stenhagen was a Swedish biochemist who was prominent among those who brought mass spectrometry into organic chemistry. In the course of his studies he also spurred instrumental developments to help carry out the investigations he envisioned and played a major role in the development of reference collections of mass spectral data.

Stenhagen was born in 1911 in Askesta, Söderala, Sweden (*1–4*). With a strong interest in technology from childhood, he began his higher education in 1929 studying electrical engineering at the Royal Institute of Technology in Stockholm, but severe tuberculosis forced him to withdraw. After time at a sanatorium in Switzerland, during which his interests shifted in the direction of medical and chemical topics, he returned to Sweden to start the study of medicine in 1933 at the Karolinska Institute. After also studying at Lund, he earned the candidate's (pre-clinical) degree in medicine at Uppsala in 1935. His strong interest in the chemistry of the tubercle bacillus brought him in contact there with docent Torsten Teorell, with whom he co-authored research papers and whose contacts led to Rockefeller Foundation support for a year in the colloid science department at Cambridge University (*5*). He returned to Uppsala where he worked with Stina Lisa Ställberg, a biochemist who had earned her candidate degree in medicine in 1939; they married in 1942. Despite his periods of illness and lack of a doctorate, publishing over twenty-five research papers (six co-authored with Ställberg) led to appointment as a docent in medical chemistry at Uppsala University in 1943. His continuing concern was with compounds associated with the tubercle bacillus; his continued high quality research brought him an honorary doctorate in 1949.

1951 was a momentous year for Stenhagen and Ställberg-Stenhagen. She too was struck by tuberculosis and also attained a position as docent at Uppsala. In addition, the couple decided that the reported capability of a modified mass spectrometer to produce structural information for hydrocarbons with molecular masses at least up to 600 warranted an effort to apply mass spectrometry in their own research. In the fall of 1951 Stenhagen and Ställberg-Stenhagen initiated discussions with Einar Hammarsten, who had been among Stenhagen's teachers at the Karolinska and who had started a mass spectrometry laboratory there primarily for work with isotopically labeled substance in biomedical studies. The result was Hammarsten's decision to have Ragnar Ryhage, who was in charge of the mass spectrometry unit, build a mass spectrometer suitable for such work (*6*,*7*, and see this volume, *Ragnar Erik Ryhage*).

In 1952 Ställberg-Stenhagen became assistant professor at the University of Gothenburg, while Stenhagen was appointed to a special research chair in medical biochemistry created for him at Uppsala. Some years later he was able to have his chair transferred to Gothenburg.

Ryhage's new instrument was ready by 1954 and a decade of Stenhagen-Ryhage collaborative research began. In over thirty papers, some only by themselves, some with various others (including seven with Ställberg-Stenhagen), they explored the structure of long-chain molecules, not only advancing the understanding of lipids and various other substances but significantly advancing the role of mass spectrometry in organic chemistry generally. They also worked on instrumental innovations, notably in early experiments with linking a gas chromatograph with a mass spectrometer (*8–15*, see also this volume, *A History of Gas Chromatography Mass Spectrometry (GC/MS)* and *The Beginnings of Organic Mass Spectrometry*).

In 1963, as Stenhagen was developing a group at the Institute for Biochemistry at Gothenburg and the collaboration with Ryhage was drawing toward a close, Stenhagen added the Ecological Station of the University of Uppsala at Skogsby on the island of Öland to his research interests, where he and Ställberg-Stenhagen focused in particular on insect pheromones (*16*).

Over the course of the 1960 s, Stenhagen and collaborators in Sweden and other countries assembled a large collection of mass spectra. He worked with Fred McLafferty and Sixten Abrahamsson to prepare a coherently organized publication of the collected data (*17*,*18*, and see this volume, *Fred W. McLafferty* and *A History of the Development of Mass Spectral Databases*). In 1973 Stina Lisa Ställberg-Stenhagen died in March and Einar Stenhagen died in December in Fässbergs near Gothenburg.

References

(*1*) Andersson, B. A. Biography: Einar August Stenhagen (1911–1973) Stina Lisa Ställberg-stenhagen (1916–1973). *Prog. Chem. Fats Other Lipids.* **1978**, *16*, 1–7.

(*2*) Abrahamsson, S. Einar Stenhagen (1911–1973). In: *Biochemical Applications of Mass Spectrometry, First Supplementary Volume*; Waller, G. R., Dermer, O. C., Eds.; John Wiley & Sons: New York, 1980, pp vii–viii.

(*3*) McLafferty, F. W. Einar Stenhagen. In: *Registry of Mass Spectral Data*; Stenhagen, E., Abrahamsson, S., McLafferty, F. W., Eds.; John Wiley & Sons: New York, 1974; p v.

(*4*) Tunlid, A. Stenhagen, Einar August. in Karlsson, Åsa, Ed., *Svenskt Biografiskt Lexikon*. Riksarkivet (National Archives of Sweden): Stockholm, 2008, vol. 168, pp. 295–301.

(*5*) Stenhagen, E.; Teorell, T. Electrophoretic Behavior in Nucleic Acid Protein Mixtures. *Nature* **1938**, *141*, 415.

(*6*) O'Neal, M. J. Jr.; Wier, T. P. Jr. Mass Spectrometry of Heavy Hydrocarbons. *Anal. Chem.* **1951**, *23*, 830–843.

(*7*) Stenhagen, E. Some Early Examples of the Application of Mass Spectrometry to the Elucidation of the Structure of Complex Molecules of Biological Origin. In: *Biochemical Applications of Mass Spectrometry*; Waller, G. R., Ed.; John Wiley & Sons: New York, 1972; pp 11–19.

(*8*) Bergstrom, H.; Ryhage, R.; Stenhagen, E. Constituents of Tall Oil. I. The Nature of "Carnauba Acid" from Pitchy Wood. *Sven. Papperstidn* **1956**, *59*, 593–598.

(*9*) Demarteau-Ginsburg, H.; Lederer, E.; Rvhage, R.; Ställberg-Stenhagen, S.; Stenhagen, E. Structure of Phthiocerol. *Nature* **1959**, *183*, 1117–1119.

(*10*) Ryhage, R.; Stenhagen, E. Mass Spectrometry in Lipid Research. *J. Lipid Res.* **1960**, *1*, 361–390.

(*11*) Andersson, C.-O.; Rvhage, R.; Ställberg-Stenhagen, S.; Stenhagen, E. Mass Spectrometric Studies. IX. Methyl and Ethyl Esters of Some Aliphatic Amino Acids. *Ark. Kemi* **1962**, *19*, 405–416.

(*12*) Stenhagen, E. Mass Spectrometry of Long Chain Esters. In: *Mass Spectrometry of Organic Ions*; McLafferty, F. W., Ed.; Academic Press: New York, 1963; pp 399–452.

(*13*) Ställberg-Stenhagen, S.; Stenhagen, E. Gas Liquid Chromatography Mass Spectrometry Combination. In: *Topics in Organic Mass Spectrometry*; Burlingame, A. L., Ed.; John Wiley & Sons: New York, 1970; pp 167–183.

(*14*) Stenhagen, E. Fatty Acids. In: *Biochemical Applications of Mass Spectrometry*; Waller, G. R., Ed.; John Wiley & Sons: New York, 1972; pp 211–228.

(*15*) Odham, G.; Stenhagen, E. Complex Lipids. In: *Biochemical Applications of Mass Spectrometry*; Waller, G. R., Ed.; John Wiley & Sons: New York, 1972; pp 229–250.

(*16*) Kullenberg, B.; Stenhagen, E. The Ecological Station of Uppsala University on Öland 1963–1973. *Zoon* **1973** (Suppl. 1), 5–8.

(*17*) Stenhagen, E., Abrahamsson, S., McLaffertv, F. W., Eds. *Atlas of Mass Spectral Data*; John Wiley & Sons: New York, 1969; vol. 3.

(*18*) Stenhagen, E., Abrahamsson, S., McLafferty, F. W., Eds. *Registry of Mass Spectral Data*; John Wiley & Sons: New York, 1974; vol. 4.

Keith A. Nier

Independent Historian, Madison, NJ, USA

David Paul Stevenson

b. January 14, 1914
d. October 24, 1999

D.P. Stevenson was an American chemist known for his work in chemical bond energetics determined through mass spectrometric measurements. His is perhaps known best for the eponymous rule that is based on his 1951 publication (*1*). Stevenson's Rule, as his observation came to be known, states that when a bond is broken via electron ionization, the charge is retained by the moiety having the lowest ionization energy (see Chapter 1 (Volume 4): *Stevenson's Rule*).

David Paul Stevenson was born in Alameda, CA, on January 14, 1914. He obtained a B.S. from the University of California (Berkeley) in 1934 and a Ph.D. in chemistry from Princeton University in 1938. Following a National Research Council Fellowship at The California Institute of Technology (1938–1939), he held fellowships from the Asphalt Institute (1939–1940) and The Westinghouse Electric and Manufacturing Company (1940–1942) (see this volume, *The Beginnings of Organic Mass Spectrometry* and *John Alfred Hipple, Jr.*). After the latter, he took a position with the Shell Development Company where he remained for the balance of his career, beginning as a chemist (1942–1952), Head of Department (1952–1957), Director of Fundamental Research (1957–1961) and, until his retirement, Head of Basic and General Scientific Research at the Emeryville, CA Research Center. He married in 1940 and together he and his wife raised two children.

In the course of his work at Shell Development, Stevenson investigated a number of fundamental phenomena that serve as the basis for our current understanding of the behavior of gas phase ions. His work focused of hydrocarbons, as might be expected for investigations being done in a petroleum company's laboratory. His studies spanned the unimolecular decompositions of ions formed in electron ionization conditions (*1–5*), isomerizations on catalytic surfaces (*6,7*) and included both experimental and theoretical studies of ion-molecule reactions (*8,9*).

References

(*1*) Stevenson, D. P. Ionization and Dissociation by Electronic Impact. Ionization Potentials and Energies of Formation of sec-Propyl and tert-Butyl Radicals. Some Limitations of the Method. *Discuss. Faraday Soc.* **1951**, *10*, 35–45.

(*2*) Stevenson, D. P. The Mass Spectra of Hydrogen and Deuterium. *J. Chem. Phys.* **1947**, *15*, 409–411.

(*3*) Schissler, D. O.; Stevenson, D. P. The Benzyl-Hydrogen Bond Dissociation Energy from Electron Impact Measurements. *J. Chem. Phys.* **1954**, *22*, 151–152.

(*4*) Stevenson, D. P. Average Energies of Molecular Ions. *Radiation Res.* **1959**, *10*, 610–621.

(*5*) Stevenson, D. P. The Franck-Condon Principle and the Ionization and Dissociation of Hydrogen by Electron Impact. *J. Am. Chem. Soc.* **1960**, *82*, 5961–5966.

(*6*) Stevenson, D. P.; Wagner, C. D.; Beeck, O.; Otvos, J. W. Isotope Effect in the Thermal Cracking of Propane-1-C13. *J. Chem. Phys.* **1948**, *16*, 745.

(*7*) Stevenson, D. P.; Morgan, J. H. Isomerization of Cyclohexane and Methylcyclopentane in the Presence of Aluminum Halides. II. Equilibrium and Side Reactions. *J. Am. Chem. Soc.* **1948**, *70*, 2773–2777.

(*8*) Stevenson, D. P.; Schissler, D. O. Reactions of Gaseous Molecule Ions with Gaseous Molecules. IV. Experimental Method and Results. *J. Chem. Phys.* **1958**, *29*, 282–294.

(*9*) Gioumousis, George; Stevenson, D. P. Reactions of Gaseous Molecule Ions with Gaseous Molecules. V. Theory. *J. Chem. Phys.* **1958**, *29*, 294–299.

Alfred L. Yergey
National Institute of Child Health and Human Development, NIH, Bethesda, MD, USA

John T. Stults

b. December 10, 1957

John Stults is an American biochemist recognized for this work in the characterization of protein post-translational modifications and in the development of methodologies for the study of proteomics.

John T. Stults was born and raised in Ohio, where he attended the College of Wooster (Ohio). He received his undergraduate degree (B.A.) in Chemistry in 1980. He received the Ph.D. in Analytical Chemistry from Michigan State University in 1985 under the guidance of Profs. Chris Enke and Jack Watson (see this volume, *Christie George Enke*). His dissertation research focused on instrumentation development. His postdoctoral research in the Biochemistry Department at Michigan State University, under the direction of Prof. Jack Watson, introduced him to peptide sequencing by tandem mass spectrometry. John was hired in 1987 to lead the research mass spectrometry group at Genentech, Inc., where he rose to the position of Senior Scientist in the Protein Chemistry Department. In 2002 he left to become Vice President of Analytical Sciences at Biospect, Inc. (later renamed Predicant Biosciences). In 2006 he returned to Genentech in the process development organization, where he is currently Director of Protein Analytical Chemistry. John is married to Kathy Stults and has two children, Michael and Carolyn.

John first encountered mass spectrometry during a summer internship in 1979 at Eastman Kodak in Rochester NY, where he worked in the laboratory of Roland Gohlke (see this volume, *Roland S. Gohlke*). This experience encouraged him to pursue graduate work in the field. His first decade in the field coincided with the development of many 'soft' ionization techniques; his publications include the use of field desorption, fast atom bombardment, ^{252}Cf-plasma desorption, electrospray, and MALDI. Much of the work early in his career at Genentech utilized a four-sector tandem mass spectrometer with a fast atom bombardment source. With this instrument he and his group developed methods for peptide sequencing and for the determination of disulfide linkages.

The concept of Peptide Mass Fingerprinting started with a need to identify proteins rapidly from bands on SDS-PAGE gels. At the time, proteins were isolated in order to obtain sequence from which cloning experiments could be performed. Frequently, abundant carrier proteins (e.g., albumin) or 'housekeeping' proteins were inadvertently isolated. Working with colleagues Bill Henzel and Colin Watanabe, in 1989 John developed a method to digest a protein, measure the peptide masses, and match the list of observed masses with predicted masses from an *in silico* digestion of a sequence database. The lack of sensitivity of FAB ionization and size of the sequence database (less than 10,000 proteins at the time) initially limited the utility of the approach; the availability of MALDI instrumentation and the dramatic increase in known sequences made the method practical for routine use by 1993. Rapid identification of proteins from gels proved to be valuable for comparative biochemical studies using 2-D gels, a field that had become known as proteomics by 1996.

Comparative proteomics studies led to the pursuit of biomarkers for early cancer detection at Predicant Biosciences, starting in 2002. This work involved development of microfluidic chip-based instrumentation intended for clinical analysis. The complexity of the serum proteome ultimately limited the success of the effort.

The methods developed during John's career, particularly Peptide Mass Fingerprinting, laid much of the foundation for current approaches used for quantitative proteomics experiments. For his work on the development of Peptide Mass Fingerprinting for protein identification, he shared the ASMS Award for Distinguished Contribution in Mass Spectrometry in 2002 with Colin Watanabe and William Henzel (see this volume, *William J. Henzel*).

The publications cited below are representative of the author's most significant contributions to the field (*1–8*).

References

(*1*) Henzel, W. J.; Billeci, T. M.; Stults, J. T.; Wong, S. C.; Grimley, C.; Watanabe, C. A Novel Approach for Identifying Two-Dimensional Gel Proteins by Molecular Mass Searching of Peptide Fragments in Protein Sequence Databases. *Proc. Natl. Acad. Sci. USA* **1993**, *90*, 5011–5015.

(*2*) Billeci, T. M.; Stults, J. T. 'Tryptic Mapping of Recombinant Proteins by Matrix-Assisted Laser-Desorption Ionization Mass-Spectrometry'. *Anal. Chem.* **1993**, *65*, 1709–1716.

(*3*) Nuwaysir, L. M.; Stults, J. T. 'Electrospray Ionization Mass Spectrometry of Phosphopeptides Isolated by

On-Line Immobilized Metal-Ion Affinity Chromatography'. *J. Am. Soc. Mass Spectrom.* **1993**, *4*, 662–669.
(*4*) Bourell, J. B.; Clauser, K. P.; Kelley, R.; Carter, P.; Stults, J. T. Electrospray Ionization Mass Spectrometry of Recombinantly Engineered Antibodies. *Anal. Chem.* **1994**, *66*, 2088–2095.
(*5*) Arnott, D.; O'Connell, K. L.; King, K. L.; Stults, J. T. 'An Integrated Approach to Proteome Analysis: Identification of Proteins Associated with Cardiac Hypertrophy'. *Anal. Biochem.* **1998**, *258*, 1–18.
(*6*) Arnott, D.; Kishiyama, A.; Luis, E. A.; Ludlum, S. G.; Marsters, J. C. Jr.; Stults, J. T. Selective Detection of Membrane Proteins Without Antibodies: A Mass Spectrometric Version of the Western Blot. *Mol. Cell Proteomics* **2002**, *1*, 148–156.
(*7*) Henzel, W. J.; Watanabe, C.; Stults, J. T. Protein Identification: The Origins of Peptide Mass Fingerprinting. *J. Am. Soc. Mass Spectrom.* **2003**, *14*, 931–942.
(*8*) Sassi, A. P.; Andel, F. III; Bitter, H. M., *et al.* An Automated, Sheathless CE-MS Platform for Discovery of Biomarkers in Human Serum. *Electrophoresis* **2005**, *26*, 1500–1512.

John T. Stults
Genentech, Inc., South San Francisco, CA, USA

Victor L'vovich Tal'roze

b. April 15, 1922
d. June 22, 2004

Victor Tal'roze, Ph.D, Dr.Sc., was a Russian chemical physicist of Jewish origin. He was one of the pioneers to study ion-molecular reactions in organic compounds and proved for the first time the existence of stable CH_5^+ ion. He also made pioneering studies in connecting liquid chromatography to mass spectrometry and was a co-inventor of the 'chemical laser' (*1*).

Tal'roze was born on April 15, 1922, in Tula, Russian Federation, in what was then the Soviet Union. After graduating from high school in 1939, he entered the Chemistry Department of Moscow State University. In 1941, he joined the Red Army and was wounded in action in the summer of 1942. While recovering, he continued his studies at Moscow State University which had been evacuated from Moscow and relocated to Sverdlovsk, but by early 1943 he was at the front again. After demobilization in 1945, Tal'roze returned to the University and started graduate diploma work at the laboratory of 'elementary processes' in the Institute for Chemical Physics USSR Academy of Sciences (head of the laboratory-academician V. Kondrat'ev). The work was dedicated to the study of the role of radicals in a chain reaction of oxygen and hydrogen in a gas phase. His immediate supervisor was V.Voevodsky, who later became one of the founders of the Siberian Branch of the USSR Academy of Sciences.

After graduating from university in July, 1947, Tal'roze was employed by the Institute of Chemical Physics, USSR Academy of Sciences, where he worked in a special department headed by M.A. Sadovskii. In this department, he took part in the development of procedures for determining the degree of conversion of nuclear fuel during full-scale test explosions on a test site. After the completion of the ground testing program in 1949, Tal'roze worked at the laboratory headed by V. Kondrat'ev and at the Institute of Chemical Physics, where he began developing procedures that would allow mass spectrometric investigation of complex reactions of free molecules. The result of these interesting pioneering studies was the construction of the first mass spectrometer in the country designed for studying elementary events of complex reactions involving ions, atoms, and free radicals. Under his supervision, commercial mass spectrometers that used a gas chromatograph for preliminary sample separation were designed and produced in the USSR for the first time. The tandem radical mass spectrometer RMS-2 designed by his group was demonstrated at the 1958 World Exhibition in Brussels (Belgium) and then at the USSR Exhibition in the United States in 1959 (see this volume, *The Development of Time-of-Flight (TOF) Mass Spectrometry*).

Gradually, having been groomed as a scientist in the school of Academician Kondrat'ev, Tal'roze created a school of his own, including not only chemical mass spectrometry but also plasma chemistry, radiation chemistry, atmospheric chemistry, quantum electronics, and many other disciplines. In 1959, the Department of Free Radicals headed by Tal'roze was established at the Institute of Chemical Physics, which was transformed into the Division of Physical Methods of Stimulation of Chemical Reactions in 1974. A representative list of his publications is given below (*2–5*).

In 1968, Tal'roze was elected a corresponding member of the USSR Academy of Sciences; from 1972 to 1987, he was deputy director of the Institute of Chemical Physics. In 1984, together with G.K. Vasil'ev, A.N. Oraevskii, and E.P. Markin, he won the Lenin Prize for a series of studies that led to the development of chain reaction and branched chain reaction chemical lasers. In 1987, on Tal'roze's own initiative, the Institute of Energy Problems of Chemical Physics was founded. He was its director until 1995.

Tal'roze successfully combined his great and fruitful research and science management activities with teaching work. In 1954, he began to teach at the Moscow Institute of Physics and Technology, where he became head of the Department of Chemical Physics in 1960 and dean of the Faculty of Molecular and Chemical Physics. He worked as dean until 1983.

Tal'roze was a member of ASMS and the Soviet and Russian Academy of Sciences. He was a permanent member of the editorial board of *Khimiya Vysokikh Energii* (High Energy Chemistry) from its foundation in 1967 and made great contributions to the establishment and development of the journal. He also worked with *Rapid Communications in Mass Spectrometry*, the *International Journal of Mass Spectrometry*, and other publications. In addition to the Lenin Prize, he received numerous honors for his military and civil services and the Thomson Medal (2003) of the International Mass Spectrometry Foundation.

His last positions were Director of the Institute for Energy Problems of Chemical Physics, Russian Academy of Sciences, and Scientific adviser of Russian Academy of Sciences. He was married to Dr. Evgenija Vladimirovna Barelko and had one daughter, Raisa Tal'roze, who is a Ph.D, Dr.Sc. in polymer chemistry at A.V. Topvhiev Institute of Petrochemical Synthesis.

References

(*1*) Nikolaev, E. Victor Talrose: An Appreciation. *J. Mass Spectrom.* **1998**, *33*, 499–501.

(*2*) Tal'roze, V. L.; Ljubimova, A. K. Secondary Processes in the Ion Source of a Mass Spectrometer. Reprinted from report of the Soviet Academy of Sciences, Volume LXXXVI, N5 (1952), 909–912. *J. Mass Spectrom.* **1998**, *33*(6), 502–504.

(*3*) Tal'roze, V. L.; Batovsky, O. M.; Vasiliev, G. K. Chemical Laser on Branched Chain Reaction of Fluorine with Hydrogen. *J. Exp. Theor. Phys.* **1969**, *9*, 341–343.

(*4*) Nikolaev, E. N.; Neronov, I.; Gorshkov, M. V.; Tal'roze, V. L. Ion Cyclotron Resonance Measurements of the Mass Difference between Tritium and Helium-3. *J. Exp. Theor. Phys.* **1984**, *39*, 534–536.

(*5*) Yakovlev, B. S.; Tal'roze, V. L.; Fenselau, C. Membrane Ion Source for Mass Spectrometry. *Anal. Chem.* **1994**, *66*, 1704–1707.

Eugene Nikolaev

Laboratory of Ion and Molecular Physics, Institute for Energy Problems of Chemical Physics, Russian Academy of Sciences, Moscow, Russia; Laboratory of Mass Spectrometry of Biomacromolecules, Institute for Biochemical Physics, Russian Academy of Sciences, Moscow, Russia; and Institute Biomedical Chemistry, Russian Academy of Medical Sciences, Moscow, Russia

Koichi Tanaka

b. Aug 3, 1959

Koichi Tanaka is a Japanese scientist who shared the Nobel Prize in Chemistry in 2002 for developing a novel method for mass spectrometric analyses of biological macromolecules. As of 2008, he is the only person to have won a Nobel Prize in a scientific field without any post-baccalaureate training.

Tanaka was born and raised in Toyama City, Japan, on Aug 3, 1959. His mother died within a month of his birth and his father shortly thereafter. He was raised by an uncle and aunt, to whom he refers as his parents. He attended Toyama Chubu High School and obtained his undergraduate training as an electrical engineer at Tohoku University. He has been married to Yuko Ikegami since 1996.

Upon completion of his degree at Tohoku University, he was immediately hired by the Central Research Laboratory of Shimadzu Corporation. At Shimadzu in February, 1985. Tanaka serendipitously developed what he called a called a 'soft laser desorption' (SLD) when he used ultra-fine metallic cobalt powder in glycerine as a matrix for forming protein ions for time-of-flight mass spectrometric analysis. Having placed the incompletely dried mixture containing the added protein on the mass spectrometer sample stage region, he attempted to speed the drying by focusing the system's laser on the target. He observed a signal at a high m/z value that was clearly different from anything else he had seen previously and recognized it for what it was – a protonated protein ion! His methodology was patented and then presented publically in May, 1987, at the Annual Conference of the Mass Spectrometry Society of Japan held in Kyoto, Japan (*1–3*). SLD is not widely used at present, having been supplanted by Matrix Assisted Laser Desorption Ionization (MALDI) (see this volume, *The Development of Matrix Assisted Laser Desorption Ionization (MALDI) Mass Spectrometry*).

References

(*1*) Tanaka, K.; Waki, H.; Ido, Y.; Akita, S.; Yoshida, Y.; Yoshida, T. Protein and Polymer Analyses up to m/z 100 000 by Laser Ionization Time-of flight Mass Spectrometry. *Rapid Commun. Mass Spectrom.* **1988**, *2*(20), 151–153.

(*2*) Tanaka, K. Koichi Tanaka – Biographical. In *The Nobel Prize in Chemistry 2002*. Nobelprize.org. At http://www.nobelprize.org/nobel_prizes/chemistry/laureates/2002/tanaka-autobio.html (accessed August 19, 2014).

(*3*) Anon. Koichi Tanaka. Wikipedia. At http://en.wikipedia.org/wiki/Koichi_Tanaka (accessed August 19, 2014).

Alfred L. Yergey
National Institute of Child Health and Human Development, NIH, Bethesda, MD, USA

Henry George Thode

b. September 10, 1910
d. March 22, 1997

Henry George Thode was a very influential Canadian chemist whose influence reached far beyond his major role in mass spectrometry. From an unpretentious farming background in Dundurn, Saskatchewan, he became a pioneer not only in sulphur geochemistry but also in nuclear chemistry. His achievements in these two fields earned him a Fellowship of the Royal Society of London in 1954. In addition to his research accomplishments, he became the driving force behind the transformation of a small Baptist-affiliated liberal arts college into one of Canada's leading research universities – McMaster University in Hamilton, Ontario. Under his guidance, McMaster developed strong science and engineering schools together with a prestigious medical school. As Vice-Chancellor and President of McMaster University he became one of Canada's most respected educators and administrators. He oversaw the installation of a nuclear reactor on campus and this facility attracted a galaxy of scientists, engineers and medicos, brought together in a multi-disciplinary research team using the reactor facilities in a variety of applications. The Nuclear Research Building was somewhat akin to the League of Nations, with people from all over the world working together on a variety of projects, attracted by Thode's reputation and the research team he had assembled. This multi-disciplinary approach was exemplified by Harry Thode himself. A Professor of Chemistry, he was also given professorial status in the School of Physics and Astronomy, and the School of Geography and Geology at McMaster.

After completing bachelors and masters degrees in chemistry at the University of Saskatchewan, Harry Thode transferred to the University of Chicago where he completed his PhD in 1934 at the age of 24 years. He introduced himself to the 1934 Nobel Prize winner, Harold Urey, at the American Chemical Society meeting in Pittsburgh in 1935 and asked if he could work with him as a Postdoctoral Fellow (see this volume, *Harold Clayton Urey*). At that time Thode was teaching at a college in Pittsburgh where it was almost impossible to carry out research. Urey was impressed with the young man, and offered him a position at Columbia University in New York. As a result of this appointment, a steady stream of papers in the design and operation of systems for the isotopic separation of nitrogen, carbon and sulphur flowed from the Columbia laboratory (*1*). Urey had shown that isotope fractionation could occur for the light elements during chemical reactions and the measurement of these fractionation effects by sensitive mass spectrometry could provide information on the mechanisms causing this fractionation.

Thode's interest in isotope separation was forged during this period at Columbia University in the knowledge that if sufficient enrichment in the isotopes of these light gases could be achieved the enriched isotopes could be used as isotopic tracers – not only in chemistry but in other scientific fields as well. After his postdoctoral fellowship with Urey was concluded, Thode spent a brief period in industry, before taking up in 1939 the position of Assistant Professor of Chemistry at McMaster University, a most unlikely institution for this precocious young man. There he set up a long fractionating column in the stairwell to separate isotopes. He also built his own gas source mass spectrometer from pyrex glass. As was the custom in those days, mass spectrometers were home-made, so of necessity Harry Thode became an expert glassblower and took great pride in his expertise.

Thode's skill in isotope separation and mass spectrometry made him of immense value to the Manhattan Project during World War II. He was seconded to the Canadian Atomic Energy Project from 1942 to 1945, where he carried out research on nuclear fission of the noble gases krypton and xenon. The task was to determine the isotopic composition of the noble gases produced by the thermal neutron-induced fission of ^{235}U. Thode was successful in measuring the relative isotopic abundances of all the fission yields for both krypton and xenon. During this work he discovered the existence of radioactive ^{85}Kr and measured its half-life. Due to security precautions during World War II, Thode's pioneering work in fission products remained unpublished until 1947 (*2*). Then followed a series of papers describing details of the fission process for krypton and xenon. Thode and his colleagues showed that fine structure occurred in the ^{235}U fission yields, which they attributed to nuclear shell structure effects (*3*). The group's work on fission products was of inestimable value to the design of nuclear reactors.

Under Thode's direction, the study of cumulative fission yields was extended to the non-gaseous elements, and to a range of fissile materials. Thode also established a clinical research team which investigated radioactive ^{131}I in the diagnosis and treatment of thyroid disorders (*4*). And he showed that the $^{10}B/^{11}B$ ratio varied in nature, thus laying the foundation for the use of boron as a tracer in geochemical processes (*5*).

Harry Thode's other great research interest was the study of variations in the isotopic composition of stable isotopes in nature, particularly sulphur (see this volume, *The Development of Mass Spectrometry in the Earth and Planetary Sciences*). He has been called the 'Father of Sulphur Isotope Geochemistry'. In 1949, he built a mass spectrometer which measured the ion beam intensities of ^{32}S and ^{34}S simultaneously, thus enabling variations in the $^{32}S/^{34}S$ ratios to be determined to a few tenths of one percent. He discovered that sulphides were depleted in ^{34}S whilst the same isotope was enriched in sulphates. This result was consistent with the predictions of equilibrium thermodynamics. Thode showed that the wide range in the isotopic composition of terrestrial sulphur was in part due to kinetic effects, particularly the low temperature, bacterial reduction of sulphates to sulphides (*6*). This important result was applicable to other isotopic systems, such as selenium (*7*).

In 1958, he investigated the origin of petroleum and the nature of hydrocarbon source rocks. His studies on ore genesis demonstrated the role of bacterial activity in the formation of the Precambrian iron formations. Profiles of the isotopic composition of sulphur from lake sediments enabled him to examine the effect of anthropogenic sulphur in the environment, especially through the effect of acid rain precipitation (*8*). Undoubtedly the culmination of his work on sulphur isotope geochemistry was his research on extra-terrestrial materials. New analytical techniques enabled small variations in ^{33}S and ^{36}S to be measured in meteorites, with implications to theories of nucleosynthesis (*9*). In the early 1970s, the arrival of lunar samples from the Apollo Mission, provided him and his colleague C.E. Rees the opportunity to investigate sulphur isotopic effects in the lunar regolith. The large difference between the isotopic composition of lunar soils and rocks helped to establish the history of the lunar regolith (*10*). It is noteworthy, that, after his retirement as President of McMaster university in 1972, he returned to the laboratory (*11,12*).

Professor Thode received many honours and awards, both as a scientist as well as an educational administrator. Apart from his Fellowships of the Royal Societies of London and Canada, he was awarded the Arthur L. Day Medal of the Geological Society of America in 1980. He received numerous honorary Doctorates from Canadian universities, and in 1967 was named a Companion of the Order of Canada. He was a sought-after member of Boards and Commissions because of his wise counsel and contagious enthusiasm.

Harry and Sadie, his wife of 62 years, had three sons. He was active in research until the end. He died in Dundas, Ontario, in 1997. He managed to successfully combine an outstanding research career alongside an effective career as a university administrator. He was a modest individual, but with powers of persuasion and a tenacious determination to achieve the very best possible. Harry Thode was undoubtedly the most influential person in the history of McMaster University, because of his great vision and unbridled optimism in the future (*13–15*).

References

(*1*) Urey, H. C.; Thode, H. G.; Gorham, J. E. The Concentration of N^{15} and S^{34}. *J. Chem. Phys.* **1938**, *6*, 296.

(*2*) Thode, H. G.; Graham, R. L. A Mass Spectrometer Investigation of the Isotopes of Krypton and Xenon from the Fission of ^{235}U by Thermal Neutrons. *Can. J. Res. A.* **1947**, *25*, 1–14.

(*3*) Thode, H. G.; MacNamara, J.; Collins, C. B. The Fission Yield of ^{133}Xe and Fine Structure in the Mass Yield Curve. *Phys. Rev.* **1950**, *78*, 129–134.

(*4*) Thode, H. G.; Jaimer, C. H. Thyroid Function Studies on Children Receiving Cobalt Therapy. *J. Amer. Med. Assoc.* **1955**, *158*, 1353–1355.

(*5*) Thode, H. G.; MacNamara, J.; Lossing, F. P.; Collins, C. B. Natural Variations in the Isotopic Content of Boron and its Chemical Atomic Weight. *J. Am. Chem. Soc.* **1948**, *70*, 3008–3011.

(*6*) Szabo, A.; Tudge, A.; MacNamara, J.; Thode, H. G. The Distribution of S^{34} in Nature and the Sulfur Cycle. *Science* **1950**, *111*, 464–465.

(*7*) Krouse, H. R.; Thode, H. G. Thermodynamic Properties and Geochemistry of Isotopic Compounds of Selenium. *Can. J. Chem.* **1962**, *40*, 367–375.

(*8*) Thode, H. G.; Monster, J.; Dunford, H. B. Sulfur Isotope Geochemistry. *Geochim. Cosmochim. Acta.* **1961**, *25*, 150–174.

(*9*) Thode, H. G.; Monster, J.; Anders, E. $^{34}S/^{32}S$ Ratios for the Different Forms of Sulphur in the Orgueil Meteorite and their Mode of Formation. *Geochim. Cosmochim. Acta.* **1965**, *29*, 773–779.

(*10*) Thode, H. G.; Rees, C. E. Measurement of Sulphur Concentrations and the Isotope Ratios $^{33}S/^{32}S$ $^{34}S/^{32}S$ and $^{36}S/^{32}S$ in Apollo 12 Samples. *Earth Planet. Sci. Lett.* **1971**, *12*, 434–438.

(*11*) Thode, H. G.; Goodwin, A. M. Further Sulphur and Carbon Isotope Studies of Late Archean Iron Formations of the Canadian Shield and the Rise of Sulphate-Reducing Bacteria. *Precambrian Res.* **1983**, *20*, 337–356.

(*12*) Thode, H. G. Sulphur Isotopes in Nature and the Environment: An Overview. In: *Stable Isotopes in the Assessment of Natural and Anthropogenic Sulphur in the Environment*; Krouse, H. R., Grinenko, V. A., Eds.; John Wiley & Sons: New York, 1991; pp 1–26.

(13) Clarke, W. B.; Crocket, J. H.; Gillespie, R. J.; Krouse, H. R.; Shaw, D. M.; Schwarcz, H. P. Henry George Thode, M.B.E.: 10 September 1910–22 March 1997. *Biog. Mems. Fell. R. Soc.* **2000**, *46*, 499–514.
(14) Shaw, D. M.; Epstein, S.; Hayes, J. M. Memorial to Henry G. Thode 1910–1997. *Geol. Soc. Am. Memorials.* **2000**, *31*, 11–15.
(15) Zack, M.; Lee, A. A.; Martin, L. *Harry Thode: Scientist and Builder at McMaster University*; McMaster University Press: Hamilton, Ontario, Canada, 2003.

John de Laeter (d.)
Curtin University, Perth, Australia

Bruce A. Thomson

b. February 25, 1950

Bruce Thomson is a Canadian physicist noted for work in the general area of ion sources and ion optics for mass spectrometry. His most widely-known contributions are the theory of the Ion Evaporation Mechanism (with PhD supervisor J. V. Iribarne) [*1*, *2*], and the developments of the Heated Nebulizer LC/MS Inlet in the early 1980s and the collision cell with axial electric fields [*3*] in the 1990s, both used in many commercial mass spectrometers. As a member of the research group with commercial mass spectrometry manufacturer Sciex (later called MDS Sciex and AB SCIEX), he has contributed to ion source and optics designs of many products.

Thomson was born in Calgary, Alberta. He obtained his PhD in Physics at the University of Toronto in 1978 under J. V. Iribarne. His thesis project of studying charge loss from evaporating charged droplets by ion mobility and mass spectrometry led to the observation and identification of organic and inorganic ions from the charged droplets. The explanation of charge loss lead to the theory of ion evaporation, which is one of the two proposed explanations for ion production in electrospray ion sources. He began work at Sciex in 1978 immediately after graduation and has continued with the same company throughout his career, as the company expanded from a small start-up to one of the major players in commercial mass spectrometry.

As a key member of the research group at Sciex, his interest in atmospheric pressure ion sources and ion evaporation lead to the first observation of a doubly charged anion by spray ionization (ASMS 1981), the first demonstration of LC/MS by ion evaporation (ASMS 1982), and the first publication on LC/MS/MS (1982, with Jack Henion and Peter Dawson). The latter work led to the development of the heated nebulizer, a method of coupling LC/MS to atmospheric pressure chemical ionization by spraying and vaporizing the liquid into a corona discharge ion source (ASMS 1983). This technique became one of the two main methods of atmospheric pressure LC/MS. Subsequent work on triple quadrupole ion optics resulted in the invention of the Linac® collision cell to eliminate crosstalk in the collision cell. Other contributions include the development of the high-pressure RF collision cell. A selection of his work is listed below (*1–6*).

He received the W. A. E. McBryde Medal in 1997 awarded by the Canadian Society for Chemistry for Contributions to Analytical Chemistry in Canada, and the Fred P. Lossing Award for Significant Contributions to Canadian Mass Spectrometry in 2004. His body of work includes approximately 45 refereed publications and 18 US patents to date.

Bruce is married with two children.

References

(*1*) Iribarne, J. V.; Thomson, B. A. On the Evaporation of Small Ions From Charged Droplets. *J. Chem. Phys.* **1976**, *64*(6), 2287–2294.

(*2*) Thomson, B. A.; Iribarne, J. V.; Dziedzic, P. J. Liquid Ion Evaporation/Mass Spectrometry for the Detection of Polar and Labile Compounds. *Anal. Chem.* **1982**, *54*, 2219–2224.

(*3*) Thomson, B. Quadrupole With Axial DC Field. US Pat. 6,111,250, Dec. 5th, 2006.

(*4*) Thomson, B. A.; Douglas, D. J.; Corr, J. J.; Hager, J. W.; Jolliffe, C. L. Improved Collisionally Activated Dissociation Efficiency and Mass Resolution on a Triple Quadrupole Mass Spectrometer. *Anal. Chem.* **1995**, *34*, 1696–1704.

(*5*) Javahery, G.; Thomson, B. A. A Segmented Radiofrequency-Only Quadrupole Collision Cell for Measurements of Ion Collision Cross Section on a Triple Quadrupole Mass Spectrometer. *J. Am. Soc. Mass Spectrom.* **1997**, *8*, 697–702.

(*6*) Thomson, B. A.; Jolliffe, C. L. Apparatus Method for MS^n in a Tandem Mass Spectrometer System. US Pat. 7,145,133, Aug. 29th, 2000.

Bruce Thomson

AB Sciex, Framingham, MA, USA

Joseph John Thomson

b. December 18, 1856
d. August 30, 1940

Sir Joseph John Thomson was an English physicist noted for both experimental and theoretical accomplishments. He won the Nobel Prize in Physics in 1906 for work on the structure of the atom. His work in the early 20th-century experiments on *positive rays* included construction of the first mass spectrometer, and the discovery of evidence for stable isotopes.

Thomson was the elder son of a Manchester bookseller. He attended Owens College (now the University of Manchester) and Trinity College, Cambridge, where he graduated in 1880 and became a lecturer and research worker at the Cavendish Laboratory. In 1884, Thomson was elected Fellow of the Royal Society, and in the same year was appointed head of the Cavendish Laboratory. He became Master of Trinity College in 1918, and remained in that post for the rest of his life. Thomson served as president of the Royal Society from 1915 to 1920. His many honors include The Nobel Prize for Physics in 1906, and his knighthood in 1908. In 1890 Thomson married Rose Elizabeth Paget, of Cambridge; they had two children. Thomson's son, George Paget Thomson, also became a Nobel laureate (*1–3*).

Thomson's early work included mathematical investigations on electrodynamics according to James Clerk Maxwell's theories (*4*). Following his appointment as head of the Cavendish Laboratory in 1884 he turned his attention to the study of electric discharges through gases. In 1897, using an apparatus capable of measuring both magnetic and electrostatic deflection, Thomson was able to show that the mass to charge ratio for the negative carriers (cathode rays) was substantially ($<1000\times$) smaller than that for the positive carriers (*5*); then in 1899 he was able to establish the charge as being the same as that measured for ions in electrolysis, and not dependent upon how they were derived, thus showing the negative charge carrying particle to have a mass less than one-thousandth that of a hydrogen atom (*6*). Thomson used the term 'corpuscle' for this particle, and started to develop the 'corpuscular theory' of the atom, but soon the term 'electron', previously coined by Stoney as a name for the unit of electric charge, was adopted (*7*).

Thomson then turned his attention to the positive rays, or *kanalstrahlen*, first noted by Goldstein in 1886, and reported by Wien (Nobel Prize, 1911) in 1898 to have masses comparable to atoms (see this volume, *Wilhelm Carl Werner Otto Fritz Franz Wien*). Thomson's 1906 apparatus consisted of a large evacuated bulb containing the discharge electrodes; the cathode having a fine-bore tube inserted; the rays passed through the tube, and were then deflected by parallel electric and magnetic fields onto a fluorescent screen. The parallel fields meant that ions of a particular mass to charge ratio should be projected in a parabola, the distance from the origin being a function of the ion's energy (see this volume, *Magnetic and Electrostatic Analyzers before 1960*).

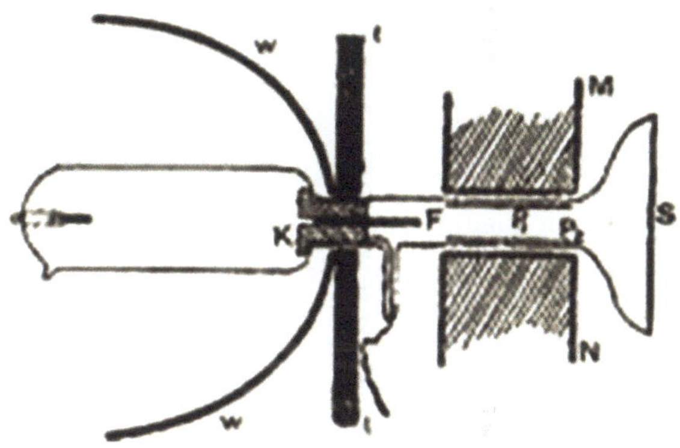

Thomson's first positive ray apparatus.

Initial results showed only a straight line from the origin to a point corresponding to the mass of a hydrogen atom, and this was present whatever the gas in the tube, supporting the view that hydrogen atoms were the universal carrier of positive charge. As the pressure was reduced a line corresponding to a hydrogen molecule (H_2) was seen, then a line corresponding to a helium atom. Improvements were made to the apparatus to allow operation at even lower pressure, and to replace the screen with a photographic plate that allowed for long exposures, and was more sensitive to higher mass ions. The expected parabolas were now evident and the straight lines reduced in brightness; moreover parabolas were seen corresponding to the masses of many of the atoms and molecules of the residual gas in the tube. The positive rays appeared to be atoms and molecules that had lost an electron (*8,9*).

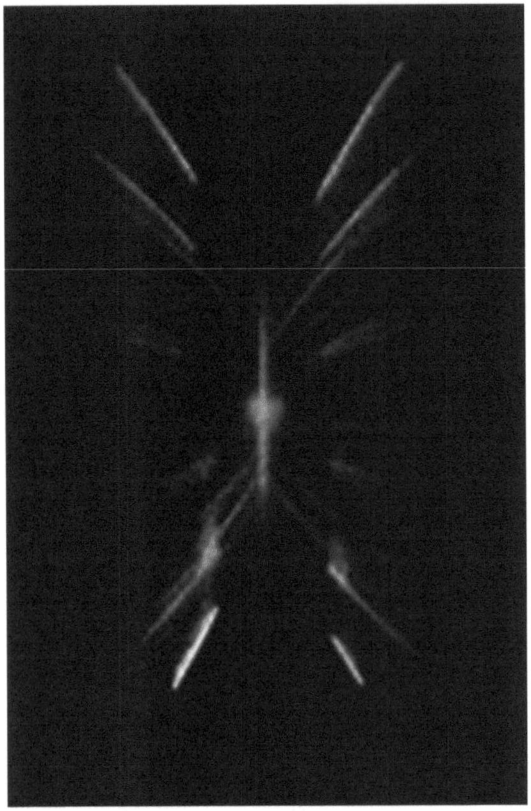

Example of a positive ray photograph.

In 1910, Thomson had been joined by Francis Aston (Nobel Prize, 1922) as his research assistant and the pair continued to work on positive rays and refine their equipment (see this volume, *Francis William Aston*). The photographs obtained are rich in information. For instance, an inference was made that the straight lines could be due to charge transfer between ions and the residual gas caused by collisions within the deflecting fields; an instrument with two magnets in series was constructed to test this. Additionally, photographs showed molecular fragments, radicals, and ions with more than one charge. In 1912, whilst experimenting with neon, a parabola was observed at mass 22 as well as the expected one at mass 20. Various possibilities, such as a doubly charged CO_2 molecule, were considered and rejected; eventually it was concluded that the line was an isotope of neon (*10*). The first stable (as opposed to radioactive) isotope had been discovered. Thomson's positive ray research is published in his book Rays of Positive Electricity (*11*).

With the outbreak of World War I, in 1914 research at the Cavendish Laboratory was suspended. In 1919, Thomson resigned his professorship after being appointed Master of Trinity College; he continued to work and publish on atomic and molecular structure (*12*).

Thomson is remembered principally as the discoverer of the electron, but for mass spectrometrists it is significant to note the closing words of his Rays of Positive Electricity: "The positive rays thus seem to promise to furnish a method of investigating the structure of the molecule, a subject certainly of no less importance than that of the structure of the atom."

References

(*1*) Davis, E. A.; Falconer, I. J. *JJ Thomson and the Discovery of the Electron*; Taylor & Francis: London, 1997.

(*2*) Griffiths, I. W. J. J. Thomson–the Centenary of His Discovery of the Electron and of His Invention of Mass Spectrometry. *Rapid Commun. Mass Spectrom.* **1997**, *11*, 2–16.

(*3*) Heilbron, J. L. Thompson, Joseph, John. In: *Dictionary of Scientific Biography*; Gillispie, C. C., Ed.; Charles Scribner's Sons: New York, 1976; vol 13, pp 362–372.

(*4*) Thomson, J. J. On the Electric and Magnetic Effects Produced by the Motion of Electrified Bodies. *Phil. Mag.* **1881**, *11*, 229–234.

(*5*) Thomson, J. J. On Cathode Rays. *Phil. Mag.* **1897**, *44*, 293–316.

(*6*) Thomson, J. J. On the Masses of the Ions in Gases at Low Pressures. *Phil. Mag.* **1899**, *48*, 547–567.

(*7*) Stoney, G. J. On the cause of double lines and of equidistant satellites in the spectra of gases. *Trans. R. Dublin Soc.* **1891**, *4*, 563–608.

(*8*) Falconer, I. J. J. J. Thomson's Work on Positive Rays, 1906–1914. *Hist. Std. Physical Biol. Sci.* **1988**, *18*, 265–310.

(*9*) Dahl, P. *Flash of the Cathode Rays: A History of J. J. Thomson's Electron*; Institute of Physics Publishing: Bristol and Philadelphia, 1997; pp. 266–292.

(*10*) Thomson, J. J.; Aston, F. W.; Soddy, F.; Merton, T. R.; Lindemann, F. A. Discussion on Isotopes. *Proc. Royal Soc. A* **1921**, *99*, 87–104.

(*11*) Thomson, J. J. *Rays of Positive Electricity and their application to Chemical Analyses*; Longmans Green and Co.: London, 1913.

(*12*) Thomson, J. J. *The Electron in Chemistry*; Franklin Institute: Philadelphia, Pa, 1923.

Anthony James Gilbert
Waters Corporation, Manchester, UK

George Robert Tilton

b. June 3, 1923
d. October 12, 2010

George Tilton was an American isotope geochemist. He was among the pioneers who developed methods for measuring isotopic ratios in meteorites and all kinds of continental rocks. He used those methods to date significant materials, to help determine when and how the solar system and earth formed, and to describe the evolution of the earth's mantle and crust. Among other notable accomplishments, he made the first zircon U/Pb age measurement.

George Tilton was born in Danville, Illinois, US, in 1923 and went to school there through high school. His college education began at Blackburn College in Illinois but was interrupted by service in the US Army during World War Two, during which he was wounded in Alsace. He completed his college work with a degree in chemistry from the University of Illinois in 1948, in which year he also married Elizabeth Jane Foster, who had been a teaching assistant in one of his classes. They would have five children, four of whom survived him.

Tilton did graduate work at the University of Chicago, in the midst of the extraordinary inter-disciplinary concentration of scientific excellence that was gathered then around the Institute for Nuclear Studies. He completed his in doctorate in 1951 and moved to the Carnegie Institution in Washington, DC. After five years in the Department of Terrestrial Magnetism he moved to the Institution's Geophysical Laboratory. In 1965 he became a professor of geo-chemistry at the University of California Santa Barbara (UCSB), where he stayed, becoming professor emeritus in 1991, and continued in research to the end of his life in 2010 (*1*,*2*).

Tilton was interested in radiochemistry when he started graduate work at Chicago. Harrison Brown steered him and fellow graduate student Clair C. Patterson to a project to use mass spectrometry to examine the uranium and lead in meteorites as a path to estimating the time of the formation of the solar system and the earth. While Patterson concentrated on determining lead contents, Tilton concentrated on uranium. This led to intensive work with several other major figures involved both with developing mass spectrometric techniques and instruments on one hand and the use of mass spectrometry in the geological and related fields, Mark Inghram perhaps most crucial among them, on the other. The work and interactions continued even as some of the group left Chicago; this research produced much improved analytic procedures as well as fundamental age determinations (see this volume, *Harrison Scott Brown, Mark Gordon Inghram II*, and *Clair Cameron Patterson*).

At the Carnegie Institution Tilton continued to refine and extend uranium-lead isotopic dating, particularly of igneous and metamorphic rocks, and to evaluate comparatively the various isotopic systems that have been used in geological dating. His research extended far beyond issues of chronology, however, and continued to expand its reach when he moved to UCSB. Together with Paul Gast and Carl Hedge he showed that the earth's mantle is not isotopically homogenous. He used isotopic analysis to explore differences among volcanic sources and to approach a time scale for the production of continental crust. He was involved in field projects world-wide, and continued his research past his academic retirement. Representative papers from his long record of research are listed below (*3–13*).

Tilton had many students, served as president of Geochemical Society, and was elected to the National Academy of Sciences among many other honors. He was a music lover who sang in the Santa Barabara Oratorio Chorale for years. He died in 2010 in Eugene, Oregon at the home of one of his children.

References

(*1*) Hopson, C. In Memorium: George Robert Tilton, 1923–2010, Professor of Geochemistry Emeritus, UCSB. *UCSB Earth Science Newsletter* **2011**, *Fall*, 8.

(*2*) Anonymous. George Robert Tilton: 1923–2010, Santa Barbara 1923–2010, *Santa Barbara Independent*, Oct 18, **2010**.

(*3*) Tilton, G. R.; Patterson, C.; Brown, H., *et al.* Isotopic Composition and Distribution of Lead, Uranium, and Thorium in a Precambrian Granite. *Geol. Soc. Am. Bull.* **1955**, *66*, 1121–1148.

(*4*) Tilton, G. R.; Davis, G. L.; Wetherill, G. W.; Aldrich, L. T. Isotopic Ages of Zircon from Granites and Pegmatites. *EOS: Trans. Am. Geophys. Union* **1957**, *38*, 360–371.

(*5*) Tilton, G. R. Volume Diffusion as a Mechanism for Discordant Lead Ages. *J. Geophys. Res.* **1960**, *65*, 2933–2945.

(*6*) Tilton, G. R.; Hart, S. R. Geochronology. *Science* **1963**, *140*, 357–366.
(*7*) Gast, P. W.; Tilton, G. R.; Hedge, C. E. Isotopic Composition of Lead and Strontium from Ascension and Gough Island. *Science* **1964**, *145*, 1181–1188.
(*8*) Tilton, G. R.; Grunenfelder, M. H. Sphene: Uranium–Lead Ages. *Science* **1968**, *159*, 1458–1461.
(*9*) Jahn, B.-M.; Vidal, P.; Tilton, G. R. Archaean Mantle Heterogeneity: Evidence from Chemical and Isotopic Abundances in Archaean Igneous Rocks. *Phil. Trans. R. Soc. A* **1980**, *297*, 353–364.
(*10*) Tilton, G. R.; Hopson, C. A.; Wright, J. E. Uranium–Lead Isotopic Ages of the Semail ophiolite, Oman, with Applications to Tethyan Ocean Ridge Tectonics. *J. Geophys. Res.* **1981**, *86*, 2763–2775.
(*11*) Tilton, G. R. Evolution of Depleted Mantle: The Lead Perspective. *Geochim. Cosmochim. Acta* **1983**, *47*, 1191–1197.
(*12*) Tilton, G. R.; Kwon, S. T. Isotopic Evidence for Crust–Mantle Evolution with Emphasis on the Canadian Shield. *Chem. Geol.* **1990**, *83*, 149–163.
(*13*) Bell, K.; Tilton, G. R. Nd, Pn and Sr Isotopic Compositions of East African Carbonatites: Evidence for Mantle Mixing and Plume Heterogeneity. *J. Petrol.* **2001**, *42*, 1927–1945.

Keith A. Nier
Independent Historian, Madison, NJ, USA

John Francis James Todd

b. May 20, 1937

John Todd is a British physical chemist widely noted for his work in the general area of ion trap mass spectrometry (*1–3*, *see this volume, The Development of the Quadrupole Mass Filter and Quadrupole Ion Trap*). In particular, he was the first to conceive the idea of combining the three-dimensional quadrupole ion trap (which he termed the QUISTOR, for 'quadrupole ion store') with a quadrupole mass filter to provide a means of studying the physical and chemical behavior of trapped ions prior to pulse-ejecting them for external mass analysis.

Todd was born in Leeds, England, the only son of Eric and Annie Lewin (nee Tinkler) Todd. He received his early education at the Leeds Grammar School, UK (1947–1955). He attended Leeds University, UK, from 1955–1964, obtaining first a BSc (First Class Honors) in Special Studies in Chemistry and then a PhD. His PhD research, conducted under the supervision of E. Collinson and F.S. (later Lord) Dainton, FRS, won him the Cohen Prize, an award given to the student with the most meritorious PhD thesis. From 1963–1965, he was a Fulbright Research Scholar and Post-Doctoral Research Fellow at Yale University, Connecticut, USA, where he worked on the chemistry of energetic ('hot') recoiling radioactive atoms, under the supervision of R. L. Wolfgang.

Returning to England in 1956, he took up an appointment as Assistant Lecturer in Chemistry at the University of Kent, Canterbury, Kent, UK. Subsequent appointments, all at Kent, include Lecturer in Chemistry, 1966–1973; Senior Lecturer in Chemistry, 1973–1990; Master of Rutherford College, 1975–1985; Reader in Physical Chemistry, 1990–1991; Professor of Mass Spectroscopy, 1991–2003; Head of Chemistry Department, 1991–1994; and finally Emeritus Professor of Mass Spectroscopy, a position he continues to hold.

Todd's interest in mass spectrometry originated in his reading of the early literature on ion/molecule reactions for their apparent relevance to his PhD research project entitled 'Chemical Changes Promoted by the Penning Effect in Argon.' The topic was inspired by the work of James Lovelock, FRS, the inventor of the argon ionization detector for gas chromatography and of the electron capture detector. His interest in mass spectrometry was extended by his having casual access to a Bendix time-of-flight instrument during his post-doctoral period at Yale and led to his first research proposal, an investigation of the lifetimes of metastable ions through the observation of the respective flight times of the precursor and product ions through the analyzer of a quadrupole mass spectrometer.

His first substantive academic appointment at Kent University in 1965 provided an opportunity to exploit the proposal through the creation of an experimental instrument in which ions could be trapped for pre-determined times within a QUISTOR and then mass-analyzed by pulse-ejecting them into a quadrupole mass analyzer. This novel system provided a means of exploring the kinetics and energetics of ion/molecule reactions (*4,5*), but was also a vehicle for characterizing the physics of the motion of ions trapped in radiofrequency quadrupole devices, both experimentally and theoretically (*6,7*).

His original conception of a combined quadrupole ion trap/quadrupole mass filter instrument in 1968 led to a series of studies in which understanding of the fundamental physics and chemistry of trapped ions was enhanced (*8*). During the period 1968 to *ca.* 1982 he was one of a very small international group of chemists (numbering about five) that kept an interest in quadrupole ion traps alive, helping to pave the way and providing a valuable body of background knowledge, for the early development and exploitation of the novel commercial ion trap mass spectrometer introduced by Finnigan MAT (later Thermo) in the 1980s (*9–11*). Since that time literally thousands of ion trap mass spectrometers of various types have been sold by several different manufacturers, and the contributions these instruments continue to make to sophisticated analytical investigations, especially those in the bioscientific and biomedical areas, is immense.

An AEI (later Kratos) MS02 double-focusing mass spectrometer procured by Todd for the department at Kent facilitated collaborative research with synthetic chemist colleagues to investigate the fragmentation reactions of organic and organometallic ions. Work with this instrument, one of the first to be equipped with negative ion capability, led to Todd's proposing a simple rationale for predicting the fragmentation routes of negative ions formed by electron attachment to different types of organic hetero-atomic compounds (*12*). Todd's research group was also involved in the construction and characterization of one of the first hybrid ion mobility spectrometer (IMS)/quadrupole mass spectrometer systems, the results from which contributed to the commercial development of IMS-based instruments for trace detection in military and civilian applications (*13*).

Todd's expertise in quadrupole ion traps has also contributed to the development of a miniature automated gas chromatography/isotope ratio mass spectrometer instrument that is currently deployed in one of the most complex and extended space missions yet attempted: after a 10-year journey, this system as part of the Rosetta Space Mission will provide an isotopic analysis of the composition of material that is older than the solar system itself when the probe reaches its target comet (*Churyumov-Gerasimenko*) in December 2014 (*14*).

Todd has been an active member of the international mass spectrometry community throughout his career, serving as Chairman of the Mass Spectroscopy Sub-Committee of Commission I.5 of the International Union of Applied Chemistry (responsible for drawing up recommendations for symbolism and nomenclature in mass spectrometry) (*15*), Chairman, British Mass Spectrometry Society, and Chairman of the Management Advisory Committee of the EPSRC National Mass Spectrometry Service Centre, Swansea University. He has been an Editor of the *International Journal of Mass Spectrometry*, and has also served on the editorial boards of a number of mass spectrometry journals and has been involved with authorities, boards, and councils promoting health, science, and education.

Todd became a Fellow of the Royal Society of Chemistry in 1974 and has been an Honorary Fellow since 2008. He became a Chartered Chemist in 1975, a Fellow of the Institute of Measurement and Control in 1982, and a Chartered Engineer in 1987. He was awarded the Thomson Medal by the International Mass Spectrometry Society in 1997 and the Aston Medal by the British Mass Spectrometry Society in 2006. In 2007, he was admitted as an Honorary Fellow of Canterbury Christ Church University, Canterbury, UK.

Todd was married to Mavis Georgina Lee in Leeds in 1963. Together they had three sons who have collectively produced six grandchildren.

References

(*1*) March, R. E.; Todd, J. F. J. *Quadrupole Ion Trap Mass Spectrometry*, 2nd ed. John Wiley & Sons Inc.: Hoboken, NJ, 2005.

(*2*) March, R. E.; Todd, J. F. J. *Practical Aspects of Ion Trap Mass Spectrometry*; Volumes 1 to 3. CRC Press: Boca Raton, FL, 1995.

(*3*) March, R. E.; Todd, J. F. J. *Practical Aspects of Trapped Ion Mass Spectrometry*; Volumes 4 and 5. CRC Press: Boca Raton, FL, 2010 and 2009.

(*4*) Bonner, R. F.; Lawson, G.; Todd, J. F. J. Ion-Molecule Reaction Studies with a Quadrupole Ion Storage Trap. *Int. J. Mass Spectrom. Ion Phys.* **1972**, *10*, 197–203.

(*5*) Lawson, G.; Bonner, R. F.; Todd, J. F. J. The Quadrupole Ion Store ('Quistor') as a Novel Source for a Mass Spectrometer. *J. Phys. E* **1973**, *6*, 357–362.

(*6*) Mather, R. E.; Todd, J. F. J. The Quadrupole Ion Store (QUISTOR) Part VI. Simultaneous Positive/Negative Ion Mass Spectrometry. *Int. J. Mass Spectrom. Ion Physics* **1980**, *33*, 159–165.

(*7*) Todd, J. F. J.; Freer, D. A.; Waldren, R. M. The Quadrupole Ion Store (QUISTOR) Part XI. The Model of Ion Motion in a Pseudo-Potential Well: An Appraisal in Terms of Phase-Space Dynamics. *Int. J. Mass Spectrom. Ion Physics* **1980**, *36*, 185–203.

(*8*) Todd, J. F. J.; March, R. E. A Retrospective Review of the Development and Application of the Quadrupole Ion Trap Prior to the Appearance of Commercial Instruments. *Int. J. Mass Spectrom.* **1999**, *190/191*, 9–35, (Contributed to a collection of invited papers in a Special Issue of the *International Journal of Mass Spectrometry* entitled 'Quadrupole Ion Trap Mass Spectrometry', in Honor of the Scientific Contributions of John F.J. Todd and Raymond E. March).

(*9*) Stafford, G. C. Jr.; Kelley, P. E.; Syka, J. E. P.; Reynolds, W. E.; Todd, J. F. J. Recent Improvements in and Analytical Performance of Advanced Ion Trap Technology. *Int. J. Mass Spectrom. Ion Processes* **1984**, *60*, 85–98.

(*10*) Louris, J. N.; Cooks, R. G.; Syka, J. E. P.; Kelley, P. E.; Stafford, G. C. Jr.; Todd, J. F. J. Instrumentation, Applications and Energy Deposition in Quadrupole Ion Trap MS/MS Spectrometry. *Anal. Chem.* **1987**, *59*, 1677–1685.

(*11*) Clarke, N. J.; Franklin, A. M.; Langford, M. L.; Penman, A. D.; Thorner, D. A.; Todd, J. F. J. Automation of Data Acquisition and Processing Involving Dynamically Programmed Scanning for a Quadrupole Ion Trap Mass Spectrometer. *Rapid Commun. Mass Spectrom.* **1995**, *9*, 723–730.

(*12*) Alexander, R. G.; Bigley, D. B.; Todd, J. F. J. Towards a Qualitative Rationalization of the Fragmentation Reactions of Organic Negative Molecular Ions. *Org. Mass Spectrom.* **1973**, *7*, 643–646.

(*13*) Proctor, C. J.; Todd, J. F. J.; Turner, R. B. Mass Analyzed Ion Mobility Studies of Nitrobenzene. *Int. J. Mass Spectrom. Ion Processes* **1984**, *60*, 137–145.

(*14*) Todd, J. F. J.; Barber, S. J.; Wright, I. P., *et al.* Ion Trap Mass Spectrometry on a Comet Nucleus: The Ptolemy Instrument and the Rosetta Space Mission. *J. Mass Spectrom.* **2007**, *42*, 1–10.

(*15*) Todd, J. F. J. Recommendations for Nomenclature and Symbolism for Mass Spectroscopy (IUPAC Recommendations 1991). *Pure Appl. Chem.* **1991**, *63*, 1541–1566.

John F.J. Todd

School of Physical Sciences, University of Kent, Canterbury, Kent, UK

Grenville Turner

b. November 1, 1936

Grenville Turner is an English physicist and geochemist who has significantly advanced the study of isotopic variations by mass spectrometry to reveal major aspects of the history of lunar, meteoritic, and terrestrial materials. He was responsible for the earliest development of the ^{40}Ar-^{39}Ar method of radiometric dating. He used it to provide a chronology of the Moon using Apollo and Luna samples. He has also worked on extinct isotopes, pre-solar isotope anomalies, combined noble gas and halogen geochemistry, and resonance ionization mass spectrometry (see this volume, *The Development of Mass Spectrometry in the Earth and Planetary Sciences*).

Turner was born on November 1, 1936, in Todmorden, West Yorkshire, UK. He attended Todmorden Grammar School, a small co-educational school, two graduates of which have won Nobel prizes in the sciences. From there he went on to a B.A. degree in 1958 from Cambridge University (St John's College) and a D.Phil. in 1962 from Oxford University (Balliol College and the Clarendon Laboratory). In Oxford he built an early stainless steel UHV mass spectrometer to study helium isotope production by 50–150 Mev protons, using the Harwell cyclotron.

In 1962 Turner joined John Reynolds' group at U.C. Berkeley to investigate xenon and other noble gases in primitive meteorites (see this volume, *John H. Reynolds*). An early result was his discovery of a component of xenon enriched in both heavy and light isotopes, the first of a group of isotope anomalies, shown later to be due to the presence of mineral grains which predated the formation of the solar system. Neutron activation was used in Berkeley to measure I-Xe ages, based on the decay of extinct 16 Ma ^{129}I. The observation by C. M. Merrihue of ^{39}Ar from (n.p) reactions on ^{39}K led Merrihue to suggest using it as a proxy for K in K-Ar dating. Turner was able to show how variations in isotope ratios during stepped heating could be understood and explained quantitatively in terms of variations in apparent age caused by partial loss by thermal diffusion of radiogenic ^{40}Ar, and could provide a powerful technique for identifying, and in many cases correcting for, effects of partial ^{40}Ar loss which were a serious problem in establishing the validity of conventional K-Ar ages. Most of the commonly used graphical and analytical methods for interpreting ^{40}Ar-^{39}Ar ages were introduced by Turner around this time. The delivery of rock samples from the Moon by Apollo 11 provided an ideal opportunity to apply the new dating method and provided the first steps in establishing a lunar chronology, in particular evidence for the late heavy bombardment, the limited duration of mare volcanism, the use of $^{38}Ar/^{37}Ar$ ratios to date recent small impact craters, and early attempts at thermochronology (*1–3*).

Attempts to apply the ^{40}Ar-^{39}Ar method to date terrestrial ore minerals (and diamonds) containing saline fluid inclusions demonstrated the potential of using ^{38}Ar produced by neutron activation of Cl as a way of distinguishing ^{40}Ar produced by in-situ decay of ^{40}K, from 'parentless' or 'excess' ^{40}Ar introduced by interaction with saline crustal fluids. More generally it opened a new way to study the sources of palaeo-fluids through their interactions with crustal rocks, and in the case of diamonds, the mantle, using the direct link between halogen and noble gas abundances.

In the 1980s Turner led the development of a new ultra sensitive instrument for xenon isotope analysis based on resonance ionization mass spectrometry and with colleagues used it for applications of I-Xe dating, martian geochemistry, and recently the discovery of xenon isotopes from now-extinct ^{244}Pu in terrestrial zircons (*4,5*).

Turner is a fellow of the Royal Society (1980), the Meteoritical Society (1980), the European Association of Geochemistry (1996), the American Geophysical Union (1998). He was a principal investigator for the NASA Apollo Program from 1968–1988. He has received the Rumford Medal from the Royal Society (1996), the Leonard Medal from the Meteoritical Society (1999), the Urey Medal from the European Association of Geochemistry (2002), and a Gold Medal from the Royal Astronomical Society (2004). Asteroid 15523 is named Grenville in his honor.

Turner is currently an Emeritus Professor at the School of Earth, Atmospheric and Environmental Sciences, University of Manchester, UK. He is married to Kathleen (Kate) Morris. They have one daughter and one son.

References

(*1*) Merrihue, C. M.; Turner, G. Potassium-Argon Dating by Activation with Fast Neutrons. *J. Geophys. Res.* **1966**, *71*, 2852–2857.

(2) Turner, G.; Miller, J. A.; Grasty, R. L. The Thermal History of the Bruderheim Meteorite. *Earth Planet. Sci. Lett.* **1966**, *1*, 155–157.
(3) Turner, G. ^{40}Ar-^{39}Ar ages from the Lunar Maria. *Earth Planet. Sci. Lett.* **1971**, *11*, 169–191.
(4) Gilmour, J. D.; Lyon, I. C.; Johnston, W. A.; Turner, G. RELAX: An Ultrasensitive Resonance Ionization Mass Spectrometer for Xenon. *Rev. Sci. Instrum.* **1994**, *65*, 617–625.
(5) Turner, G.; Harrison, T. M.; Holland, G.; Mojzsis, S. J.; Gilmour, J. Extinct ^{244}Pu in Ancient Zircons. *Science* **2004**, *306*, 89–91.

Grenville Turner
School of Earth, Atmospheric, and Environmental Sciences, University of Manchester, Manchester, UK

Harold Clayton Urey

b. April 29, 1893
d. January 5, 1981

Harold Urey was an American chemist whose work profoundly affected many other fields of science (*1–3*). Best known for his Nobel-Prize-winning discovery of deuterium, made in 1931 (with the Prize following in 1934), Urey also strongly influenced the spread of mass spectrometry.

Harold Urey was born in Walkerton, Indiana, where his father was a teacher and a clergyman. After high school, Harold taught in small schools in Indiana and Montana until he entered the University of Montana, in 1914. He finished his undergraduate work in three years, with majors in biology and chemistry. After the US entered World War I, Urey went to work as a chemist in Philadelphia. In 1919, he returned as an instructor in chemistry to the University of Montana. In 1921 he obtained a fellowship and started graduate work in chemistry at age 28 at the University of California at Berkeley, CA. He studied math and physics as well as chemistry and received his Ph.D. in 1923. While some of his dissertation work was published in the *Journal of the American Chemical Society* (*4*), much of it was the basis for a paper in the *Astrophysical Journal* (*5*), and his next year was spent at Niels Bohr's Institute of Theoretical Physics, learning quantum theory in the middle of its creation. In 1924 he returned to the US and obtained a position in chemistry at the Johns Hopkins University. He and Frieda Daum married in 1926; they had four children. In 1929 Urey began a sixteen year period at Columbia University.

Urey and physicist Arthur Ruark published in 1930 one of the earliest texts in English to introduce quantum theory in chemistry (*6*). Urey's attention increasingly turned to the study of isotopes, which in turn soon led to the discovery of deuterium and the fame, funding, and influence that stemmed from that. Isotopes thenceforth were central to his research for almost half a century. He eagerly explored ways in which isotopes could be differentiated, including, of course, early mass spectrographs.

Urey's first researches involving mass spectrometry appeared in 1934; they were collaborations with physicist Walker Bleakney of Princeton University, a pioneer in this area (*7,8*, and see this volume, *Walker Bleakney*). One of these already manifested Urey's interest in topics outside the common concerns of chemistry as it dealt with isotope abundances in certain meteorites.

Through the 1930s, Urey influenced the advance of mass spectrometry in several interrelated ways. These included investigation of multiple modes of isotope separation and development of ways to prepare small amounts of rare stable isotopes of hydrogen, nitrogen, oxygen, and carbon for use in research, which required mass spectrometry for analysis. He led his graduate student David Rittenberg to take up mass spectrometry, got him hired as a general experimenter in Columbia's department of biochemistry, and spurred Rittenberg's path-breaking collaborations with Rudolf Schoenheimer, providing isotopes for use in, and arranging funding for, their research (see this volume, *David Rittenberg* and *Rudolf Schoenheimer*). This, along with related work by Mildred Cohn and Urey, went a long way toward establishing the power of isotopic tracer techniques in biochemical and biological research and initiated a major transformation in the biochemical understanding of cell metabolism. In promoting such work Urey helped move mass spectrometry into laboratories in the life sciences where its use become more and more widespread over subsequent decades (*9*).

Having become a the leading expert on methods of isotope enrichment and separation, during the period just before and at the beginning of World War II Urey was drawn into American studies of uranium fission and isotopes. He was soon at the head of the investigations of various ways to produce uranium enriched in ^{235}U and thus arranging for mass spectrometers and people to operate them for isotopic analysis of inputs, intermediates, and outputs of each possible method. As the scale of the Manhattan Project grew, some areas were split off and Urey became the overseer of the work to develop the gaseous diffusion process – and for that even more mass spectrometry was called for (*10,11*). The sheer number of mass spectrometers built to implement these decisions, along with the design improvements made in them and the number scientists and technicians familiarized with them, transformed the world of mass spectrometry. Mass spectrometers developed for the Manhattan Project's gaseous diffusion plant at Oak Ridge, Tennessee, with the people who created and ran them, led directly to postwar deployment and improvement of high precision devices that could be applied widely not only in atomic and nuclear studies but in organic as well as inorganic chemistry (see this volume, *Alfred Otto Carl Nier*).

When World War II ended, Urey sought to avoid all military-related research. He took a new position in the Institute for Nuclear Studies (later the Fermi Institute) at the University of Chicago and selected, from his many interests, the development of the earth, the moon, and the solar system in general as the future direction of his work. At Chicago, and later in San Diego, Urey was at the heart of a large interdisciplinary group of scientists, attracting brilliant graduate students and postdoctoral fellows, who were in the middle of the metamorphosis of the disparate fields of geology, geo-physics, climatology, astro-physics, meteoritics, solar system astronomy, and more into an integrated area of earth-and-planetary sciences (*12,13*). Urey both exemplified and promoted these developments.

Urey generalized the understanding of isotope concentration processes into an overall analysis of the thermodynamics of isotopically varied substances (*14*). Urey recognized the implication that natural processes operating over time on earth and other bodies would produce isotopically varied traces which, if accurately measured, could inform us about past developments. The first example was the investigation of temperature and thus climate conditions and changes in the Cretaceous period (*15,16*). This required a notable improvement in the design of mass spectrometers for isotope ratio measurement; this dual inlet system is the one innovation in instrument design actually linked, though only in part, to Urey himself (*17*). This aspect of instrumentation has been very influential (*18*). The example of the research was far more influential, opening the field for quantitative studies of paleotemperatures and leading the way by example in approaching all manner of topics in geo- and planetary-history and dynamics through mass spectrometric investigation of isotopic abundances.

Urey quickly expanded his sights to the moon, the planets, the stars, regarding all as part of a field he named cosmochemistry. Hans Suess and he codified available information in an improved table of elemental and isotopic abundances in the earth, meteorites, planets, and stars and showed its relationship to nuclear structure (*19*). This was crucial to the development of explanations of nucleo-synthesis and thus to understanding not only the development of our solar system but the universe as a whole. It pointed in turn to the importance of further extension of and improvements in mass spectrometric determinations of isotopic abundances on this and other planets and in interplanetary and interstellar material. Urey naturally became a strong supporter of space exploration.

Urey faced mandatory retirement in 1958 at the University of Chicago; rejecting this option, he moved to the newly created University of California at San Diego. Before long a considerable number of other significant figures from Chicago and elsewhere relocated to this new center of research that Urey helped promote, and mass spectrometry was further advanced as a result. A large proportion of the young scientists around Urey during his years in Chicago and in San Diego went on to employ and improve these instruments in study after study and to train further investigators, who have helped make these tools of inquiry ubiquitous in departments and programs in the earth and planetary sciences (see this volume, *The Development of Mass Spectrometry in the Earth and Planetary Sciences*).

Urey remained active in research until his death in his 87th year; he published over a hundred papers after his move to San Diego. In addition to the Nobel Prize he received far more honors than can be listed here, including recognition from the National Academy of Sciences, the Royal Society of London, the American Chemical Society, the Geological Society of America, the Geochemical Society, the Meteoritical Society, and the Royal Astronomical Society. And though only a small amount of his work was directly concerned with design and operation of mass spectrometers, his early use of them and his powerful promotion of their use by others mark him as one of the most influential pioneers in the world of mass spectrometry.

References

(*1*) Arnold, J. R.; Bigeleisen, J.; Hutchison, C. A., Jr. *Harold Clayton Urey: April 29, 1893–January 5, 1981. Biographical Memoirs v. 68*; National Academies Pr.: Washington, DC, 1995. 363–411.

(*2*) Cohen, K. P.; Runcorn, S. K.; Suess, H. E.; Thode, H. G. Harold Clayton Urey, 29 April 1893–5 January 1981. *Biogr. Mems. Fell. Roy. Soc.* **1983**, *29*, 622–659.

(*3*) Miller, S. L.; Oró, J.; Harold, C. Urey 1983–1981. *J. Mol. Evol.* **1981**, *17*, 263–264.

(*4*) Urey, H. C. The Heat Capacities and Entropies of Diatomic and Polyatomic Gases. *J. Am. Chem. Soc.* **1923**, *45*, 1445–1455.

(*5*) Urey, H. C. The Distribution of Electrons in the Various Orbits of the Hydrogen Atom. *Astrophys. J.* **1924**, *59*, 1–10.

(*6*) Ruark, A. E.; Urey, H. C. *Atoms, Molecules and Quanta*; McGraw-Hill: New York, 1930.

(*7*) Rittenberg, D.; Bleakney, W.; Urey, H. C. The Equilibrium Between the Three Hydrogens. *J. Chem. Phys.* **1934**, *2*, 48–49.

(*8*) Manian, S. H.; Bleakney, W.; Urey, H. C. The Relative Abundance of the Oxygen Isotopes O16:O18 in Stone Meteorites. *J. Am. Chem. Soc.* **1934**, *56*, 2601–2609.

(*9*) Ratner, S. Applications of Stable Isotopes to the Study of Intermediary Metabolism: Historical Survey. In: *Biochemical Applications of Mass Spectrometry*;

Waller, H. C., Ed.; John Wiley & Sons: New York, 1972; pp 1–10.
(*10*) Rhodes, R. *The Making of the Atomic Bomb*; Simon and Schuster: New York, 1986.
(*11*) Nier, A. O. C. Some Reminiscences of Mass Spectrometry and the Manhattan Project. *J. Chem. Educ.* **1989**, *66*, 385–388.
(*12*) Doel, R. E. The Earth Sciences and Geophysics. In: *Science in the Twentieth Century*; Krige, J., Pestre, D., Eds.; Harwood Academic Pubs.: Amsterdam, 1997; pp 391–416.
(*13*) Doel, R. E. *Solar System Astronomy in America: Communities, Patronage, and Interdisciplinary Research, 1920–1960*; Cambridge Univ. Pr: New York, 1996.
(*14*) Urey, H. C. The thermodynamic Properties of Isotopic Substances. *J. Chem. Soc.* **1947**, 562–581.
(*15*) Urey, H. C.; Lowenstam, H. A.; Epstein, S.; McKinney, C. R. Measurement of Paleotemperatures and Temperatures of the Upper Cretaceous of England, Denmark, and the Southeastern United States. *Bull. Geol. Soc. Amer.* **1951**, *62*, 399–416.
(*16*) Epstein, S.; Buchsbaum, R.; Lowenstam, H. A.; Urey, H. C. Carbonate Water Isotopic Temperature Scale. *Bull. Geol. Soc. Amer.* **1951**, *62*, 417–426.
(*17*) McKinney, C. R.; McCrea, J. M.; Epstein, S.; Allen, H. A.; Urey, H. C. Improvements in Mass Spectrometers for the Measurement of Small Differences in Isotope Abundance Ratios. *Rev. Sci. Instr.* **1950**, *21*, 724–730.
(*18*) Brand, W. A. High Precision Isotope Ratio Monitoring Techniques in Mass Spectrometry. *J. Mass. Spectrom.* **1996**, *31*, 225–235.
(*19*) Suess, H. E.; Urey, H. C. Abundances of the Elements. *Rev. Mod. Phys.* **1956**, *28*, 53–74.

Photo source: http://www.nobelprize.org/nobel_prizes/chemistry/laureates/1934/urey-bio.html

Keith A. Nier
Independent Historian, Madison, NJ, USA

Marvin L. Vestal

b. September 8, 1934

Marvin L. Vestal, PhD, is an American engineer and physicist whose scientific interest has been to develop practical mass spectrometers for biological applications. He has also developed theoretical models for ionization processes and mass analyzers. He built the first commercial Chemical Ionization mass spectrometer, and developed and commercialized thermospray liquid chromatography-mass spectrometry (LC-MS) and high performance matrix assisted laser desorption time of flight (MALDI-TOF) and tandem time of flight (TOF-TOF) (see this volume, *The Development of Time-of-Flight (TOF) Mass Spectrometry*).

Vestal was born on September 8, 1934, in Pendleton, Indiana, USA. He graduated from Purdue University with B.S. (1958) and M.S. (1960) in Engineering Sciences. During his undergraduate years he took a break to join the US Army Signal Corp. Vestal worked part-time at Johnston Laboratories in Lafayette, Indiana, from 1958 while he continued to study for his M.S. degree in Engineering Sciences. He continued with Johnston Labs after their move to Baltimore until the founding of Scientific Research Instruments in the same city where he worked until going back to school in 1970, obtaining his Ph.D. in 1975 from University of Utah in Chemical Physics.

In the early 1960s Vestal collaborated with Henry Rosenstock to develop the Coincidence Time-of-Flight mass spectrometer for which Vestal invented a secondary electron multiplier that improved the performance of this instrument (see this volume, *Henry Meyer Rosenstock*). He also worked with Wahrhaftig to correct a major problem with the original quasi-equilibrium theory developed by Rosenstock and Wahrhaftig and applied the theory to calculations on fragmentation of 'large' molecular ions such as propane (see this volume, *Austin L. Wahrhaftig*). Vestal cofounded Scientific Research Instrument Corporation, where he built the first commercial chemical ionization source in 1968 (see this volume, *A History of Chemical Ionization (CI) in Mass Spectrometry*). In the early 1970s as a PhD student at the University of Utah, he built a crossed-beam apparatus to study ion-molecule reaction, the first tandem quadrupole mass spectrometer to study photodissociation mechanisms. He also began to design and build an LC/MS instrument.

In 1980, Vestal and co-workers developed the Thermospray technique (see this volume, *The Development and Demise of Thermospray Ionization Mass Spectrometry*), which not only provided an innovative interface between an LC and a mass spectrometer but also served as a new soft ionization method for polar and non-volatile molecules. In 1983, Vestal founded his own company, Vestec, to produce commercial LC-MS instruments based on the Thermospray interface. Soon, this technique became the first practical LC-MS interface, enabling direct analysis of biological samples and drawing enormous attention from the life-sciences community. Vestal's studies of ion evaporation/desorption mechanisms involved in Thermospray ionization provided a fundamental understanding of common traits in spray ionization. Furthermore, his successful applications of Thermospray to biologically important and relevant nucleosides and peptides heightened interest in the biomedical communities and prompted further development of the interface. Soon thereafter, other soft ionization techniques such as electrospray and MALDI were introduced. In particular, Electrospray ionization (ESI) invented by John Fenn provided a far more stable LC-MS interface, which was critically important to the broad user base created by the commercial Thermospray LC-MS instruments (see this volume, *The Development of Electrospray Ionization (ESI) for Mass Spectrometry*, and *John Bennett Fenn*). Although Thermospray 'passed the torch' to ESI, it may not be an overstatement to say that Vestal set the stage for practical LC-MS in life sciences and paved the way for ESI.

Following the invention of MALDI by Karas and Hillenkamp (see this volume, *The Development of Matrix Assisted Laser Desorption Ionization (MALDI) Mass Spectrometry, Franz Hillenkamp*, and *Michael Karas*), Vestal's focus became developing MALDI instruments with TOF analyzers. He and his colleagues at Vestec built the first commercial MALDI-TOF instrument in 1990, and Vestec's acquisition by PerSeptive Biosystems (later merged to Applied Biosystems Incorporation, ABI) expanded its commercial viability. A remaining problem, however, was that MALDI-TOF had low performance: mass resolving power was marginal, accuracy of mass measurements was coarse, and the lack of MS-MS functionality was a substantial limitation.

Vestal's most significant accomplishment in this area was the development of a comprehensive theoretical model of the various components of a TOF analyzer with a view toward optimizing performance of a complete system for particular applications. The first application of this theoretical approach was the design of a family of MALDI-TOF instruments employing delayed extraction which led to a series of MALDI-TOF instruments with mass resolving power exceeding 10,000 and mass accuracy with sub-10-ppm mass errors for the first time. More than half of the MALDI-TOF instruments in use today are based on Vestal's designs.

By refining his theory and extending it to the design of tandem TOF, Vestal successfully constructed MALDI-TOF-TOF instruments, which were introduced commercially by Applied Biosystems as the 4700 Proteome Analyzer, and later as the 4800 TOF-TOF; these instruments are used today in most proteomics centers. Both MALDI-TOF MS and MS-MS systems have had an enormous impact on many important areas of research that include proteomics, glycomics, cell signaling, structural biology, organelle imaging, and polymer science. There are few scientists and innovators who have had significant impact on their field both through research and commercial technology development. Perhaps no one has had a more profound dual impact on modern mass spectrometry than Vestal during a career that spans five decades. After 'retirement' from ABI in 2004, Vestal started a new company, Virgin Instruments, at which he continues to develop MALDI-TOF-TOF instruments, taking advantage of recent advances in related technology and his theoretical predictions.

Vestal is a member of the American Society for Mass Spectrometry. He has received the Lester W. Strock Award (1985), the Field and Franklin Award for Distinguished Contribution in Mass Spectrometry (2005), and the ASMS Distinguished Contribution in Mass Spectrometry Award (2010). He served on the editorial board of *Molecular & Cellular Proteomics* from 2002–06.

Vestal is married to fellow mass spectrometrist Dr. Christina Vestal and serves as CSO of Virgin Instruments.

The publications cited below are representative of the author's most significant contributions to the field (*1–5*).

References

(*1*) Blakley, C. R.; Carmody, J. J.; Vestal, M. L. A New Soft Ionization Technique for Mass Spectrometry of Complex Molecules. *J. Am. Chem Soc.* **1980**, *102*, 5931–5933.

(*2*) Vestal, M. L. Studies of Ionization Mechanisms Involved in Thermospray LC-MS. *Int. J. Mass Spectrom. Ion Phys.* **1983**, *46*, 193–196.

(*3*) Vestal, M. L. High Performance Liquid Chromatography-Mass Spectrometry. *Science* **1984**, *226*, 275–281.

(*4*) Vestal, M. L.; Juhasz, P. Resolution and Mass Accuracy in Matrix-assisted Laser Desorption Ionization Time-of-Flight. *J. Am. Soc. Mass Spectrom.* **1998**, *9*, 892–911.

(*5*) Vestal, M. L. Modern MALDI Time of Flight Mass Spectrometry. *J. Mass Spectrom.* **2009**, *44*, 303–317.

Marvin Vestal
SimulTOF Instruments, Framingham, MA, USA

Austin L. Wahrhaftig

b. May 5, 1917
d. November 11, 1997

Austin L. Wahrhaftig was an American chemist known for his pioneering work developing quasi-equilibrium theory (QET) of fragmentation patterns in molecular mass spectrometry and for innovative developments in mass spectrometry instrumentation.

Wahrhaftig was born on May 5, 1917, in Sacramento, CA, USA (*1*). He received an A.B. in Chemistry (with great distinction) from the University of California, Berkeley, in 1938 and a Ph.D. in Chemistry (Physical) from the California Institute of Technology, Pasadena in 1941. His dissertation was on the absorption spectrum of chlorine fluoride as determined via molecular spectroscopy (*2*). He remained at Caltech as a postdoctoral research fellow until 1945, working on various war projects. The experience he gained during those years included instrument design and construction, electronics, photography, engineering, and a wide assortment of minor applications of physical chemistry and mathematics. He then spent a year working on the research and development of evaporated coatings, electronic equipment, and interferometric devices with Dr. W Ewart Williams, and a year as a university fellow at Ohio State University.

From 1947 until his death Wahrhaftig was associated with the University of Utah at one point serving as Associate Chair of the Chemistry Department. He also twice held a Visiting Professor position at La Trobe University, Bundoora, Melbourne, Australia in 1972 and 1980. He was also heavily involved in the design of new physical facilities for the university chemistry department. His last position was Professor Emeritus of Chemistry at the University of Utah, until his death on November 11, 1997.

At Utah, with funding from the Atomic Energy Commission, he built a strong experimental mass spectroscopy research program that produced many high-impact publications. Wahrhaftig's interest in innovative instrumentation extended over his whole career (*3*). Still, he became best known for work in the development and testing of theory, particularly the statistical theory of how molecules fragment in a mass spectrometer after ionization by electron impact, which was inaugurated in classic paper by two of his graduate students (Henry Rosenstock and Merrill Wallenstein), himself, and Henry Eyring (*4–7*). That theory has become known as the quasi-equilibrium theory or QET (see this volume, *The Beginnings of Organic Mass Spectrometry* and *Henry Meyer Rosenstock*).

In addition to membership in various professional scientific societies, Wahrhaftig served on the Board of Directors for the ASTM Committee E-14 on Mass Spectrometry (1962–64) and the American Society of Mass Spectrometry (1970–72).

Wahrhaftig was married to Ruby M. Wahrhaftig, née Ruby Dixon Elledge; they had no children. When his arthritis permitted, he enjoyed skiing, mountain hiking, some river running, and traveling in the U.S. and abroad.

References

(*1*) (Anon.) Austin Levy Wahrhaftig. *Deseret News*, **1997**, *Nov. 16*.

(*2*) Wahrhaftig, A. L. The Absorption Spectrum of Chlorine Fluoride. *J. Chem. Phys.* **1942**, *10*, 28.

(*3*) Randal, L. G.; Wahrhaftig, A. L. Direct Coupling of a Dense (Supercritical) Gas Chromatograph to a Mass Spectrometer Using a Supersonic Molecular Beam Interface. *Rev. Sci. Instrum.* **1981**, *52*, 1283–1295.

(*4*) Rosenstock, H. M.; Wallenstein, M. B.; Wahrhaftig, A. L.; Eyring, H. Absolute Rate Theory for Isolated Systems and the Mass Spectra of Polyatomic Molecules. *Proc. Natl. Acad. Sci.* **1952**, *38*, 667–678.

(*5*) Rosenstock, H. M.; Wahrhaftig, A. L.; Eyring, H. Metastable Ions in Mass Spectra. *J. Chem. Phys.* **1955**, *23*, 2200.

(*6*) Vestal, M.; Wahrhaftig, A. L.; Johnson, W. H. Improved Rate Expression in the Quasi-Equilibrium Theory of Mass Spectra. *J. Chem. Phys.* **1962**, *37*, 1276–1283.

(*7*) Barfield, A. F.; Wahrhaftig, A. L. Determination of Appearance Potentials by the Critical Slope Method. *J. Chem. Phys.* **1964**, *41*, 2947–2948.

Peter B. Armentrout
The University of Utah, Salt Lake City, UT, USA

Harold Williams Washburn

b. June 23, 1902
d. November 1, 1986

Harold Washburn was an American physicist who led in the creation of the first commercial mass spectrometer and in its application to analysis in the petroleum refining industry. His mix of scientific, technical, and business skills greatly influenced the development of the analytical instrumentation industry and the spread of mass spectrometry applications through the 1940s and 1950s.

Harold Washburn was born in Jacksonville, Oregon, in 1902. He earned undergraduate and MS degrees in electrical engineering from MIT and a PhD in physics from the University of California, Berkeley in 1932 under the direction of E. O. Lawrence. His dissertation involved both the creation of an instrument and its use in research, a combination that, on a larger scale, characterized much of his career (*1*). He was an instructor at MIT for a year and chief physicist for Western Geophysical Co. for 1933–1937 before becoming vice president in charge of research at the newly formed Consolidated Engineering Corporation (CEC, later Consolidated Electrodynamics Corp.).

Washburn and the company's president, Herbert Hoover Jr., assembled a productive scientific and engineering team that identified an appropriate mass spectrometer design. Based on that design, they produced a modified instrument that could be manufactured and marketed successfully. When the originally forecast use for the machine – oil exploration – turned out not to be viable, Washburn identified an alternative market in hydrocarbon mixture analysis for oil refineries and successfully launched the business of making and selling mass spectrometers to that market (*2*, see this volume, *Magnetic and Electrostatic Analyzers before 1960* and *Mass Spectrometry Manufacturers and Markets from the Beginning to 1960*). Success in this area required demonstration of the potential for mass spectrometry in this kind of analysis, as distinct from its previous common uses in atomic and nuclear physics (*3*,*4*, and see this volume, *The Beginnings of Organic Mass Spectrometry*). As competition developed, Washburn and his team produced improved techniques (including early use of computers in data analysis), more models for more uses, and of course many patents. Some further publications are listed below (*5–10*).

Washburn was honored with the American Chemical Society's Beckman Award for chemical instrumentation in 1956. He left CEC in 1960 when the company was acquired by Bell and Howell Corp. He took a position with the Jet Propulsion Laboratory, California Institute of Technology in connection with the Ranger space program, and he ended his career as professor emeritus of Electrical Engineering at California State University, Long Beach. He died in 1986 in Long Beach, California (*11*).

References

(*1*) Washburn, H. W. Electro-Optical Shutter. *Phys. Rev.* **1932**, *39*, 688–701.

(*2*) Grayson, M. A., Ed. *Measuring Mass: From Positive Rays to Proteins*; Chemical Heritage Press: Philadelphia, PA, 2002.

(*3*) Hoover, H., Jr.; Washburn, H. W. A Preliminary Report on the Application of the Mass Spectrometer to Problems in the Petroleum Industry. *Am. Inst. Mining Met. Engrs. Tech. Pub.* **1940**, *No. 1205*.

(*4*) Washburn, H. W.; Wiley, H. F.; Rock, S. M. The Mass Spectrometer as an Analytical Tool. *Ind. Eng. Chem., Anal. Ed.* **1943**, *15*, 541–547.

(*5*) Washburn, H. W.; Wiley, H. F.; Rock, S. M.; Berry, C. E. Mass Spectrometry. *Ind. Eng. Chem., Anal. Ed.* **1945**, *17*, 74–81.

(*6*) Berry, C. E.; Wilcox, D. E.; Rock, S. M.; Washburn, H. W. Computer for Solving Linear Simultaneous Equations. *J. Appl. Phys.* **1946**, *17*, 262–272.

(*7*) Washburn, H. W.; Berry, C. E. Effects of High Initial Energies on Mass Spectra. *Phys. Rev.* **1946**, *70*, 559.

(*8*) Washburn, H. W. Mass Spectrometry. In: *Physical Methods in Chemical Analysis*; Berl, W. G., Ed.; Academic Press: New York, 1950; vol. 1, pp 587–639.

(*9*) Robinson, C. F.; Washburn, H. W.; Berry, C. E.; Perkins, G. D. A Mass Spectrometer for Monitoring Continuous Processes. *Instruments* **1951**, *24*, 25–29.

(*10*) Washburn, H. W.; Berry, C. E.; Hall, L. G. Measurement of Deuterium Oxide Concentration in Water Samples by the Mass Spectrometer. *Anal. Chem.* **1953**, *25*, 130–134.

(*11*) Bowker, R. R. *American Men and Women of Science*, 19th ed. New Providence, New Jersey, US. *s.v.* "Washburn, Harold." 1995.

Keith A. Nier
Independent Historian, Madison, NJ, USA

Gerald J. Wasserburg

b. March 25, 1927

Source: http://www.archives.caltech.edu/pictures/GJW57.1-1.jpg.

Gerald J. Wasserburg (always known as Jerry) is an American geochemist and geo- and astro-physicist. He is known for his work in the fields of isotope geology, cosmochemistry, and meteoritics. His efforts were crucial to ensuring that the best science would be done with the first lunar samples returned to earth, and in the course of that preparation he helped drive high-precision high-sensitivity mass spectrometry to new levels.

Jerry Wasserburg was born in New Brunswick, New Jersey, US, in 1927. He was fascinated by minerals as a child but not academically inclined as a youth. After getting into the army by lying about his age, he served in the European theater during World War II. He completed high school, entered night school at Rutgers University in New Brunswick, became seriously interested in geology, and transferred to the University of Chicago in 1948. At Chicago he earned his BS in physics in 1951. He got a job running a mass spectrometer for Harold Urey's team, married Naomi Orlick (a graduate student in physiology), and in 1954 completed a Ph.D. in geology with Urey and Mark Inghram as advisors (*1*, *2*, and see this volume, *Mark Gordon Inghram II* and *Harold Clayton Urey*). Getting a degree in geology with advisors in chemistry and physics exemplified both the interdisciplinary character of the Institute for Nuclear Studies at Chicago and the revolution that was beginning to transform geology – a transformation Wasserburg's subsequent career significantly advanced (see this volume, *The Development of Mass Spectrometry in the Earth and Planetary Sciences*). After one post-doctoral year at Chicago he became an assistant professor at the California Institute of Technology in Pasadena, California, in 1955. He remained at Caltech until he retired in 2001 from the John D. MacArthur Professorship of Geology and Geophysics. Since adding "Emeritus" to his title he has continued his research, directing most of his attention to astrophysical problems (*3–5*).

Wasserburg's research has produced significant advances in a large number of major subject areas. Some illustrative papers are listed below; only a suggestion can be given of their range and importance here (*6–20*). He made major contributions in regard to several lines of isotopic dating, working over the course of his long career on the ^{40}K–^{40}Ar, U–Pb, ^{87}Rb–^{87}Sr, ^{230}Th, ^{40}K–^{39}Ar, ^{187}Re-^{187}Os systems, and particularly the ^{147}Sm-^{143}Nd system, which, once it could be detected at all, became a cornerstone of the understanding of the chemical evolution of the deep interior of the Earth. He and his numerous colleagues have used measurements of variations in the occurrences of these isotopes to discover features of and determine the timing of many aspects of the formation of the solar system generally, of the meteoric and planetary members of it, and of the various components and stages in the consolidation and differentiation of the earth.

Wasserburg's most influential measurements have been made with the special instrument he designed and assembled in the 1960s, partially in anticipation of the analysis of lunar samples, which was given the name "Lunatic I" while its home was named "the Lunatic Asylum" of the Charles Ames Laboratory of Geological Sciences at Caltech. This was a custom-built, high precision, high-sensitivity machine with computer controlled field scanning, switching, and data acquisition, built and usually operated together with D. A. Papanastassiou. Wasserburg was also largely responsible for creating functional facilities and procedures for receiving and handling lunar samples and distributing them to researchers worldwide.

Wasserburg and colleagues were able to determine the age of the moon and the later age of lunar maria and developed evidence indicative of a period of major impact events half a billion years after initial formation. With the same instrument and in the same years as analyzing lunar material, Wasserburg made fundamental investigations of Allende meteorite samples that he had collected himself very soon after its fall in 1969. Some of the inclusions proved to be older than any previously measured materials, predating the formation of the earth.

The ability to measure extremely tiny isotopic variations enabled Wasserburg and co-workers to discover traces of the extinct radioactive nuclides ^{26}Al and ^{107}Pa and to work out their implications for solar system and planetary history.

The combination of results from the several isotopic systems and their differential appearance in diverse sources around the world has made it possible to determine significant aspects of and the timing of the differentiation and dynamics of the earth's core and mantle, the formation of continental crust, the circulation of the earth's water, and paleoclimatology.

His meteoritic studies have revealed evidence that the pre-solar cloud was not fully homogeneous and that it was influenced by more than one supernova. Wasserburg has recently extended his research to the chemical evolution of different types of stars at different times in galactic and cosmic development.

Jerry and Naomi Wasserburg have been married for over half a century; they have two sons and several grandchildren. Wasserburg has been honored in ways too numerous to catalog, including election to the National Academy of Sciences and receipt of several medals for science and public service from NASA. He received the Arthur L. Day Medal from the Geological Society of America and the Leonard, Goldschmidt, and Bowie Medals, the highest awards of the Meteoritical and Geochemical Societies and the American Geophysical Union, respectively. In 1986 Wasserburg and Claude Allègre shared the Crafoord Prize, Nobel equivalent for the fields in covers (see this volume, *Claude Jean Allègre*). And in 1991 he received an honor, the very existence of which demonstrates the wide impact of the kind of science he has pursued – the Royal Astronomical Society awards a Gold Medal for Geophysics and Planetary Sciences.

References

(*1*) Wasserburg, G. J. Isotopic Adventures – Geological, Planetological, and Cosmic. *Annu. Rev. Earth Planet. Sci.* **2003**, *31*, 1–74, Available at: http://www.annualreviews.org/doi/abs/10.1146/annurev.earth.31.100901.141409.

(*2*) Marvin, U. B. Oral Histories in Meteoritics and Planetary Science: XII. Gerald J. Wasserburg. *Meteorit. Planet Sci.* **2004**, *39*(Suppl.), A177–A197, Available at: http://onlinelibrary.wiley.com/doi/10.1111/j.1945-5100.2004.tb00355.x/pdf.

(*3*) Janos, L. Gerald Wasserburg: Timekeeper of the Solar System. In: *A Passion to Know: 20 Profiles in Science*; Hammond, A. L., Ed.; Charles Scribner's Sons: New York, 1984; pp 48–59.

(*4*) Pounds, K. A. The Address by the President (Professor K. A. Pounds) on the Presentation of the Gold Medal to Professor G. J. Wasserburg on Friday 1991 May 10. *Quart. J. Roy. Astr. Soc.* **1991**, *32*, 351–352.

(*5*) Jacobsen, S. B. Citation: 2008 William Bowie Medal Winner, Gerald J. Wasserburg. *Eos, Trans. Am. Geophys. Union* **2009**, *90*, 5, Available at: http://sites.agu.org/honors/winners/gerald-j-wasserburg-3/?sub=citation.

(*6*) Wasserburg, G. J. ^{40}Ar-^{40}K Dating. In: *Nuclear Geology: A Symposium on Nuclear Phenomena in the Earth Sciences*; Faul, H., Ed.; John Wiley & Sons: New York, 1954; pp 341–349.

(*7*) Butler, W. A.; Jeffery, P. M.; Reynolds, J. H.; Wasserburg, G. J. Isotopic Variations in Terrestrial Xenon. *J. Geophys. Res.* **1963**, *68*, 3283–3291.

(*8*) Wasserburg, G. J. Geochronology and Isotopic Data Bearing on Development of the Continental Crust. In: *Advances in Earth Science*; Hurley, P. M., Ed.; MIT Press: Cambridge, MA, USA, 1966; pp 431–459.

(*9*) Wasserburg, G. J.; Papanastassiou, D. A.; Nenow, E. V.; Bauman, C. A. A Programmable Magnetic Field Mass Spectrometer with On-line Data Processing. *Rev. Sci. Instrum.* **1969**, *40*, 288–295.

(*10*) Wasserburg, G. J.; Huneke, J. C.; Burnett, D. S. Correlation between Fission Tracks and Fission-Type Xenon from an Extinct Radioactivity. *Phys. Rev. Lett.* **1969**, *22*, 1198–1201.

(*11*) Gray, C. M.; Papanastassiou, D. A.; Wasserburg, G. J. The Identification of Early Condensates from the Solar Nebula. *Icarus* **1973**, *20*, 213–239.

(*12*) Jessberger, E. K.; Huneke, J. C.; Wasserburg, G. J. Evidence for a ~4.5 Aeon Age of Plagioclase Clasts in a Lunar Highland Breccia. *Nature* **1974**, *248*, 199–202.

(*13*) Lee, T.; Papanastassiou, D. A.; Wasserburg, G. J. ^{26}Al in the Early Solar System: Fossil or Fuel? *Astrophys. J. Lett.* **1977**, *211*, L107–L110.

(*14*) Kelly, W. R.; Wasserburg, G. J. Evidence of the Existence of ^{107}Pd in the Early Solar System. *Geophys. Res. Lett.* **1978**, *5*, 1079–1082.

(*15*) Jacobsen, S. B.; Wasserburg, G. J. The Mean Age of Mantle and Crustal Reservoirs. *J. Geophys. Res.* **1979**, *84*, 7411–7427.

(*16*) Chen, J. H.; Wasserburg, G. J. The Isotopic Composition of Uranium and Lead in Allende Inclusions and Meteorite Photosphates. *Earth Planet. Sci. Lett.* **1981**, *52*, 1–15.

(*17*) Wasserburg, G. J. Isotopic Abundances: Inferences on Solar System and Planetary Evolution. *Earth Planet. Sci. Lett.* **1987**, *86*, 129–173.

(*18*) Anbar, A. D.; Papanastassiou, D. A.; Wasserburg, G. J. Determination of Iridium in Natural Waters by Clean Chemical Extraction and Negative Thermal Ionization Mass Spectrometry. *Anal. Chem.* **1997**, *69*, 2444–2450.

(*19*) Qian, Y. Z.; Wasserburg, G. J. Determination of Nucleosynthetic Yields of Supernovae and Very Massive Stars from Abundances in Metalpoor Stars. *Astrophys. J* **2002**, *567*, 515–531.

(*20*) Allègre, C. *Isotope Geology*; Cambridge University Press: Cambridge, etc., 2008.

Keith A. Nier
Independent Historian, Madison, NJ, USA

Wilhelm Carl Werner Otto Fritz Franz Wien

b. January 13, 1864
d. August 30, 1928

Wilhelm Wien, who identified himself as Willy, was a German physicist noted for both experimental and theoretical accomplishments. He won the Nobel Prize in Physics in 1911 for his work on heat radiation, particularly development of the Wien displacement law (*1*). The results and the techniques of his late 19th- and early 20th-century experiments on *kanalstrahlen*, or channel rays, became the foundation of mass spectrometry.

Wien was born in Gaffken, Germany (now Russia). His family were minor landholders in what then was East Prussia. He attended the universities of Göttingen and Berlin and earned his doctorate in 1886 at Berlin under the direction of Herman von Helmholtz. After helping manage the family estate for a few years, Wien became an assistant at the Physikalische Technische Reichsanstahlt in 1890. His next position was as associate professor at the Technische Hochschule in Aachen in 1896. In 1898, Wien married Luise Mehler, of Aachen, with whom he had four children. Wien subsequently was called to professorships at the universities of Giessen, Würzburg, and Munich in 1899, 1900, and 1920, respectively (*2*,*3*). He died in Munich in 1928.

At the end of his years in Berlin he had worked with a cathode ray apparatus as an X-ray source. In Aachen, Wien took up research on both cathode rays and *kanalstrahlen*. Many researchers were studying cathode rays in this period since electric discharge in rarified gasses was considered to be a very promising area in which to get at the fundamental characters of matter, electromagnetism, radiation, and their relationships. Wien's experiments supported his expectation that cathode rays were negatively charged particles; he reached this result just after J. J. Thomson did and he arrived at a similar value for the ratio of mass to charge for these particles.

Wien was apparently the only one in the 1890s carrying out comparable research on another kind of ray that could be produced in a discharge (or cathode ray) tube, namely *kanalstrahlen* or channel rays. These had been discovered by Eugen Goldstein in 1886 but little had learned about them since then. Goldstein established that when small channels (*kanalen*) had been bored through a metal plate cathode and the discharge tube was operating, luminous lines projected out from the openings on the opposite side from the space in which cathode ray phenomena were observed. Like cathode rays, their direction was basically perpendicular to the plate surface and they could cause glass to fluoresce. In contrast, however, rays from the channels were luminous themselves, even without hitting a surface, and they came in several versions, differing in color depending on the identity of the residual gas in the tube where the discharge took place. Further, Goldstein found that magnets able to readily deflect cathode rays did not measurably deflect these *kanalstrahlen* (*4*).

Wien, after concluding that cathode rays are negatively charged particles, demonstrated that *kanalstrahlen* are positively charged particles. This seemed likely since they apparently originated in the gas in the discharge tube and then moved toward and through the negatively charged perforated cathode. Wien first showed that an electrode, isolated from the discharge circuit and cathode rays, accumulated positive charge when channel rays were directed on it. The channel rays thus were like cathode rays, though producing opposite electrification.

To identify the rays unambiguously as particles carrying charge rather than something inducing charge at the target, Wien then demonstrated that they were appropriately deflected by electrostatic and by magnetic forces, although limitations in vacuum technology made this very difficult to show. Philipp Lenard had shown that cathode rays could pass through a very thin metal foil. This allowed their investigation in a space more highly evacuated than the discharge tube. Wien did not find a way to do the same with channel rays, and a pressure in the discharge tube suitable for it to serve as a source of channel rays was not low enough to facilitate easy maintenance on the other side of the cathode of sufficient electrostatic potential between plate electrodes to deflect the rays that came through to that part of the tube. Nevertheless, using a 2.5 cm thick iron disk cathode with a single 2 mm hole drilled through it, he was able for short times to obtain both a clear channel ray beam and a good enough vacuum to sustain a constant 2000 V potential difference between plates 1.7 cm apart. This could be sustained just long enough to pass the beam between the electrodes and measure the direction and amount of electrostatic deflection before its passage increased the conductivity of the residual gas enough to cause the potential to discharge.

The same vacuum limitation ruled out placement of a deflecting magnet a significant distance from the cathode opening, yet a very powerful magnet close to the cathode might disrupt the discharge source on the other side. Wien arranged iron shielding around the discharge tube section of his apparatus to allow having the poles of a powerful electromagnet outside the glass tube but immediately adjacent to the cathode plate. He was thus able to produce and determine the direction and amount of magnetic deflection of channel rays.

He reported the results both briefly and more thoroughly in papers in 1898 (*5,6*). Channel rays consisted of positively charged particles. From his measurements of the deflections and field strengths, Wien calculated the velocity of the rays as around a hundred times less than that of cathode rays and the ratio of mass to charge (m/e) as very much greater, about that of a hydrogen ion in electrolysis. The fluorescing spot at the end of the observation space where the beam hit the glass was slightly spread in both the electrical and magnetic cases, so Wien pointed out that the beam contained a mixture of rays, somewhat differently deflected.

In 1902 Wien reported further experiments with somewhat different apparatus, capable of working with higher vacua and higher potentials. He reinforced the general conclusion that channel rays are positively charged atomic-scale particles produced during electric discharge in rarefied gases and he showed that rays produced by different gases have different ratios of mass to charge (*7–10*). He also showed that the presence of the least amounts of contaminants could significantly affect the phenomena and that it took a great deal of care to get these at all under control.

Wien continued research on ion beam phenomena over the remainder of his life (*11*). He also directed the initial work of several students who took up research in this field and his research stimulated various other scientists as well, among whom may fairly be counted J. J. Thomson. He systematized the accumulating results over the decades in substantial monographs in major reference Handbooks on radiation research and experimental physics (*12*).

Although he devoted so many years to ion physics, he has been renowned mainly, in the words of fellow Nobel Prize-winner Max von Laue, for the "immortal glory" of having "led us to the very gates of quantum physics," and he is mentioned in histories of physics almost exclusively for his theoretical studies (*1*). From the perspective of the world of mass spectrometry, however, additional accomplishments come to the fore.

Wien was the first to demonstrate that electric and magnetic fields could differentiate different kinds of ions by their mass-to-charge ratios and thus opened the very possibility of mass spectrometry. Any apparatus with electrostatic and magnetic fields perpendicular both to each other and to the general direction of particle motion is referred to as a Wien filter; Wien first made use of such an arrangement in the investigation of cathode rays that preceded his initial study of *kanalstrahlen.*The importance of vacuum conditions and of control of residuals and contaminants exemplified in his studies are points that have been relearned endlessly. Yet Willy Wien's most influential contribution to mass spectrometry, after providing its foundation, was his influence on the research direction of Arthur Dempster, a Canadian graduate student under his guidance at Würzburg from 1912 to 1914, from whose work almost all later mass spectrometry has stemmed (see this volume, *Arthur Jeffrey Dempster*).

References

(*1*) Dahlgren, E. W. The Nobel Prize in Physics 1911. In: *Nobel Lectures, including Presentation Speeches and Laureates' Biographies, Physics 1901–1921*; Elsevier Publishing Co. Amsterdam, 1967.

(*2*) Kangro, H. Wien, Wilhelm Carl Werner Otto Fritz Franz. In: *Dictionary of Scientific Biography*; Gillispie, C. C., Ed.; Charles Scribner's Sons: New York, 1976; vol. XIV, pp 337–342.

(*3*) Wien, W. *Aus dem Leben und Wirken eines Physikers*; Barth: Leipzig, 1930.

(*4*) Goldstein, E. Über eine noch nicht untersuchte Strahlungsform an der Kathode inducirter Entladung. *Sitzungsberichte der Königlichen Akademie der Wissenschften zu Berlin.* **1886** [29 July], *39*, 691. Translation in Magie, W. F., ed. *A Source Book in Physics*; Harvard Univ. Pr.: Cambridge, Mass., 1935 & 1963, 576–578.

(*5*) Wien, W. Die elektrostatische und magnetische Ablenkung der Kanalstrahlen. *Verhandlungen der Deutsche Physikalische Gesellschaft zu Berlin* **1898**, *17*, 10–12.

(*6*) Wien, W. Untersuchungen über die elektrische Entladung in verdünnten Gasen. *Annalen der Physik und Chemie.* **1898**, *3. Folge 65*, 440–452. Excepted and translated in: Magie, W. F., ed. *A Source Book in Physics*; Harvard Univ. Pr.: Cambridge, Mass., 1935 & 1963, 597–600.

(*7*) Wien, W. Elektrische Entladung in verdünnten Gasen. *Annalen der Physik und Chemie* **1902**, *4. Folge 8*, 244–266.

(*8*) Wien, W. Über die Natur der positiven Elektronen. *Annalen der Physik und Chemie* **1902**, *4. Folge 9*, 660–664.

(*9*) Falconer, I. J. Thomson's Work on Positive Rays, 1906–1914. *Historical Studies in the Physical and Biological Sciences* **1988**, *18*, 265–310.

(*10*) Dahl, P. *Flash of the Cathode Rays: A History of J. J. Thomson's Electron*; Institute of Physics Publishing: Bristol and Philadelphia, 1997. pp. 266–269.
(*11*) Wien, K. 100 Years of Ion Beams: Willy Wien's Canal Rays. *Braz. J. Phys.* **1999**, *29*, 401–414.
(*12*) Wien, W. Kanalstrahlen. In: *Kanalstrahlen und Ionisation bei hohen Tempersturen: Handbuch der Radiologie, IV*; Marx, E., Ed.; Akademische Verlagsgesellschaft: Leipzig, 1917; pp 1–210.

Photo source: http://www.nobelprize.org/nobel_prizes/physics/laureates/1911/wien-bio.html

Keith A. Nier
Independent Historian, Madison, NJ, USA

Charles Lee Wilkins

b. August 14, 1938

This is a photograph of Wilkins (standing) with Robert White, his student, working with a permanent magnet FTMS system for analytical chemistry applications in 1980. (Used with permission of Elsevier from Figure 2, International Journal of Mass Spectrometry, 287, 1–6 (2009).)

Charles Wilkins is an American chemist widely noted for work in the general areas of Fourier transform mass spectrometry and hyphenated analysis systems. He is recognized for his contributions to design of the second Fourier transform mass spectrometer and to the first such system especially designed for analytical chemistry applications of FTMS. The latter resulted in early pioneering papers emphasizing the value of the technique for high performance analytical mass spectrometry.

Charles Wilkins was born in Los Angeles, CA, USA, where he also attended high school in Los Angeles. He obtained an undergraduate B.S. degree from Chapman College in Orange, California. He went on to the University of Oregon, where he received the Ph.D. in Chemistry under the direction of Professor Lloyd Dolby in 1966. Following a postdoctoral appointment at the University of California, Berkley, with Professor Andrew Streitweiser, Dr. Wilkins accepted an Assistant Professor appointment at the University of Nebraska, Lincoln, in 1967, becoming Professor in 1976. He then moved to University of California, Riverside in 1981 as Professor of Chemistry was promoted to Distinguished Professor in 1997 and in 1998 became Distinguished Professor of Chemistry, University of Arkansas in Fayetteville, AR.

Professor Wilkins interest in mass spectrometry began early in his academic career and, in collaboration with Dr. Michael Gross, was one of the earliest proponents of the technique that later became known as Fourier transform mass spectrometry (also known as Fourier transform ion cyclotron resonance) (see this volume, *A History of Ion Cyclotron Resonance (ICR) and Fourier Transform (FTICR) Mass Spectrometry* and *Michael L. Gross*). Together they and their students fabricated the first analytical chemistry FTMS instrument, which later led to a commercial version marketed by Nicolet Analytical Instruments (Madison, WI). Wilkins also has made significant contributions to research in the areas of chemical pattern recognition, graph theoretic applications in chemistry, liquid chromatography- NMR, and the development of hyphenated chemical analysis systems. Some papers representative of this work are listed below (*1–6*).

Wilkins has been recognized by a number of awards for his work in mass spectrometry including the 1997 Field and Franklin Award for Outstanding Achievement in Mass Spectrometry, the 1994 Pittsburgh Analytical Chemistry Award, the 1994 Tolman Medal of the Southern California American Chemical Society, the 2002 Eastern Analytical Symposium Award for Outstanding Achievement in the Fields of Analytical Chemistry and the 2013 Award in Chemical Instrumentation the Analytical Chemistry Division of the American Chemical Society. He was named a Fellow of the American Society for the Advancement of Science in 1996, a Fellow of the Society for Applied Spectroscopy in 2006, and a Fellow of the American Chemical Society in 2010. In 2006 he was also made an Honorary Member of the Society for Applied Spectroscopy. His academic honors have included the 2004 University of Oregon Department of Chemistry Alumni Achievement Award, the 2003 Arkansas Alumni Award for Excellence- Research, and the 2005 University of Arkansas Fulbright College Master Researcher Award.

He married Susan J. Otos in 1967 and a son (Mark R. Wilkins) was born to them in 1971. Following their divorce in 1996, Charles married fellow chemist, Ingrid Fritsch, with whom he has two children, Connor W. Fritsch (born 2000) and Eric F. Wilkins (born 2005).

References

1. Wilkins, C. L. Fourier Transform Mass Spectrometry. *Anal. Chem.* **1978**, *50*, 493A–498A.
2. Wilkins, C. L.; Brissey, G. M.; Giss, G. N.; Steiner, S. Direct-Linked Gas Chromatography-Fourier Transform Infrared-Mass Spectrometer Analysis System. *Anal. Chem.* **1981**, *53*, 113–117.
3. Wilkins, C. L.; Gross, M. L. Fourier Transform Mass Spectrometry for Analysis. *Anal. Chem.* **1981**, *53*, 1661A–1668A.
4. Wilkins, C. L.; Giss, G. N.; White, R. L.; Brissey, G. M.; Onyiriuka, E. Mixture Analysis by Gas Chromatography/Fourier Transform Infrared Spectrometry/Mass Spectrometry. *Anal. Chem.* **1982**, *54*, 2260–2264.

5. Ijames, C. F.; Wilkins, C. L. First Demonstration of High Resolution Laser Desorption Mass Spectrometry of High Mass Organic Ions. *J. Am. Chem. Soc.* **1988**, *110*, 2687–2688.
6. So, H.; Wilkins, C. L. On the First Observation of Carbon Aggregate Ions >C600+ by Laser Desorption Fourier Transform Mass Spectrometry. *J. Phys. Chem.* **1989**, *93*, 1187–1189.

Charles L. Wilkins
Department of Chemistry and Biochemistry, University of Arkansas, Fayetteville, AR, USA

Dudley Howard Williams

b. May 25, 1937
d. November 3, 2010

Dudley Williams was an English chemist best known for his work on Vitamin D and the antibiotic Vancomycin, identifying the active (metabolized) form of the former and the mechanism of action of the latter (*1*). This work benefited substantially from the new and creative ways in which Williams used mass spectrometry and NMR spectroscopy to unravel difficult and sometimes confusing chemistry. His championship of these techniques to elucidate organic structures, forms, and energetics revolutionized the way organic chemists of the 1960s, 1970s, and 1980s approached their work (*2*).

Williams was born on May 25, 1937, in Leeds, UK. He was raised and educated there, attending the Pudsey Grammar School and Leeds University. For his work at the university on the synthesis and chemistry of Vitamin D, Williams received a PhD in 1961. He spent the next three years in the US in Carl Djerassi's lab at Stanford University as a Fullbright Scholar and later a Research Associate (see this volume, *Carl Djerassi*). Returning to Cambridge University in 1964, Williams became a Lecturer in Chemistry at Churchill College. He became a Reader in Organic Chemistry in 1974, and a Professor of Biological Chemistry in 1996, a position from which he retired in 2004. In 1988, he became Deputy Director of the Cambridge Centre for Molecular Recognition, a position he held until his retirement.

Out of Williams' exceedingly productive time at Stanford and early days at Cambridge came seven books (*3–9*) that were intended to serve as guides to the organic chemist, providing a new approach to organic chemistry that involved heavy reliance on MS and NMR. About *Mass Spectrometry of Organic Compounds* (*9*) reviewers of such note as Seymour Meyerson and F. W. McLafferty wrote,

> ...brings together in remarkably clear and concise form essentially everything that is known about the ways in which organic molecules break up under electron impact. The material is presented in the language of the organic chemist. It is carefully organized around functional groups for easy reference, and the pieces are tied together by numerous beautiful examples in which a well-defined reaction sequence is shown to operate in many classes of compounds containing similar controlling structural elements (*10*, see this volume, *Seymour Meyerson*).

and

> ...I feel that the overall impact of this book provides the strongest case to date for the merit of this approach. The presentation of such a large collection of facts in logical mechanistic terms instead of on an empirical basis makes the data much easier to comprehend and assimilate for the modern organic chemist. ... We have found the compilation invaluable in our laboratory and can recommend it highly to anyone using this technique (*11*, see this volume, *Fred W. McLafferty*).

Praise like this is indicative of the reception the books enjoyed and indeed continued to enjoy: in 1982, the Institute for Scientific Information listed *Mass Spectrometry of Organic Compounds* as a 'citation classic,' noting that it had been cited in more than 2700 publications between 1967 and 1982. Further, "The approach to rationalize the fragmentation behaviour of organic molecules ionized by electron impact by formulating bond fission and rearrangement processes triggered by a localized charge allowed systemization of the vast experimental material and thus enabled the organic chemist to use mass spectrometry efficiently in his work" (*12*).

Williams practiced what he preached with regard to his belief in and reliance on top-of-the-line technology. After his exposure to the high quality instruments in the Djerassi lab at Stanford, he made it a condition of employment at Cambridge that top quality NMR instruments and mass spectrometers be purchased, so that the laboratories there would have technical capabilities equivalent to their American counterparts. For many years, he continued to develop new techniques to make NMR and mass spectrometry into ever more powerful tools for determining chemical structures and for probing chemical properties (*13*).

The focus in Williams' early work was on the chemistry of molecules, notably the study of what could be learned about them from various aspects of their mass spectrometry (*14*). He published a well-received series of papers in the 1970s in which he exposed details of potential energy surfaces by analyzing the consequences of kinetic energy release in metastable fragmentations (e.g., *15,16*).

In 1970, Williams and colleagues showed how the inactive form of Vitamin D found in foods is metabolized, first in the liver and then in the kidney, to the biologically active form crucial for calcium absorption (*17*). This work has had profound effects on the treatment of disorders involving calcium resorption by bone. During the same period, Williams and colleagues began working on the structure and shape of vancomycin, as well as its interaction with peptides on the cell walls of bacteria. Williams famously estimated the time required to complete the vancomycin work at about 6 months. The research effort ultimately occupied several decades, laying the foundation on which drugs to fight methicillin-resistant *Staphylococcus aureus* (MRSA) infections were developed (*1,18,19*). At about the same time, Williams' laboratory developed a low resolution mixture analysis approach to the sequencing of peptides (*20*). This method was made all the more useful in the 1980s, as Williams was quick to appreciate, when the advent of fast atom bombardment (FAB) and later electrospray ionization (ESI) mass spectrometry lifted the volatility barrier for macromolecules (see this volume, *Static and Continuous Flow Fast Atom Bombardment (FAB)* and *The Development of Electrospray Ionization (ESI) for Mass Spectrometry*).

As Williams' attention turned more and more toward biology and biochemistry, he continued to use the vancomycin experience as a vehicle for fundamental thinking about molecular shape and flexibility of molecules and about the thermodynamics of their solvation, binding, and interactions. With increased focus on the structures and mechanisms of action of antibiotics, as well as their interactions with DNA, he continued to rely on what he termed his favorite tools, mass spectrometry and NMR (*21–23*). His work was extremely influential. The author of more than 350 publications, he was over a number of years (1965–1978) the most cited scientist at Cambridge, the most cited chemist in Britain, and the most cited organic chemist outside the US (*24*).

Williams was an elected member of many professional organizations and received many awards for his work, among them election to the Royal Society of Chemistry in 1969 and receipt of the Tilden Medal and Lectureship in 1983 and the 1990 Bader Award for Organic Chemistry from that society. He was also elected a Fellow of the Royal Society in 1983 and received its 1984 Award for Structural Chemistry, in addition to the Leo Friend Award of the American Chemical Society in 1996.

Devoted to what he called his academic family – the students and post-docs who passed through his lab and contributed greatly to the work accomplished there – Williams reveled in the accomplishments of their careers. Williams encouraged them "to think laterally and imaginatively, and to have the courage to work in new areas." He regarded provocative and testable ideas that might turn out to be wrong to be "more important than the pursuit of boring details" (*13*).

Williams' interests outside chemistry were many and varied: he was an accomplished pianist, a fierce squash and badminton competitor, an intrepid skier, and an avid walker, particularly in the Lakes. He was a renowned raconteur and conversationalist, able to liven the deadliest committee meeting discussion with his acerbic wit and clever turn of phrase – as well as it divert it with a philosophical excursion into an interesting but nonetheless unrelated topic. Williams succumbed to a very aggressive form of liver cancer, diagnosed only a month before his death, leaving behind his wife of 47 years, Pat, and two sons (*1,2,25*).

References

(*1*) Watts, G. Obituary for Dudley Williams. *Lancet* **2011**, *377*, 120.

(*2*) Sanders, J. K. M. Dudley Williams Obituary. *The Guardian 24 November*, **2010**,

(*3*) Budzikiewicz, H.; Djerassi, C.; Williams, D. H. *Interpretation of Mass Spectra of Organic Compounds*; San Francisco: Holden-Day, 1964. 464 pp.

(*4*) Budzikiewicz, H.; Djerassi, C.; Williams, D. H. *Structure Elucidation of Natural Products by Mass Spectrometry. Vol. 1 Alkaloids*; Holden-Day: San Francisco, 1964. 540 pp.

(*5*) Budzikiewicz, H.; Djerassi, C.; Williams, D. H. *Structure Elucidation of Natural Products by Mass Spectrometry. Vol. 2 Steroids, Sugars, and Terpenes*; Holden-Day: San Francisco, 1964. 540 pp.

(*6*) Bhacca, N. S.; Williams, D. H. *Application of NMR in Organic Chemistry: Illustrations from the steroid field*; Holden-Day: San Francisco, 1964. 198 pp.

(*7*) Fleming, I.; Williams, D. H. *Spectroscopic Methods in Organic Chemistry*; McGraw-Hill: London, 1966.

(*8*) Fleming, I.; Williams, D. H. *Spectroscopic Problems in Organic Chemistry*; McGraw-Hill: London, 1967.

(*9*) Budzikiewicz, H.; Djerassi, C.; Williams, D. H. *Mass Spectrometry of Organic Compounds*; Holden-Day: San Francisco, 1967. 690 pp.

(*10*) Meyerson, S. Review of *Mass spectrometry of organic compounds* by Budzikiewicz, Djerassi, and Williams. *J. Chem. Educ.* **1968**, *45*(A440), A438.

(*11*) McLafferty, F. W. Review of *Mass Spectrometry of Organic Compounds*, Budzikiewicz, Djerassi, and Williams. *J. Am. Chem. Soc.* **1968**, *90*, 3900.

(*12*) Garfield, E. This Week's Citation Classic, Budzikiewicz H, Djerassi C & Williams D H. *Mass Spectrometry of Organic Compounds*. http://garfield.library.upenn.edu/classics1982/A1982PB23800001.pdf (accessed November 22, 2014).

(*13*) Daley, D. J. Obituary for Dudley Williams. *Rapid Commun. Mass Spectrom.* **2011**, *25*, 1117–1118.

(*14*) Bowers, M. T. Editor's (Personal) Forward in Special Issue Honoring Dudley Williams. *Int. J. Mass Spectrom* **2004**, *236*, vii–viii.

(*15*) Williams, D. H.; Hvistendahl, G. Kinetic Energy Release in Relation to Symmetry-Forbidden Transitions. *J. Am. Chem. Soc.* **1974**, *96*, 6753–6755.
(*16*) Williams, D. H.; Hvistendahl, G. Kinetic Energy Release as a Mechanistic Probe. The Role of Orbital Symmetry. *J. Am. Chem. Soc.* **1974**, *96*, 6755–6757.
(*17*) Lawson, D. E.; Franser, D. R.; Kodicek, E.; Morris, H. R.; Williams, D. H. Identification of 1,25-Dihydroxycholecalciferol, a New Kidney Hormone Controlling Calcium Metabolism. *Nature* **1971**, *230*, 228–230.
(*18*) Smith, K. A.; Williams, D. H.; Smith, G. A. Structure Studies on Antibiotic Vancomycin; the Nature of the Aromatic Rings. *J. Chem. Soc., Perkin Trans. I* **1974**, *20*, 2369–2376.
(*19*) Williams, D. H.; Bardsley, B. The Vancomycin Group of Antibiotics and the Fight against Resistant Bacteria. *Angew. Chem. Int. Ed.* **1999**, *38*, 1172–1193.
(*20*) Morris, H. A.; Williams, D. H.; Midwinter, G. G.; Hartley, B. S. A Mass-Spectrometric Sequence Study of the Enzyme Ribitol Dehydrogenase from *Klebsiella aerogenes*. *Biochem. J.* **1974**, *141*, 701–713.
(*21*) Giovannini, M. G.; Poulter, L.; Gibson, B. W.; Williams, D. H. Biosynthesis and Degradation of Peptides Derived from *Xenopus laevis* Prohormones. *Biochem. J.* **1987**, *243*, 113–120.
(*22*) Beauregard, D. A.; Williams, D. H.; Gwynn, M. N.; Knowles, D. J. C. Dimerisation and Membrane Anchors in the Extra-Cellular Targeting of Vancomycin Group Antibiotics. *Antimicrob. Agents Chemother.* **1995**, *39*, 781–785.
(*23*) Williams, D. H.; Maguire, A. J.; Tsuzuki, W.; Westwell, M. S. An Analysis of the Origins of a Cooperative Binding Energy of Dimerization. *Science* **1998**, *280*, 711–714.
(*24*) Anonymous. Biograpy of Dudley Howard Williams. *Int. J. Mass Spectrom* **2004**, *236*, ix–xliii.
(*25*) Robinson, C. V. Obituary. Dudley Williams – May 25, 1937–November 3, 2010. *J. Am. Soc Mass Spectrom* **2011**, *22*, 791–792.

P. Jane Gale
Waters Corp., Milford, MA, USA

Vicki Hopper Wysocki

b. March 1, 1960

Vicki Wysocki is an American chemist best known for her work on the gas-phase structure and fragmentation mechanisms of peptides and peptide fragment ions. She received her B.S. in Chemistry from Western Kentucky University in 1982 and her Ph.D. in Chemistry from Purdue University in 1987 under the direction of Professor R. Graham Cooks (see this volume, *R Graham Cooks*).

Dr. Wysocki spent time as a post-doctoral fellow at Purdue University in 1987 and also completed a National Research Council Postdoctoral Fellowship at the Naval Research Laboratories (1988–1989). Following her postdoctoral appointments, Dr. Wysocki joined the faculty of the Department of Chemistry at Virginia Commonwealth University in 1990 and was promoted to Associate Professor in 1994. She joined the University of Arizona in 1996 and was promoted to Professor in 2000. Most recently she was Chair of the Department of Chemistry and Biochemistry at the University of Arizona. Leaving Arizona as Professor Emeritus in August, 2012, Professor Wysocki joined the Ohio State University as an Ohio Eminent Scholar.

The mobile proton model for which Wysocki and Gaskell shared the 2009 ASMS Distinguished Contribution Award provides specific predictions regarding peptide fragmentation that enable mass spectra to be interpreted and simulated and serves as a foundation on which more advanced models of fragmentation can be built. Based in part on data-mining of large MS/MS datasets, it also underpins increasingly sophisticated database search algorithms for protein identification.

Wysocki and her group continue to focus on developing a better understanding the hydrogen-bonding interactions and fragmentation patterns of activated protonated peptides with the goal of providing additional 'rules' that will enable faster and more efficient ways to determine protein sequences. They are also investigating mechanisms of action of microorganisms in an effort to identify microorganism biomarkers that could be used in developing diagnostic tests. In addition, they are working on instrument development for improved structural characterization of large non-covalent protein complexes by surface-induced dissociation and ion mobility.

Dr. Wysocki serves on several editorial boards and was the recipient of the 1992 American Society for Mass Spectrometry Research Award. In 2009 she and Simon Gaskell shared the Distinguished Contribution to Mass Spectrometry Award given by the American Society for Mass Spectrometry for their work (collaborative and independent) on the mobile proton model (see this volume, *Simon James Gaskell*).

The publications cited below are representative of the author's most significant contributions to the field (*1*–*6*).

References

(*1*) Wysocki, V. H.; Burinsky, D. J.; Cooks, R. G. Competitive Dehydration and Deamination of Amino Alcohols and Amino Acids in the Gas Phase. *J. Org. Chem.* **1985**, *50*, 1287–1291.

(*2*) Schey, K. L.; Kenttämaa, H. I.; Wysocki, V. H.; Cooks, R. G. Low-Energy Collisional Activation of Polyatomic Ions with Different Target Gases. *Int. J. Mass Spectrom. Ion Process* **1989**, *90*, 71–83.

(*3*) Wysocki, V. H.; Ross, M. M. Charge-Remote Fragmentation of Gas-Phase Ions: Mechanistic and Energetic Considerations. *Int. J. Mass Spectrom. Ion Process* **1991**, *104*, 179–211.

(*4*) McCormack, A. L.; Jones, J. L.; Wysocki, V. H. Surface-Induced Dissociation of Multiply-Protonated Peptides. *J. Am. Soc. Mass Spectrom.* **1992**, *3*, 859–862.

(*5*) Gamage, C. M.; Fernandez, F. M.; Kuppannan, K.; Wysocki, V. H. Submicrosecond Surface-Induced Dissociation of Peptide Ions in a MALDI TOF MS. *Anal. Chem.* **2004**, *76*(17), 5080–5091.

(*6*) Zhou, M; Dagan, S; Wysocki, V. Protein Subunits Released by Surface Collisions of Non-Covalent Complexes Are Revealed by Ion Mobility-Mass Spectrometry to have Native-like Compact Structures. *Angew. Chem.* **2012**, *51*, 4336–4339.

Stephanie M. Cologna
National Institute of Child Health and Human Development, National Institutes of Health, Bethesda, MD, USA

Alfred L. Yergey
National Institute of Child Health and Human Development, NIH, Bethesda, MD, USA

Richard A. Yost

b. May 31, 1953

Richard A. Yost, Ph.D., is an American analytical chemist who is recognized internationally as a leader in the field, particularly in tandem mass spectrometry (MS/MS). He is best known for his work in inventing the triple quadrupole mass spectrometer as a graduate student with Chris Enke (see this volume, *Christie George Enke*).

Yost was born on May 31, 1953, in Martins Ferry, Ohio. He received his BS degree in Chemistry *summa cum laude* with Honors in 1974 from the University of Arizona. He was an NSF Graduate Fellow (1975–79) and an ACS Analytical Division Fellow (1978–79). He received his Ph.D. in Analytical Chemistry in 1979 from Michigan State University, performing graduate research with Professor Chris Enke. He then joined the faculty of the University of Florida as an Assistant Professor where he has remained.

Yost's professional activities have focused on research and teaching in analytical mass spectrometry, particularly tandem mass spectrometry (MS/MS). His group's research has reflected a balance between fundamental studies, instrumentation development, and applications in analytical chemistry. His design and development of early versions of several instruments, notably the triple quadrupole mass spectrometer, the ion trap tandem mass spectrometer, and the laser microprobe tandem mass spectrometer contributed to their later commercialization. He has done innovative work in fast gas chromatography using short capillary columns under vacuum, quantitative MALDI tandem mass spectrometry, imaging biomolecules in intact tissue, and high-field ion mobility (FAIMS), as well as in the application of novel mass spectrometric and ion mobility methods and techniques to areas such as biomedical, pharmaceutical, environmental, petrochemical, and forensic chemistry Yost and his research group currently focus their efforts on imaging mass spectrometry and high-field ion mobility spectrometry.

Yost is a member of the American Society for Mass Spectrometry and has served as secretary and treasurer for the organization. In 1993, he and Chris Enke shared the ASMS Award for Distinguished Contribution to Mass Spectrometry for development of the triple quadrupole mass spectrometer.

Yost is married and has one daughter and two sons. He has served as a Scout leader for over 30 years, and has received the Silver Beaver and Silver Fox Awards of the Boy Scouts of America.

He currently holds the positions of Robin & Jean Gibson Professor, College of Liberal Arts & Sciences, and has served as Head of Analytical Chemistry, Department of Chemistry, at the University of Florida as well as being a member of the Florida Board of Governors (Regents) and as the Chair of the Advisory Council of Faculty Senates of Florida.

The publications cited below are representative of the author's most significant contributions to the field (*1–5*).

References

1. Yost, R. A.; Enke, C. G. Triple Quadrupole Mass Spectrometry for Direct Mixture Analysis and Structure Elucidation. *Anal. Chem.* **1979**, *51*, 1251A–1264A.
2. Johnson, J. V.; Yost, R. A. Tandem Mass Spectrometry for Trace Analysis. *Anal. Chem.* **1985**, *57*, 758A–768A.
3. Lee, M.; Yost, R. A. *Biomed. Environ. Mass Spectrom.* **1988**, *15*, 193–204.
4. Johnson, J. V.; Yost, R. A.; Kelley, P. E.; Bradford, D. C. *Anal. Chem.* **1990**, *62*, 2162–2172.
5. Garrett, T. J.; Yost, R. A. Analysis of Intact Tissue by Intermediate-Pressure MALDI on a Linear Ion Trap Mass Spectrometer. *Anal. Chem.* **2006**, *78*, 2465–2469.

Richard A. Yost
Department of Chemistry, University of Florida, Gainesville, FL, USA

LIST OF CONTRIBUTORS

Contributors are listed in alphabetical order together with their affiliations. Titles of articles which they have authored follow in alphabetical order, along with the respective page numbers. Where articles are co-authored, this has been indicated by an asterisk preceding the article title.

ENKE, CHRISTIE G. (Department of Chemistry and Chemical Biology, University of New Mexico, Albuquerque, NM, USA, and Department of Chemistry, Michigan State University, East Lansing, MI, USA)
Christie George Enke: 64

FENSELAU, CATHERINE (Department of Chemistry & Biochemistry, University of Maryland, College Park, MD, USA)
Catherine Clarke Fenselau: 75

FINNIGAN, ROBERT (Thermo Electron Corporation [ret.])
**Evan Charles Horning and Marjorie Groothuis Horning*: 119

FRANZ, ANDREAS H. (Department of Chemistry, University of the Pacific, Stockton, CA, USA)
Curt Brunnée: 29; *Josef Heinrich Elisabeth Mattauch*: 147

FUTRELL, JEAN H. (Physical Sciences Division, Pacific Northwest National Laboratory, Richland, WA, USA)
Jean H. Futrell: 83

GALE, P. JANE (Waters Corp., Milford, MA, USA)
**Michael Barber*: 11; *Alfred Benninghoven*: 15; *John Herbert Beynon*: 17; **Brian T. Chait*: 37; **Michael L. Gross*: 91; **Michael Guilhaus*: 93; **Alexander George Harrison*: 97; *Zdeněk Herman*: 105; *John Leonard Holmes*: 114; **Richard Edward Honig*: 117; **Evan Charles Horning and Marjorie Groothuis Horning*: 119; **Donald Frederick Hunt*: 122; *Chava Lifshitz*: 134; *Helmut Schwarz*: 188; *David Smith*: 194; *Dudley Howard Williams*: 235; **Introduction to Part B: Criteria for Inclusion in these Biographies of Notable People in Mass Spectrometry*: 1

GASKELL, SIMON J. (Queen Mary's College, London, UK)
**Michael Barber*: 11; *Simon James Gaskell*: 85

GILBERT, ANTHONY JAMES (Waters Corporation, Manchester, UK)
Joseph John Thomson: 213

GRAYSON, MICHAEL A. (Washington University in St. Louis, St. Louis, MO, USA [ret.])
Kenneth Tompkins Bainbridge: 9; *Klaus Biemann*: 20; **John Bennett Fenn*: 72; *Fred W. McLafferty*: 152

HACHEY, DAVID L. (Vanderbilt University, Nashville, TN, USA)
**Rudolf Schoenheimer*: 185

HALLIDAY, ALEX N. (University of Oxford, Oxford, UK)
Alexander Norman Halliday: 95

HENZEL, WILLIAM J. (Genentech, Inc., South San Francisco, CA, USA [ret.])
William J. Henzel: 101

HERCULES, DAVID M. (Department of Chemistry, Vanderbilt University, Nashville, TN, USA)
David Michael Hercules: 103; *Andrew Gans Sharkey, Jr.*: 192

HOLDEN, NORMAN E. (National Nuclear Data Center, Brookhaven National Laboratory, Upton, NY, USA)
Angus Ewan Cameron: 33

HOLMES, JOHN L. (Chemistry Department, University of Ottawa, Ottawa, ON, Canada)
Frederick Pettit Lossing: 136

HOUK, R. SAM (Chemistry Department, Iowa State University, Ames, IA, USA)
Robert Samuel Houk: 121

KEBARLE, PAUL (University of Alberta, Edmonton, AB, Canada)
Paul Kebarle: 129

Klein, E. Roseland (Baylor College of Medicine, Houston, TX, USA)
**Rudolf Schoenheimer*: 185

Koenig, Simone (University of Muenster, Muenster, NRW, Germany)
Michael Karas: 128

Labowsky, Michael (Ansama Research, LLC, Wayne, NJ, USA)
**John Bennett Fenn*: 72

Laeter, John de (d.) (Curtin University, Perth, Australia)
William Compston: 43; *Henry George Thode*: 209

Levsen, Karsten (Fraunhofer Institute for Toxicology and Experimental Medicine (ITEM), Hannover, Germany)
Hans-Dieter Beckey: 13

Makarov, Alexander A. (Thermo Fisher Scientific GmbH, Bremen, Germany)
Alexander A. Makarov: 138

March, Raymond E. (Trent University, Peterborough, ON, Canada)
Raymond E. March: 141

Markey, Sanford P. (Laboratory of Neurotoxicology, National Institute of Mental Health, Bethesda, MD, USA)
Henry Marshall Fales: 70; *John Michael Hayes*: 99; **Donald Frederick Hunt*: 122

Marshall, Alan G. (National High Magnetic Field Laboratory, Florida State University, Tallahassee, FL, USA)
Alan G. Marshall: 143

McLafferty, Fred W. (Chemistry Department, Cornell University, Ithaca, NY, USA)
Roland S. Gohlke: 89

Morris, Howard R. (Division of Molecular Biosciences, Imperial College, London, UK)
Howard Redfern Morris: 159

Nibbering, N.M.M. (d.) (Laser Lab Amsterdam, Vrije Universiteit, HV Amsterdam, The Netherlands)
Seymour Meyerson: 155; *Nicolaas Martinus Maria Nibbering*: 165

Nier, Keith A. (Independent Historian, Madison, NJ, USA)
Arthur John Ahearn: 3; *Claude Jean Allègre*: 4; *Walker Bleakney*: 23; *Harrison Scott Brown*: 27; **Brian T. Chait*: 37; *Robert N. Clayton*: 38; *Willi Dansgaard*: 49; *Vernon Hamilton Dibeler*: 57; *Samuel Epstein*: 67; *Heinz Ewald*: 69; *Russell Elwell Fox*: 81; *Johannes Geiss*: 87; *Richard Franz Karl Herzog*: 107; *Heinrich Hintenberger*: 110; *John Alfred Hipple, Jr.*: 112; *Keith R. Jennings*: 126; *Charles Kemball*: 131; *Fred Loomis Mohler*: 157; *Alfred Otto Carl Nier*: 167; *Wolfgang Paul*: 172; *Rowland Ivor Reed*: 174; *Sibyl Martha Rock*: 179; *Ragnar Erik Ryhage*: 183; *Nicholas John Shackleton*: 190; *Einar August Stenhagen*: 201; *George Robert Tilton*: 215; *Harold Clayton Urey*: 221; *Harold Williams Washburn*: 227; *Gerald J. Wasserburg*: 228; *Wilhelm Carl Werner Otto Fritz Franz Wien*: 230; **Introduction to Part B: Criteria for Inclusion in these Biographies of Notable People in Mass Spectrometry*: 1

Nikolaev, Eugene (Laboratory of Ion and Molecular Physics, Institute for Energy Problems of Chemical Physics, Russian Academy of Sciences, Moscow, Russia; Laboratory of Mass Spectrometry of Biomacromolecules, Institute for Biochemical Physics, Russian Academy of Sciences, Moscow, Russia; and Institute Biomedical Chemistry, Russian Academy of Medical Sciences, Moscow, Russia)
Boris Alexandrovich Mamyrin: 139; *Victor L'vovich Tal'roze*: 206

Parr, Albert (National Institute of Science and Technology, Washington, DC, USA)
**Mark G. Inghram*: 124; **Henry Meyer Rosenstock*: 181

PRICE, P. BUFORD (Physics Department, University of California, Berkeley, CA, USA)
John H. Reynolds: 175

REINHARDT, CARSTEN (Chemical Heritage Foundation, Philadelphia, PA, USA)
Carl Djerassi: 58

ROBINSON, CAROL VIVIEN (Physical and Theoretical Chemistry Laboratory, Oxford, UK)
Carol Vivien Robinson: 178

SHEIL, MARGARET M. (University of Melbourne, VIC, Australia)
**Michael Guilhaus*: 93

STAFFORD, GEORGE C. JR. (Agilent Technologies, Santa Clara, CA, USA)
George C. Stafford, Jr.: 196

STOCKBAUER, ROGER (Louisiana State University, Baton Rouge, LA, USA)
**Mark G. Inghram*: 124; **Henry Meyer Rosenstock*: 181

STULTS, JOHN T. (Genentech, Inc., South San Francisco, CA, USA)
John T. Stults: 204

TAKAYAMA, MITSUO (Yokohama City University, Mass Spectrometry Laboratory, Yokohama, Kanagawa, Japan)
Hisashi Matsuda: 145

THOMSON, BRUCE (AB Sciex, Framingham, MA, USA)
Bruce A. Thomson: 212

TODD, JOHN F.J (School of Physical Sciences, University of Kent, Canterbury, Kent, UK)
John Francis James Todd: 217

TRAEGER, JOHN C. (La Trobe University, Victoria, Australia)
James Douglas Morrison: 161

TURNER, GRENVILLE (School of Earth, Atmospheric, and Environmental Sciences, University of Manchester, Manchester, UK)
Grenville Turner: 219

VESTAL, MARVIN (SimulTOF Instruments, Framingham, MA, USA)
Marvin L. Vestal: 224

WILKINS, CHARLES L. (Department of Chemistry and Biochemistry, University of Arkansas, Fayetteville, AR, USA)
Charles Lee Wilkins: 233

WYSE JACKSON, PATRICK N. (Department of Geology, Trinity College, Dublin 2, Ireland)
Clair Cameron Patterson: 170

YERGEY, ALFRED L. (National Institute of Child Health and Human Development, NIH, Bethesda, MD, USA)
**Brian T. Chait*: 37; *Norman David Coggeshall*: 40; *R. Graham Cooks*: 45; **Michael L. Gross*: 91; **Alexander George Harrison*: 97; *Franz Hillenkamp*: 108; *Frederick W. Lampe*: 132; *James A. McCloskey*: 150; *David Paul Stevenson*: 203; *Koichi Tanaka*: 208; **Vicki Hopper Wysocki*: 238; **Introduction to Part B: Criteria for Inclusion in these Biographies of Notable People in Mass Spectrometry*: 1

YOST, RICHARD A. (Department of Chemistry, University of Florida, Gainesville, FL, USA)
Richard A. Yost: 239

FSC®
www.fsc.org
MIX
Papier | Fördert gute Waldnutzung
FSC® C083411

Kolibri 360 GmbH
Ferdinand-Jühlke-Straße 7
99095 Erfurt, Deutschland
produktsicherheit@kolibri360.de